GENERAL VIEW OF THE AGRICULTURE OF THE COUNTY OF SUSSEX

GENERAL VIEW of the AGRICULTURE of the COUNTY OF SUSSEX

A Reprint of the Work Drawn up for the
Consideration of the Board of Agriculture
and Internal Improvement

by

The Rev ARTHUR YOUNG

DAVID & CHARLES REPRINTS

7153 4782 9

This book was first published in 1813
This edition published in 1970

Printed in Great Britain by
Clarke Doble & Brendon Limited Plymouth
for David & Charles (Publishers) Limited
South Devon House Railway Station
Newton Abbot Devon

GENERAL VIEW

OF THE

AGRICULTURE

OF THE

COUNTY OF SUSSEX.

DRAWN UP FOR THE CONSIDERATION OF

THE BOARD OF AGRICULTURE

AND INTERNAL IMPROVEMENT.

BY THE REV. ARTHUR YOUNG.

LONDON:

PRINTED FOR SHERWOOD, NEELY, AND JONES,
PATERNOSTER-ROW :

SOLD BY G. AND W. NICOL, PALL-MALL; J. DONALDSON, T. POL-
LARD, AND W. PAINE, BRIGHTON; MESSRS. LEE, LEWES;
P. HUMPHREY, AND W. SEAGRAVE, CHICHESTER;
AND MRS. SPOONER, WORTHING.

1813.

ADVERTISEMENT.

———◆———

THE desire that has been generally ex-
pressed, to have the AGRICULTURAL SURVEYS
of the KINGDOM reprinted, with the additional
Communications which have been received since
the ORIGINAL REPORTS were circulated, has induced the BOARD OF AGRICULTURE to come to a
resolution to reprint such as appear on the whole
fit for publication.

It is proper at the same time to add, that the
Board does not consider itself responsible for
every statement contained in the Reports thus
reprinted, and that it will thankfully acknowledge
any additional information which may still be
communicated.

————————

N. B. *Letters to the Board, may be addressed
to Sir* JOHN SINCLAIR, *Bart. the President,
No.* 32, *Sackville-Street, Piccadilly, London.*

CONTENTS.

AGRICULTURAL SURVEY

OF

SUSSEX.

CHAP. I.

GEOGRAPHICAL STATE AND CIRCUMSTANCES.

SECT. I.—SITUATION AND EXTENT.

SUSSEX is a maritime county, bounded on the west by Hampshire, on the north by Surrey, on the north-east and east by Kent, and on the south by the British Channel.

In contains, according to the mensuration in Templeman's Tables, 1416 square miles, and 1,140,000 acres: the extent, by the same authority, is 65 miles, and the breadth 26. But, according to this calculation, the real length is considerably under-rated, whilst the breadth is increased; which we find to be the case by later, and more accurate surveys. Another calculation reduces the number of acres to 908,952; both of which are confessedly erroneous.

The length of this county, as measured from Emsworth to Kent-ditch, extends 76 miles, and the medium breadth falls short of 20. The superficial contents

amount

amount to 933,360, and each parish averages 2982 acres*.

SECT. II.—DIVISIONS.

THE artificial divisions of the county are comprehended in six rapes. those of Chichester, Arundel, and Bramber, form the western division, and in which the quarter-sessions are held at Chichester. Midhurst, Petworth, and Horsham; Lewes, Pevensey, and Hastings rapes, form the eastern quarter of the county, for which the quarter-sessions are always held at Lewes. The number of parishes in the county are 313.

SECT. III.—CLIMATE.

THE climate in the western part of the maritime district is very warm, and highly favourable to the powers of vegetation†. But upon the bleak situation of

---

* In the account annexed to the Poor Returns, drawn up under the inspection of the Right Hon. George Rose, the number is 935,040.

† As Mr. Young justly remarks, the climate of the South Downs is warm, and in some respects favourable to vegetation. In the severest frosts we had in the year 1789, I exposed a thermometer at Willingdon-mill, one of the highest points on the hills, after sun set: it stood three degrees of Fahrenheit's scale lower than one in the village of Jevington, and three degrees higher than the thermometers were reported to have stood in London at the same time.

When I say the climate is in some respects favourable to vegetation, I mean, it hastens both the birth and maturity thereof, but no plants whatever attain that rank luxuriancy commonly to be observed in particular spots in most countries. This influence affects the animals as well as vegetables indigenous on the hills; and the hares and partridges are apparently smaller than those of some other parts of England.— *Rev. Mr. Sneyd.*

the

the South Down hills, exposed to the south-west, the winds have b en known to strip the thatch off corn-stacks, and the covering from all thatched buildings; and it has sometimes happened, that farmers have suf-fered considerable losses by the violence of these wes-terly gales in harvest, blowing the standing corn out of the ear, and doing other damage. When impreg-nated with saline particles*, occasioned by the west and south-west winds beating the spray against the beach, all the hedges and trees on the windward side are destroyed; and, generally speaking, the foliage wears the aspect of its wintery dress. The hedges seem to be cut by the spray, as if it were artificially; and in very exposed situations it penetrates the houses, though built with brick, even at a considerable dis-tance from the coast. The consequence of this has been, that the greatest part of the buildings in the dis-trict are situated in hollow protected situations, in

---

* This is so generally received an opinion, that it is perhaps presump-tion to contradict it; but I greatly doubt if the spray of the sea does the injury here ascribed to it. It must necessarily gain a considerable height above the level of the sea, to be carried far inland. Now, it is well known, sea-salts are not exhaled by the sun; and strong winds are ob-served to depress and bear to the ground all light bodies, such as smoke, steam, and the like. As to the spray produced by the sea, driven vio-lently by the south-west wind on the beach, it must needs mount per-pendiculasly about 150 feet before it could surmount the cliff; whereas an easterly or south-east wind, which makes a more broken sea, and consequently more spray, has no cliff to surmount between Beachy-head and Hastings; therefore would extend its influence farther and more powerfully: yet the foliage immediately exposed thereto is never injured thereby, though but at a short distance from the shore, whilst all the injury is done from the south-west, where, as I before remarked, we have the cliff, which seems to present an insurmountable barrier against those injuries we observe some how affected, and which perhaps is caused by the force of the wind *solely* obstructing by its agitation the sourse of those juices, which should nourish the leaves,—*Rev. Mr. Sneyd.*

order

order to shelter them from these distressing consequences.

———◆———

SECT. IV.——SOIL AND SURFACE.

THE investigation of the nature and properties of the
varieties of soil, in this or any other county, so as
accurately to chalk out the line where one soil ends
and another begins, can be thoroughly made only by
those who have a most exact and intimate knowledge
of the county. In attempting to give the Board this
information, it appeared that the variations would be
more clearly traced out, and more accurately defined,
by a map of the soil, than any other mode that could
be adopted ; sensible, however, at the same time, that
it will be but imperfect, and liable to errors which are
unavoidable.

The different soils of chalk, clay, sand, loam, and
gravel, are found in this county.

The first is nearly the universal soil of the South
Down hills* ; the second, in general, of the Weald † ;
                                                      the

———————————————

* This, strictly speaking, is not the case: the pure, native, untouched
soil of the Downs is chiefly a rich, light, hazel mould, whose immediate substratum is a loose chalk. These become mixt by the plough;
and the more frequently the earth is turned, the more predominant the
chalk becomes.

There is also a very considerable portion of the hills between Cuck
mare river and East Bourne, whose soil is a strong red loam. There is a
vein of this sort near four miles long, east and west, and full three-
fourths of a mile, north and south, running from the western extremity
of Excit-hills to Willingdon-mill. This soil is very deep, some feet even
on the tops of the hills : it is rather what is called cold land, but when
mended with chalk, becomes extremely productive.—*Rev. Mr. Sneyd.*

† The *Weald* is an indefinite expression for a country, the limits of
which are unknown. In a legal acceptation, it means the woodland
                                                                districts

the third principally occupies the north side of the county; the fourth is found on the south side of the hills; and the last lies between the rich loam of the coast and the chalk.

The soil of the South Downs varies according to its situation. On the summit is usually found (more especially in the eastern parts) a very fleet earth; the substratum chalk, and over that a surface of chalk rubble, covered with a light stratum of vegetable calcareous mould. Sometimes along the summit of the Downs there is merely a covering of flints, upon which the turf spontaneously grows. Advancing down the hills, the soil becomes of a deeper staple, and at the bottom is every where a surface of very good depth for ploughing. Here the loam is excellent, nine or ten inches in depth, and the chalk hardish and broken, and mixed with loam in the interstices, to the depth of some feet, which must make it admirable land for sainfoin.

West of the river Arun, the soil above the chalk is very gravelly, intermixed with large flints. Between the rivers Adur and Ouse, a substratum of reddish sand is discovered; the usual depth of the soil above the chalk, varies in almost every acre of land, from one inch to a foot. The general average between Eastbourne and Shoreham, does not exceed five inches. West of Shoreham the staple is deeper, and between Arundel and Hampshire the soil is deeper still*.

At

---

districts in the counties of Sussex, Kent, and Surrey, in which woodlands pay no tithe; but as a district relative to soil, it is extremely various, containing, besides the predominant clay, much sand, &c.

* It is the remark of a Nobleman in this county, that the surface of these hills being usually very steep to the north, the hard chalk, so favourable

At the northern extremity of these chalk hills, and usually extending the same length as the Downs, is a slip of very rich and stiff arable land, but of very inconsiderable breadth: it runs for some distance into the vale, before it meets the clay. The soil of this narrow slip is an excessively stiff calcareous loam on a clay bottom: it adheres so much to the share, and is so very difficult to plough, that it is not an unusual sight to observe ten or a dozen stout oxen, and sometimes more, at work upon it. It is a soil that must rank amongst the finest in this or any other county, being pure clay and calcareous earth: to the eye it appears whitish, from the mixture of chalk. Some of it that appears of a blacker nature, is less mixed with that substance: it is generally deep, and under it is a pure clay.

South of these hills is an extensive arable vale of singular fertility. This maritime district, extending from Brighthelmstone to Emsworth, 36 miles, is at first of a very trifling breadth, between Brighton and Shoreham. The nature of this soil, which is proba-

---

vourable for all the purposes of the farmer, is at hand to assist his industry in the cultivation of the strong retentive soil of the Weald, which lies at the northern extremity of these hills; whilst the surface to the south gradually and almost imperceptibly unites itself to the rich district on the coast. where the soft chalk, or chalk marl, is found equally propitious to the pursuits of the farmer, which shews (to make use of his Lordship's words) how beneficially Nature has distributed her gifts, in adapting to every s il a manure so suitable and near at hand.

Directly opposite to the South Down hills, to the north, are the Surrey hills, falling abruptly to the southward, and sloping gradually to the north; and between these two lines of hills is the Weald of Sussex and Surrey, where the Sussex marble (which is nothing else than a concretion of shells) is to be found. The position and formation of these opposite hills is such, that, in the opinion of his Lordship, they appear as if torn asunder by some violent commotion of Nature.

bly

bly equal to any in the kingdom, is a rich loam, either upon a reddish brick earth, or gravel; the general depth of the upper soil varying from ten to sixteen inches. Proceeding westward, gravel is generally found under the surface. This maritime district is in parts stiff, but more usually light, intermixed with sand, and beneath which is sand. Between Brighton and Shoreham, the general breadth of this uncommonly rich vale falls short of one mile; between the rivers Adur and Arun it is increased to three miles, and from the Arun to the borders of Hampshire, it becomes still wider; from three to seven miles. In the south-west angle the land is stiffer and more retentive, and in Selsea peninsula, more argillaceous; and the farmers here not having the same opportunities of marling as their brethren on the eastern side of Pagham-harbour, the soil is not equal to it in fertility.

Between this maritime district and the South Downs runs a vein of land, not equal to the foregoing in richness, but admirable land for the turnip husbandry. It is provincially called *shravey**, stony or gravelly, the flints (where they have not been picked off the land) lying so thick, as effectually to cover the ground; and it is curious to observe how vegetation flourishes through such beds of stones. The general opinion is, that if the farmers were to put themselves to the trouble and expense of picking them off the land, the soil

---

* This term is applied by the natives of the South Downs more generally to those spots on the sides of steep hills, where the turf has slipped away and exposed the soil. These scars or holes are termed *shraves*. I am at a loss for the true derivation of the word, but think it probably comes from the Saxon Schƿamme, which signifies a scar, slash, or trench. —*Rev. Mr. Sneyd.*

The Earl of Egremont observes, that is a common provincial word for stony land, or any soil mixed with sandstone, &c.

would

would be most materially injured. Some, indeed, who
have tried this experiment, are thoroughly convinced
of the loss thereby sustained, the land having never
since produced such fine crops of corn as before; but
this remark applies only to some places where the stones
are so numerous.

In the line from Chichester to Emsworth, north of
the road, we meet with the same kind land for turnips
and barley. The declivity of Hanbrook-common is
wet and springy to the south, but on the north it is
dry and gravelly. This common is a light gravelly
or stony loam upon a gravel bottom: a brick earth,
18 inches in thickness, frequently intervenes between
the upper soil and the gravel. It has been for some
time in contemplation to apply to the Legislature to
enclose this common. Some of those who live in the
neighbourhood of it, would, if it were enclosed, freely
give 30*l.* per acre for the best of it; at present it is not
worth one shilling.

The soil of the Weald is generally a very stiff loam
upon a brick clay bottom, and that again upon sandstone. Upon the range of hills running through the
county in a north-west direction, the soil is different. It is here either sandy loam upon a sandy grit-
stone, or it is a very poor black vegetable sand on a
soft clay marl. A great proportion of these hills is
nothing better than the poorest barren sand. St. Leonard's Forest contains 10,000 acres of it, and Ashdown
18,000 more, besides many thousand acres more in
various other parts of the county.

The depth of the sand on those rabbit-warrens is various—full 12 inches in many places: the soft clay,
which in its outward appearance resembles marl, is
much deeper. In the neighbourhood of Handcross,
upon

upon St. Leonard's, this substratum is several feet in
depth, as may be seen on the declivity of a new road
lately made by Mr. Marcus Dixon. An extensive
tract of this unimproved sandy soil, stretching into
Kent on one side, and, with some intersection of cultivation, into Hampshire on the other, and calling loudly
for improvement, occupies chiefly the northern division of the county. I do not affirm that this unproductive soil is united from one end of the county to
the other, since it is broken into and intersected by
interventions of the clay district; but it is usually
to be met with running east and west at the north
side of the county. It is commonly understood to
form a part of the Weald, which in its utmost extent
comprehends all that district of Sussex at the foot of
the South Down hills, or within two or three miles of
them. In its more appropriate signification, it has reference to the deep and heavy clay loam district, being bounded to the west by the Arun.

Respecting the surface of this tract of land, the
sands produce the birch, hazel, beech, and some other
under-growth, of which some profit is annually made.

So predominant is the timber and wood of one sort
or another in the Weald, that when viewed from the
South Downs, or any eminence in the neighbourhood, it presents to the eye hardly any other prospect but a mass of wood. This is to be ascribed to
the great extent and quantity of wood; preserved by a
custom of a nature so extraordinary, that it is not a
little surprising no steps have been taken to put an end
to it.

When this country was first improved by clearing, it was a common practice to leave a *shaw* of
wood several yards in width, to encompass each
distinct

distinct enclosure, as a nursery for the timber, &c.
The size of these enclosures being small, must of course
contribute to render the general aspect of it woody.
Anterior to the Conquest, the Weald was a continued
forest, extending from the borders of Kent to the con-
fines of Hampshire, across the whole county of Sussex;
and the names of a variety of parishes situated in this
line, and evidently derived from Saxon original, attest
this fact to the present day. In truth, the forest now
remaining occupies a considerable portion of Sussex.

Besides the soils already treated of, there is a large
tract of marsh land adjacent to the sea-coast between
the eastern extremity of the South Downs and Kent.
The soil is a composition of rotten vegetables, intermixed with sand and other matter, collected from the
floods and filth which settle on the surface. In Lewes
Level this vegetable mould is at least twelve inches in
thickness.

In Pevensey Level it is many feet deep, and under
it a very heavy black silt, intermixed with various
sorts of shells. Water-logs, stumps of trees, and
timber, have been dug from Pevensey Level; trees,
each containing one load, cubic measure, have been
taken from Lewes marshes.

SECT. V.—MINERALS.

RESPECTING the minerals of Sussex, it is not
inferior to many in the production of this most valuable material. Limestones of every description are
to be met with in the most eastern parts of the Weald.
The Sussex marble, when cut into slabs for orna-
menting chimney-pieces, &c. is equal to most in
beauty

beauty and quality, when highly polished. The
Earl of Egremont has several chimney-pieces at Petworth, formed of it. It is an excellent stone for
square building, and for paving is not to be exceeded. It affords a very valuable manure, equal,
and by some thought to be superior, to chalk, and
cheaper to those who live near the place where it is dug.
It is found in the highest perfection upon an estate of
the Earl of Egremont's, at Kirdford, from 10 to 20
feet under ground, where it is in flakes nine or ten
inches in thickness. Much of it was used in the
Cathedral at Canterbury, the pillars, monuments,
vaults, pavement, &c. of that venerable structure,
being built of this article, called there the *Petworth marble.* The Archbishop's chair is an entire
piece.

Besides the limestones of this district, I shall set
down a short account of what I had a more immediate
opportunity of seeing, by observing the gradations in
the earth, and mineral beds of ironstone and limestone,
to the depth of 120 feet, at Ashburnham-furnace.

The received opinion of the range of the limestone
in this neighbourhood is, that it runs eight miles from
east to west, and one from north to south. How far
this opinion of the limited continuation of limestone is
well founded, has not as yet been decided. The soil
tending immediately to sand, is of the hazel kind:
that tending to marl, connected either with iron or
limestone, is formed of a more tenacious and closer texture ; and every where the substrata bear a strict analogy to the surface. The limestone and ironstone generally rise very near the surface; often within three
feet: the depth to which the limestone continues, has
not as yet been discovered, having never in this country

try been drawn deeper than 120 feet, where it is firmer, and superior to that at any other depth.

The appearance of the ironstone more than 40 feet under the surface, is different; certainly not so good, being coarser, and seems more dull, and works heavier in the furnace; and the very best of the veins are frequently intersected with stripes, the thickness of a quill, filled with a soft marley matter; and the marl-beds which the iron lies in, wear a bluer appearance than where it is good; but the beds of limestone have no such resemblance at any depth. It is a curious fact, and worthy the attention of men conversant in matters of this sort, to account for the difference, which perhaps may not be very difficult, upon fully considering the component parts of each substance. The fact certainly is, that ironstone diminishes in goodness from depth, and limestone does not; neither the grey, which is composed of shells, and the exuvia of marine animals; nor the blue, which is a perfectly indurated calcareous ma 1. As it is now sufficiently proved that there are under-stones, that, with clearing and burning, will make equally as good lime as the top-bed, or *great blue* (as it is provincially called), from which one stratum is at the distance of 21 feet; so that instead of two to two feet and a half of blue stone generally drawn and used, there is now produced, without spoiling any more surface, upwards of seven feet. This fact shews that the perseverance of the Earl of Ashburnham, in drawing the deep under-stones at his works, and thereby setting an example which other limestone-drawers are now following, has been truly useful: for that part of Sussex must have ceased to avail itself of the advantage of lime as a manure without some change of this sort.

The

The alternate order of sandstone and ironstone is every where found through the Weald in all directions. The sandstone, marl, and ironstone, all dip into the hill.

Under this, at a considerable depth, the various sorts of limestone are discovered in the order in which they are set down, with the thickness and shale of each different sort.

	Thickness.			Shale.		
	Ft.	In.		Ft.	In.	
The first limestone,	3	3	....	8	0	grey.
second ditto,	0	9	....	9	0	ditto.
third ditto,	4	0	....	39	0	ditto.
fourth ditto,	0	8	....	3	0	ditto.
fifth ditto,	0	8	....	2	0	ditto.
sixth ditto,	8	3	....	4	0	ditto.
seventh ditto,	2	0	....	1	6	blue.
eighth ditto,	0	6	....	0	4	ditto.
ninth ditto,	0	9	....	1	3	ditto.
tenth ditto,	1	2	....	0	4	ditto.
eleventh ditto,	0	8	....	1	1	ditto.
twelfth ditto,	1	1	....	1	6	ditto.
thirteenth ditto,	0	6	....	8	0	ditto.
fourteenth ditto,	2	3	....	......		

The great blue by far the best.

This last stone is fine enough to set a razor.

This is the succession in which they are found.

The Sussex limestone, upon trial, has been discovered to be superior both to the Maidstone and Plymouth stone, and it is now supposed that for cement, none equal to it is found in the kingdom.

II. Iron.

II. Ironstone.

This mineral abounds in an eminent degree in Sussex; and it is to the ferruginous mixture with which the soil of this county is in many places so highly impregnated, that is to be ascribed the sterility of so large a portion of it.

At Penhurst, in the neighbourhood of Battel, the soil is gravelly to an indeterminate depth. At the bottom of the Earl of Ashburnham's park, sandstone is found, solid enough for the purpose of masonry. Advancing up the hill, the sand-rock is 21 feet in thickness, but so friable, as easily to be reduced to powder. On this immediately a marl sets on, in the different depths of which the ironstone regularly comes on in all the various sorts, as follows :

1. Small balls, provincially, *twelve-foots*, because so many feet distant from the first to the last bed.
2. Grey limestone ; what is used as a flux.
3. Foxes.
4. Rigget.
5. Balls.
6. Cabalia balls.
7. White-burn.
8. Clouts.
9. Pitty.

This is the order in which the different ores are discovered. Advancing on, I crossed a valley where the mineral bed seems entirely broken, and the sandstone sets on. At the distance of something above a mile, the ironstone is again seen. Another intervention of sand, and then, at low water, when the tide goes out, the

the beds of ironstone appear regularly on the shore : an indisputable proof that, however the appearance of the surface may vary, the substrata continue the same.

In taking the range northwardly from the bottom of Ashburnham-park, for twelve miles at least, the strata are nearly the same, there being no material inequality of surface which does not partake of sandstone, marl, ironstone, and sand again at the top. Sand being the general cap to the hills, the cultivated soil of these districts is made up so largely of it; even the loamy and marly soils, after rain, very evidently discover it in small glittering particles, which, in process of time, have been washed from their native beds.

III. Chalk—Marl—Fullers'-Earth.

Beside the minerals above-mentioned, a vast range of hills, the composition of which is *chalk*, occupy a considerable part of the county, adjoining the coast. *Marl* is dug up on the south side of these hills, in various places. *Fullers'-earth* is found at Tillington, and consumed in the neighbouring fulling-mills ; and red-ochre at Graffham, and in various places adjoining the sea, as Chidham, &c. much of which goes to London.

SECT. VI.—RIVERS.

THE chief rivers are, the Ouse, the Adur, and the Arun ; they rise in the northern parts of the county, and after dividing the chalk-hills into four or five parts, empty themselves into the Channel ; the first at Newhaven, the second near Shoreham, and the third at at Little Hampton. Although comparatively small, they render the greatest benefit to the county at large, by furnishing points of connexion for the canals already finished, or in agitation. Assisted by the public-spirited and enterprising conduct of one or two Noblemen, Sussex, on the completion of those canals, will not be inferior to other counties in the advantages of inland navigation ; but as this subject comes under the article *Canals*, I shall have occasion to speak of it more at large under that head.

CHAP.

17

CHAP. II.

STATE OF PROPERTY.

SECT. I.—ESTATES, AND THEIR MANAGEMENT.

IN so large, populous, and cultivated a county, estates must necessarily vary: the largest does not exceed 7500*l.* a year. In this, as in all other counties, gentlemen of property have stewards, or superintendants, to examine the state and condition of their lands. Most proprietors hold land in their own occupation; and the increasing attention observable in the better cultivation of this county, affords an agreeable spectacle, not only of rational amusement and satisfaction, but it is also eminently useful in a national light, insomuch that all the great improvements in our agriculture have been patronized, propagated, and encouraged by gentlemen of large landed property and scientific exertions.

In this class, it is impossible for the Author not to mention the Earl of Egremont. To do justice to the exertions of this distinguished nobleman, is far above the reach of my humble capacity. Suffice it to say, that his Lordship's estates are conducted upon a great scale, in the highest style of improvement. Every attention is here given to the suggestion of whatever hints have a probability of being turned to the use and advantage of his country. The Duke of Richmond has made great and beneficial exertions. The Earl of

SUSSEX.]   Chichester

Chichester and Lord Sheffield have practised with great success ; but the number of those gentlemen who have thus promoted the good of their country, is too great to repeat all their names. The following pages will contain details sufficient to establish the fact, that the land-owners of Sussex have not been behind the general spirit of this agricultural age.

CHAP. III.

BUILDINGS.

SECT. I.—HOUSES OF PROPRIETORS.

MANY of the noblemen's and gentlemen's seats are raised upon a splendid, no less than a rational, plan, and eminently contribute to the ornament and embellishment of the county. Without specifying each individual residence, it may be observed, that few districts have to boast more elegant structures.

SECT. II.—FARM-HOUSES AND OFFICES.

WHEREVER the quarries are conveniently situated, stone is the usual material for farm-buildings and offices, no less than for gentlemen's seats; and as an excellent building-stone is found under a very considerable proportion of Sussex, it is a valuable circumstance to have materials for building of such a quality.

On the South Downs, and in the neighbourhood, another material, equally good, is made use of in the construction of houses, which are flints, and a better it is impossible to meet with: farm-houses, barns, stables, out-houses, and, in general, all the buildings

in

in this district, are formed of flint. Tile is much used as a facing for houses, especially in situations exposed to the inclemency of the west or south-west winds.

I do not know whether this tile-facing for houses is used beyond the limits of Sussex and Hampshire; but it is very prevalent in Sussex, and in open and exposed situations effectually checks the fury of the storms, and preserves the inside of the house air-tight and dry: they are very common all over the county.

Under this head, certain beneficial practices in the construction and arrangement of farm-buildings and offices, deserves to be particularly noticed, and cannot be too forcibly recommended to a more extended practice.

The pleasing manner which the farmers adopt throughout a great part of this county, and especially in the western division, of stacking their corn on circular stone piers, cannot be admired too much. It requires some art and attention in the construction of these stacks, and nice management to adjust them in their truest proportion. I take this to be the best method of preserving wheat; and it is no small recommendation, that it most effectually prevents all vermin from lodging in the sheaves, and hereby obviating incalculable losses to the owner.

In the fatting of oxen, it is not unusual to find excellent contrivances to save labour in attendance: stalls, or sheds of flint, requisite for the number of cattle, are frequently contrived (as at Mr. Thomas Ellman's, of Shoreham), with keelers in each stall for watering, with troughs of communication to convey the water from a pump in the farm-yard to the general trough at the outside of the building, which is again conveyed to each stall; so that all the trouble of

tieing

tieing and untieing, and driving to water, is avoided.
Each stall is sufficient to contain two oxen, five feet
room being allowed to each.

Sheep-yards, or standing folds, are very judiciously
constructed on the South Downs. Mr. Ellman has
one which contains an area of 50 yards by 20, which is
sufficient for 750 sheep, at the rate of one yard and a
half for each; so arranged as to contain sheds all
around nine or ten feet in width, and across the centre,
if the flock is numerous. A rack for hay is placed
against the wall which surrounds the whole, and another, a double one, ought to stand along the central
shed, for the sheep to feed from in each division of the
yard. These practices, which are in the economy of
a well-ordered farm, deserve universal imitation*.

---

* I shall take the liberty of observing to the Board of Agriculture
(and worthy it is of their consideration), that great improvements may
be made both to landlords and tenants, by placing barns, &c. conveniently on large farms. All buildings necessary to a farm, undoubtedly
ought to be placed in the most eligible spots, on large Down farms, for
the tenant to cultivate them to advantage. The inconvenience and
extra expenses it occasions to the farmer, when all the farm-yard buildings (which too often is the case) surround the house, is incredible.
It behoves then every landlord who is possessed of any of those inconvenient farms, to have them inspected (not only for his benefit do I wish
it, but for a public good), and order buildings to be removed, or new
ones erected, as convenient as possible for the farmer's use.

The most advantageous way to lay barn-floors, to prevent rats and
mice from undermining them, is to get flints or hard stones; break them
fine, in the same manner as they do on turnpikes, and lay them twenty
inches thick; consolidate them with a heavy rammer; and at each side
build a foundation-wall with grey lime, to lay the ends of the planks
even. This method, if done well, preserves the timber, and is a preventative against vermin.—*Anon.*

SECT. III.——COTTAGES.

THE miserable construction of cottages in many
parts of the kingdom, and the too great exclusion of
comfort, are circumstances which ought to be remedied.   No signs of prosperity like new-built cottages :
the dwellings of the poor are, in most counties, but
mud-cabbins, with holes that expose the inhabitants
to the rigour of the climate.   In the Weald of Sussex
they are in general warm and comfortable, and many
of them built of stone; and on the Downs with flints.
Certainly the lower class of people are here in much
more eligible circumstances, than in many other parts
of England which might be named.

CHAP. IV.

MODE OF OCCUPATION.

—————

SECT. I.——SIZE OF FARMS.

THIS most important division of the rural economy of the county is exceedingly variable. It is
usually governed by the soil. Farms here, as elsewhere, are to be found more extensive, and the management in general highly superior on dry soils,
to what is usually the case on wet ones. This is
precisely the fact with respect to the county now
under consideration. Compare the Weald with the
South Downs, and this circumstance will be sufficiently manifest.

In the Weald, although farms sometimes rise to
200*l*. a year and upwards, yet of this magnitude they
are not often to be met with ; and in a general inquiry,
a far greater number fall very considerably below this
calculation, insomuch that the average size in this
district is under 100*l*. a year. On the South Downs
they rise much higher. Many farmers occupy the
greatest part, if not the whole of their respective parishes, as in Buttolphs, Kingston, Coombs, Bramber, NorthStoke, Bletchington, Falmer, Piddinghoe,
and many others in the neighbourhood of Lewes,
East Bourne, and Brighton. Many of these have
marsh-land annexed to their farms, for the convenience of maintaining and fattening their oxen, the
work

work for the most part depending upon their labour.
A farm of 1200 acres at East Bourne has 200 acres
of marsh ; another of 1260 has 300. Farms in this
district average 350*l.* per annum. In the triangle
formed by Shoreham, Lewes, and East Bourne,
they rise much higher, and on the western side of
the Downs they fall lower. In the maritime district they vary from 70*l.* to 150*l.* Three farms
out of five are under 100*l.* rent. In the peninsula of
Selsea, rented at 1800*l.* and containing more than
2000 acres, farms vary from 50*l.* to 400*l.* Upon the
large gravelly soil situated between the maritime district
and the South Downs, they average at 200*l.* In the
hundred of West Bourne, they are met with sometimes
unusually small. The hamlet of Prinstead contains
nine farms, each not exceeding 50*l.* per annum.
And within a circuit of five or six miles round West
Bourne, they fall short of 100*l.* per annum. Between
Nutbourn-turnpike and Emsworth there are 1500
acres divided into 14, on which 50 horses are kept.
If that tract of land was in three, instead of fourteen, the rent might be 1200*l.* instead of 1000*l.* ;
there would be 500*l.* worth more of cattle and sheep
kept there than at present ; 500*l.* per annum more
corn raised ; and 36 horses kept instead of 50, with
much more employ for labourers. This is an exact
representation of many other small-farm districts, as
well as this.

The proper size of a farm is a point upon which
a variety of opinions have been entertained : some as-
serting that farms should be limited by law, and over-
grown ones divided ; whilst others, on the contrary,
contend, that large farms only should be encouraged. As no doubt exists in my own mind as
                                                        to

to which the preference should be given (though absolute freedom is the only thing to be contended for),
I shall merely consider the arguments advanced by
the advocates who contend for the superiority of
small farms over large occupations. The arguments
on this side of the question are, that industry is re-
warded, merit encouraged, markets plentifully supplied, and population increased, by the little occupier.
All which appear more specious than solid. Respect
ing the encouragement of industry in a small farm,
by holding out a reward to those labourers who are sufficiently industrious and active in their occupation, to
be enabled to lay by their gains for investment in a farm,
the present situation of little farmers in many counties
has been sufficiently discouraging, to afford the smallest prospect of successful industry in that manner.
From the observation which I have made in the
county under consideration, and which holds out a
striking instance of the comparative superiority of
great over small farms in every point of view, I hold
the active and industrious labourer to be more easy
in his circumstances, and the domestic economy of
his family far better arranged for promoting his hap-
piness, than he could possibly expect in the other situation, to which his ambition might possibly prompt
him. No class of men, such as the labourer converted
into a farmer, work more intensely, and none fare so
hardly. Surely, therefore, at such a crisis as the
present*, when, from the high and increasing price in
all the necessary articles of living, and the still more
formidable increase of parochial assessments, which
fall with such distress upon the small occupier, it

---

* Written at a time of scarcity.

must

References.

A *Three divisions of the Floor laid down on the*
   Timbers
B *The Timbers shewn that support the Floor*
C *The place to heap the Corn in time of Thrashing*

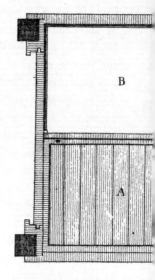

J. Upton del.^t Petworth 1795.

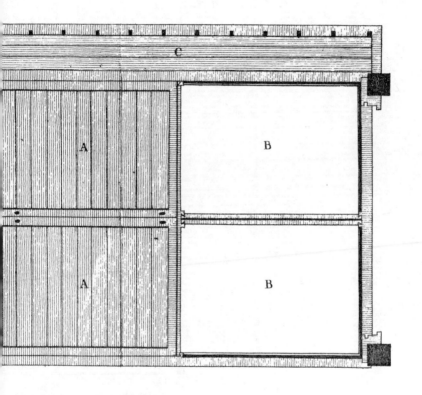

A

A

B

B

C

PLAN of the FLOOR.

S.J. Neele sculp.

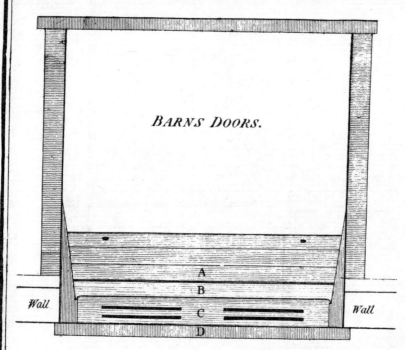

BARNS DOORS.

Wall *Wall*

A
B
C
D

References.

A _ *Two divisions of the Floor turned up, in one of which is described by dotted lines, a Rack to feed Cattle, fix't to the three moveable divisions on that side, and let down at pleasure.*

B _ *The back where the Corn is heap't against*

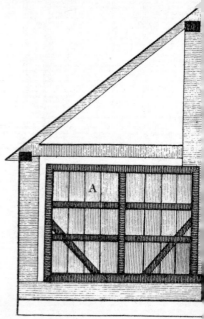

A

J. Upton del.t Petworth 1795.

SECTION *of the inside of the* BARN.

ound and floor

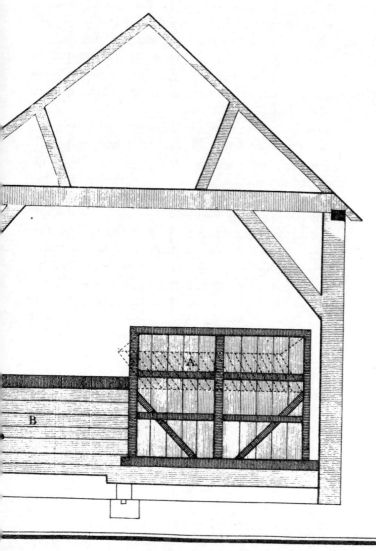

A

B

must appear little short of absolute ruin, to encourage laborious industry, by holding out the superior advantages of small farms. Markets may be, and perhaps are, more plentifully provided with a few articles; and so far some of the conveniences of life may be afforded to sale at a cheaper rate to those whose easy circumstances, or affluence, it is of little consequence to encourage. In the present inquiry, it is the laborious classes of life that are supposed to be chiefly affected in the markets, and to the benefit of which these farms are supposed so highly to contribute. But a great proportion of the commodities of markets, as butter and eggs, pigs and poultry, cannot be said to enter into the composition of a labourer's diet. But the great hinge upon which this system revolves, is the increased population to which it is thought to give birth. The union of small properties, it is said, has a tendency to depopulate: but within the last 20 or 30 years this evil of engrossing land has increased; consequently we ought to expect that population would be checked in proportion as this evil has increased; yet every one, I presume, is by this time convinced, and concurs in acknowledging, that the population of England, within the last forty years, has increased rapidly. The fact is, that in proportion to the paucity of families occupying, will be good management, and the greater the surplus of free hands for employment in trade and manufacture. But small farms, so far from being favourable to population, are directly the reverse; for the greater number of horses that are required for the cultivation of a little farm, decides the question. And since, upon the same ground that a certain proportion of horses are maintained, an equal number of men might subsist;

subsist ; it follows, that large farms are more favourable to population.

But, without doubt, the wisest measure to be em-
braced, is to leave the size of farms to find their own
level, unshackled by laws, unlimited in extent, for
capitals of all sorts to find employment.

SECT. II.—RENT.

RENT, of course, varies in proportion to the quality of the land. In the Weald it averages at 9s. per
acre (but in a great part of the Weald, is from 12s.
to 20s. per acre), excepting the north and north-
western parts, comprehending a considerable portion of poor, and frequently wet sandy land, which is
lett at 7s. and 8s. per acre ; whilst good loamy clay
on the eastern side rises to 15s. At the foot of the
South Downs, not included in this district, there is a
slip of excellent arable, which, taken by itself, is
rented from 20 to 25s. per acre. But this is generally
included with Down farms. A great quantity of waste
land, not less than 100,000* acres, in this part of
Sussex, is lett from 1s. to 1s. 6d. Of this, St. Leonard's and Ashdown forest contain at least 30,000
acres. With respect to the rental of the South Downs,
we find that farms are occupied at a rate much lower
on these hills than on the cold wet soils on the Weald,
when the nature of the soil, situation, &c. are considered. Some farmers on the Downs rent their
farms at a valuation under what the same lands would

---

* This might probably be made worth five times as much.—*Anon.*

yield

yield in many other parts of England. This practice deserves consideration, as low rents do not always generate exertion and activity*.

The native down, or sheep-walk, is rented at various prices—from 1s. to 8s. 6d. A very large tract of the hills between Newhaven and Shoreham, averages at 5s. 9d. and the arable at 11s.; very rich at 20s. Between Lewes and East Bourne, the Down is 2s. 6d.; arable, 10s. 6d. Between East Bourne and Shoreham, 4s. 1d. On the light gravelly soils, the rent is 12s. 6d.; where the quality is better, as in

Prinstead manor, ...	24s.
Chidham ditto, ...	20
Horney Bickley, ...	20
Bosham, ...	20
West Bourne, ...	20
Funtingdon, ...	20
East Mardin, ...	10
Racton, ...	16
Stoughton, ...	12†

In the maritime district, rents rise from 20s. to 30s.

---

* I am glad to find this idea in the minds of so many of the Reporters.—*William Dann, Gillingham.*

† The average price per acre on these parishes, I presume; the calculation to be made on statute acres. The reason of my observing this is, because, to my knowledge, the acres are very unequal as to the number of rods they contain. In the parishes of Prinstead and Bosham, the size of acres are from 107 to 212 rods; and in Chidham, Funtingdon, West Bourne, and many other adjoining parishes, 107 rods to the acre. In the parishes (many of them) between Arundel and Chichester, there is no regular measure, for in some farms the acres are from 110 to 120, or 130 rods, to the acre. I believe, if the size of all acres were made statute measure, it would be more satisfactory to the tenant, and of no injury to the landlord.—*Anon.*

Statute measure.—*A.Y.*

This

This land is almost entirely arable; and property is much divided, almost all the farmers enjoying land of which they are the proprietors.

Excluding the rents of pasture in the vicinity of towns, with all grass-land which enjoys any local or particular advantages, grass in the Weald averages at 13 or 14s. per acre, but it is seldom lett by itself. On the western side of the county, where the admirable practice of irrigating is understood and practised, meadow rents as high as 40s. per acre; in East Lavant, at 25s.; in West Bourne, 35s.; in West Hampnet, meadow, which before watering rented at 5s. per acre, is now lett at 40s. and has been valued as high as 60s.

The river Lavant, from the spring-head at East Dean to Chichester, irrigates between four and five hundred acres.

A large tract of marsh-land adjoining the coast, varies from 20s. to 40s. per acre. Some small parcels rise as high as 50s. and even 60s. Pevensey-level averages at 30s.; Winchelsea, 25s.; Brede, 35s.; Pett, 25s.; Lewes and Lawton, the same; Beeding, 30s.; Arundel rape, 25s. The rental of the parishes of Pevensey and Westham, amounts to 7510l. almost entirely grazing land. Pevensey parish contains only four arable acres; about two-thirds of it is occupied by the parishioners, and the other third by graziers living at a distance.

I shall conclude this account of the rent of land by the following statement of the rent, produce, and division of the land.

Down-

	Acres.			£.				
Down land,	68,000	at	7s.*	is	23,800	at	3 rents	71,400
Rich arable,	100,000	—	20s.	—	100,000	—	5 —	500,000
Marsh,	30,000	—	25s.	—	37,500	—	2½ —	75,000
Waste,	110,000	—	1s. 6d.	—	8250	—	1½ —	12,375
Arable and pasture in the Weald, }	425,000	—	12s.	—	255,000	—	3 —	765,000
Woods, &c.	170,000	—	8s.	—	68,000	—	2 —	136,000
	£. 903,000			£. 492,550				£.1,559,775

The remainder is composed of water, roads, build‑
ings, &c. ; so that the general rent is 492,500*l.* or 10*s.*
per acre, including all sorts of land; and the produce
1,559,775*l.*

SECT. III.—TITHES.

THE mode of collecting tithes is variable. In the
western parts of the county, the composition which
generally takes place, is at the average rate of 4*s.* 6*d.*
in the pound. The lay impropriators compound by
the acre. Wheat, 4*s.* 6*d.* ; barley, oats, and pease,
2*s.* 6*d.* ; pasture and meadow, 2*s.* per acre. These
tithes, on the whole, are allowed to be moderate and
very fair.

In other parts of Sussex, tithes are higher, and fall
with greater weight upon the occupier. About

---

* Down-land at 7*s.* is surely too high.—*Anon.*

When I speak of Down-land, it is to be understood that I take into
the estimate all the land on the Downs, arable as well as native down;
and when it is considered what a considerable portion of these hills is
under the plough, and that the pure down of itself rents in a variety
of places at from 4*s.* to 7*s.* and the arable from 10*s.* to 15*s.*, I think it
not far from the truth, in setting the average at 7*s.*—*A. Y.*

Cuckfield,

Cuckfield, wheat from 5s. to 6s.; barley, 2s. 6d. to 3s. In many places they are taken in kind, as Hailsham, &c.

In the level of Westham, tithe on grazing land is 2s.; upon full rents* of arable, 1s. 4d.†

In Battel, the composition for wheat is 4s. and Lent corn 2s. per acre. A very considerable part of the parish is tithe-free, being abbey-lands, the possession of Sir Godfrey Webster. By a return of the corn tithes of the above extensive parish, transmitted by Sir Godfrey to the President of the Board of Agriculture, some tolerable estimate may be formed, not only in the parish of Battel, but in other parts of the county, " of the comparative progress of improvement, and the additional benefits which result from moderate compositions."

Corn Tithes of Battel for Thirty-seven Years, 1758 to 1794.

		Acres.	Average.
1758 to 1764, ...	wheat,	1039	148¼
1771, ...	——	1260	180
1778, ...	——	1655	236½
1785, ...	——	1502	214¼
1792, ...	——	1583	226
1793, ...	——	....	249
1794, ...	——	....	238

* If this was general, there would be no complaint respecting the payment of tithes.—*William Dann.*

† Throughout the levels of Westham, both great and small tithes are due in kind. The great tithes are in lay hands, and taken in kind; the vicarial are mine, and lett to the landholders, not by their rents, but compounded for by the acre.—*Rev. Mr. Snryd.*

Lent

Lent Corn.

			Acres.		Average.
1758 to 1764,	....	wheat,	1682	....	240⅓
1771,	....	——	1915	....	273½
1778,	....	——	2132	....	304½
1785,	....	——	2018	....	288½
1792,	....	——	2257	....	322½
1793,	....	——	....	....	364
1794,	....	——	....	....	403

The mode, as at present adopted, of collecting tithes, although perhaps levied with as little hardship upon the occupier as the nature of the case admits, is, without any doubt, exposed to the strongest objections. These have of late been so much and so ably discussed, that a repetition of the complaints would be needless. Certainly tithes are a heavy deduction from the profit of farming, and an onus of no inconsiderable weight upon improvement. An arrangement of such a nature as to embrace equally the interest of the farmer and clergyman, is the object so much to be wished for.

SECT. IV.——RATES.

The rates for the maintenance of the poor in Sussex, collectively taken, are not comparatively so high as in other counties where manufactures prevail. But the increase of them in almost every district of the kingdom, is truly alarming, and operates as a most discouraging check to agricultural exertion, at a time when the comforts accruing to the poor are inversely

as

as the increase of rates. From an inspection of the rate-books in various parts of the county, it esta- blishes the fact, of a considerable increase having al- most invariably arisen. But this is to be understood as relating to those parishes where houses of indus- try have not been set up; since, where these have been established (although very recently founded), the contrary has followed. In eleven parishes united at Sutton, in the lower rape of Arundel, though the junction was formed as late as 1794, the rates have diminished. It is in some measure to be attributed to the good or bad management of those who are en- trusted with the superintendance of the poor, that much of the expenses may be said to be increased or diminished; and until gentlemen of liberal education and independent fortune, in their respective parishes attend more closely to the concerns of the poor, they may surely be said to connive at the evil. But this burthen, so alarming in its magnitude, and so dis- tressing in its consequences, lies deeper than this. The system of the poor-laws perhaps needs revision, before any radical remedy will succeed. It is a growing evil, which should be timely curbed by legis- lative interference. Temporary laws enacted upon the spur of the moment, for the purpose of ward- ing off present inconveniences, and removing the evil day out of sight, cannot fail of proving unsuc- cessful. By the multiplication of acts, difficulties are entailed, the whole system becomes complex, and the execution sometimes impracticable.

That the reader may in some measure be made ac- quainted with the progressive increase of the rates, I shall set down a few extracts, as specimens for the county at large.

SUSSEX.]   In

In Battel parish, containing between 1800 and 1900 people, and rapidly increasing, the rise of rates has been in proportion:

In 1769, the collection was .... £.656
1788, ... 1071
1789, ... 1113
1790, ... 927

At present they are considerably augmented, being 6s. in the pound.

In Selsea, the rates in 1786 were set at 4s. and produced 356l.; the next year, at 5s. 1d. 4s. 6d. 4s. 9d. 3s. 7d. 3s. 8d.; and in 1792, 3s. 3d. This diminution is entirely ascribed to the very excellent management of the overseers. In Petworth, the rates for 1791 were 3s. 2d. in the pound; the next year, 4s. 6d.; and in 1793 and 1794, 3s. 6d.; which, if we take into consideration the scarcity of all the articles of living, is certainly low; and this in a large and highly populous place. The conduct of this parish, in all that respects the economy of their poor, is excellent; and although they are contracted for at a regular stipend, yet the situation of the paupers is in every respect the reverse of that consequence which is so generally understood to flow from this method of farming the poor. No mismanagement results in this parish from such a conduct. The governor's salary is fixed by agreement. A sack manufactory, which promises success, has been lately established.

LAND TAX.

Sussex, land at 4s. in the pound.

Arundel, Upper Division, 25.

	£	s.	d.
Arundel,	201	2	0
Angmering,	296	1	0
Barnham,	127	16	0
Binstead,	83	14	0
Burpham,	91	11	0
Climping,	209	5	11
Eastergate,	85	11	6
Ferring,	118	11	4
Ford,	57	4	0
Goreing,	212	7	2
Little Hampton,	128	17	8
Kingston,	43	8	0
Leominster,	302	15	2
Madehurst,	76	10	0
Middleton,	50	0	4
Phelpham,	212	4	8
Poleing,	77	3	8
Preston, East,	63	2	0
Rushington,	138	2	0
Stoke, North,	48	4	6
Stoke, South,	91	18	4
Tortington,	103	9	0
Walberton,	130	10	0
Warningcamp,	55	5	4
Yapton,	218	0	0
Total, £.	3312	14	7

Arundel, Lower Division, 34.

	£.	s.	d.
Amberley,	150	0	8
Bignar,	87	6	4
Billingshurst,	427	16	4
Bury,	210	9	4
Burton,	59	18	6
Chiltington, West,	130	17	4
Coates.	29	13	0
Diddlesfold,	18	11	0
Duncton,	57	0	4
Egdean,	34	1	4
Fittleworth,	149	14	0
Greatham,	58	1	8
Hardham,	98	19	4
Houghton,	50	2	8
Kindford,	619	16	8
Lavington, Bar,	85	12	0
Lavington, Wod,	138	13	6
Lurgershall,	265	3	0
North Chappel,	175	1	8
Parham,	47	18	4
Petworth,	594	17	0
Pulborough,	468	7	8
Rudgwick,	373	17	8
Slinfold,	319	14	8
Stopham,	51	7	4
Storrington,	189	9	0
Sutton,			

	£.	s.	d.
Sutton,	108	7	8
Tillington,	323	6	4
Walthamcold,	71	6	4
Weggonholt,	69	8	0
Wisborough-green,	527	6	0
Total,	£.5992	12	0

Bramber, Upper Division, 31.

	£.	s.	d.
Alborne,	142	0	0
Ashington,	90	0	0
Ashurst,	164	10	0
Beeding, Upper,	299	6	0
Beeding, Lower,	142	4	0
Brambec,	50	0	0
Broadwater,	234	0	0
Buttolphs,	47	2	6
Chiltington, East,	119	18	8
Clapham,	87	2	0
Coombs,	53	0	0
Durrington,	56	12	0
Edburton,	60	0	0
Findon,	120	0	0
Heene,	51	10	0
Henfield,	446	13	4
Kingston-by-sea,	61	8	0
Lanceing,	158	0	0
Patching,	65	0	0
Old Shoreham,	89	0	0
New Shoreham,	130	0	0
Steyning,	361	6	8
Sompting,	156	0	0

	£.	s.	d.
Southweek,	72	16	0
Sullington,	116	0	0
Tarring,	180	0	0
Thakeham,	200	18	0
Washington,	164	13	4
Wiston,	176	4	0
Woodmancoat,	152	0	0
Worminghurst,	89	12	0
Total,	£.4336	16	6

Bramber, Lower Division, 11.

	£.	s.	d.
Cowfold,	295	14	0
Grinstead, West,	446	10	8
Hitchingfield,	119	10	2
Horsham parish,	701	17	11
———— borough,	185	13	3
Ifield,	302	0	2
Nuthurst,	184	1	0
Rusper,	186	8	0
Shermanbury,	153	19	0
Shipley,	497	9	5
Warnham,	353	4	3
Total,	£.3436	7	10

Chichester, Upper Division, 45.

	£.	s.	d.
Aldingbourne,	277	10	0
Appledram,	96	12	0
Binderton,	92	13	0

Birdham,

	£.	s.	d.
Birdham,	182	12	0
West Bourne,	363	4	3
South Bersted,	341	14	0
St. Bartholomew's,	63	12	6
Bosham,	332	8	0
Boxgrove,	235	19	4
Chichester,	643	19	8
Compton,	60	0	3
Chidham,	126	0	0
East Dean,	123	15	0
West Dean,	209	6	0
Donnington,	106	5	0
Earnly,	103	14	0
Eartham,	74	10	0
Fishbourne,	96	18	0
Funtington,	227	6	0
West Hampret,	152	0	0
Hunston,	102	13	0
Itchenor,	33	16	0
Mid Lavant,	89	10	9
E. & W Lavant,	148	4	0
North Maiden,	51	18	0
Upper Maiden,	151	16	3
East Maiden,	58	17	7
Merston,	96	2	0
N. Mundham,	203	8	0
Oving,	351	4	3
Pagham,	427	8	0
St. Pencrass,	53	4	0
Racton,	66	9	10
Rombaldweek,	89	2	9
Selsea,	255	4	0
Siddlesham,	338	8	0
Singleton,	179	9	2
Slindon,	127	14	0

	£.	s.	d.
West Stoke,	52	4	0
Stoughton,	176	8	8
Tangmer,	97	8	0
Thorney,	69	2	0
E. Whittering,	115	6	0
W. Whittering,	188	8	0
Upper Waltham,	39	16	0

Total, £.7429 4 8

Chichester, Lower Division, 24.

	£.	s.	d.
Bepton,	94	16	0
Budington,	30	13	5
Chithurst,	45	14	0
Cocking,	188	5	6
Didling,	54	1	0
Eastbourne,	335	0	0
Elsted,	108	14	0
Fernhurst,	190	16	0
Graffham,	84	10	0
Hasting,	545	10	0
Heyshott,	142	6	4
St. John's,	25	7	6
Iping,	94	2	0
Lodsworth,	152	18	4
Lynch,	67	18	9
Lynchmore,	104	0	0
Midhurst,	209	15	0
Rogate,	218	16	0
Selham,	73	5	3
Stedham,	154	7	0
Terwick,	41	18	0
Trayford,			

	£.	s.	d.
Trayford,	91	3	6
Trotton,	201	15	0
Woolbeding,	119	5	0
Total, £.	3374	17	7

Hastings Rape, 40.

	£.	s.	d.
Ashburnham,	233	8	11
Battel,	482	8	0
Bexhill,	378	9	9
Beckley,	276	4	0
Bodiam,	115	16	2
Breed,	336	2	0
Brightling,	226	9	10
Burwash,	528	5	8
Castle parish,	46	11	8
Catsfield,	113	9	0
Crowhurst,	116	9	1
Dallington,	143	1	1
Etchingham,	221	8	6
Ewhurst,	321	8	5
Fairlight,	93	3	7
Guestling,	169	5	6
East Guildford,	350	4	6
Heathfield,	322	9	6
Hollington,	113	13	0
Hove,	211	5	1
Hurstmonceux,	342	14	7
Icklesham,	278	15	11
Iden,	276	7	10
St. Leonard's,	42	15	5
Munfield,	203	12	3
Nenfield,	127	12	0

	£.	s.	d.
Northyam,	254	9	8
Ore,	107	5	4
Peasmarsh,	261	19	8
Penhurst,	76	19	2
Pett,	187	2	8
Playden,	80	9	8
Salehurst,	587	1	6
Seddlescomb,	147	11	9
Ticehurst,	492	3	3
Udimer,	215	11	0
Watlington,	91	5	7
Warbleton,	250	7	6
Westfield,	244	10	9
Wartling,	326	19	7
Total, £.	9395	8	4

Lewes, Upper Division, 38.

	£.	s.	d.
Bareomb,	366	9	2
Brighthelmstone,	263	8	6
Chailey,	251	8	10
Chillington,	141	9	8
Clayton,	181	6	0
Ditcherling,	308	0	8
Falmer,	159	3	4
Fulking,	115	12	6
Hamsey,	224	0	0
Hangleton,	110	16	0
Hove,	59	16	10
Hurstperpoint,	499	10	0
Iford,	116	1	10
Keymer,	236	5	0

Kingston,

	£.	s.	d.
Kingston,	106	17	0
Meeching,	81	0	10
Newick,	123	16	8
Newtimber,	108	5	0
Ovindean,	60	0	0
Patcham,	262	8	0
Pidinghoe,	93	19	2
Piccomb,	91	17	6
Plumpton,	162	3	6
Portslade,	124	10	4
Poynings,	130	12	8
Preston,	53	17	0
Rodmell,	130	17	8
Rottendean,	141	7	0
Street,	93	8	0
St. Ann's, P. M.	151	17	6
St. Michaels,	157	11	8
St. John's,	150	1	8
All Saints,	113	9	0
Southees,	61	1	6
Southover,	142	1	6
Telescomb,	43	2	8
Westmiston,	123	12	0
Vivilsfield,	212	18	0

Total, £. 5954 4 2

Lewes, Lower Division, 9.

	£.	s.	d.
Ardingly,	201	8	3
Balcomb,	187	5	6
Bolney,	215	0	0

	£.	s.	d.
Crawley,	51	14	1
Cuckfield,	755	1	0
Hoathley, West,	284	16	8
Slaugham,	170	19	2
Twineham,	173	18	0
Worth,	480	7	6

Total, £. 2520 10 2

Pevensey, Upper Division, 32.

	£.	s.	d.
Alciston,	95	0	0
Allfriston,	106	8	0
Arlington,	402	8	0
Bedingham,	144	5	0
Berwick,	92	16	0
Bishopston,	71	16	0
Bletchington,	36	1	0
Bourne, East,	365	8	0
Chalvington,	52	8	0
Dean, East,	62	16	0
Dean, West,	122	4	0
Denton,	41	4	0
Firle, West,	292	0	0
Folkington,	74	4	0
Friston,	68	0	0
Glynd,	152	13	0
Hailsham,	243	4	0
Hellingly,	325	17	0
Heyton,	21	14	0
Jevington,	70	18	0
Laughton,	368	4	0
Littleington,	32	18	0

Lulling-

	£.	s.	d.
Lullington,	42	8	0
Ringmer,	425	16	0
Ripe,	158	4	0
Selmeston,	177	12	0
South Malling,	212	8	0
Stanmer,	62	4	0
St. Thomas,	113	17	0
Tarning Nevil,	50	14	0
Willingdon,	237	14	0
Wilmington,	117	12	0

Total, £.4840 15 0

Pevensey, Lower Division, 20.

	£.	s.	d.
Buckstead,	439	16	0
Chidingly,	231	4	0
Fletching,	417	9	0

	£.	s.	d.
Frant,	308	6	0
Frantfield,	374	0	0
Grinstead, East,	857	2	0
Hoathly, East,	153	18	0
Hartfield,	377	12	0
Horsted, Little,	123	10	0
Horsted Keyns,	179	11	8
Isfield,	115	14	8
Lamberhurst,	154	4	0
Linfield,	396	11	0
Maresfield,	210	4	0
Mayfield,	674	18	0
Rotherfield,	638	17	0
Uckfield,	155	8	0
Wadhurst,	600	4	0
Waldron,	254	4	8
Withyam,	306	8	0

Total, £.6969 2 0

Cinque Ports.

	£.	s.	d.
Rye, ...	473	18	0
Seaford, ...	141	0	0
Pevensey, ...	1088	10	0
Winchilsea, ...	405	0	0

Hastings, as under:

	£.	s.	d.
All Saints, ...	93	8	0
Castle Parish, ...	170	0	0
St. Clement's, ...	114	18	0
Hastings, total,	£.378	6	0

Charge,

Charge, ...£.60,050 4 10
Collector and clerks, ...   1125 18 9
Deserters, ... £.25
Militia, ... 227 } 347 0 0
Hemp, &c.   95

                           £.58,086 1 1

Rapes.	Parishes.
Arundel, Upper, ...	25
——— Lower, ...	31
Bramber, Upper, ...	31
——— Lower, ...	24
Chichester, Upper, ...	45
——— Lower, ...	24
Hastings, ...	40
Lewes, Upper, ...	38
——— Lower, ...	9
Pevensey, Upper, ...	32
——— Lower, ...	20

Cinque Ports.

Rye, ...	1
Seaford, ...	1
Pevensey, ...	1
Winchilsea, ...	1
Hastings, ...	3

                           326

Houses.

Houses.

	£.	s.	d.
Houses and windows, 1793 (commutation),	7782	16	1½
New inhabited,	10,365	6	2
Inhabited duty 1793,	1105	8	7
	£. 19,253	10	10½

Horses.

	£.	s.	d.
Horses,	1875	10	0
Additional,	411	15	0
	£. 2287	5	0

Carriages, Four Wheels.

	£.	s.	d.
Four wheels,	2478	0	0
Additional,	280	0	0
	£. 2758	0	0

Carriages, Two Wheels.

	£.	s.	d.
Two wheels,	1186	10	0
Carriages,	£. 3944	10	0
Ten per cent. assessed taxes,	£. 1681	13	6

SECT.

SECT. VI.—LEASES.

THE term of leases every where varies. They are granted for seven, fourteen, and twenty-one years. It sometimes happens that none are allowed, and the tenant depends upon the good faith and honour of his landlord. Leases are unquestionably the greatest possible encouragement to agricultural improvement, and when exertions are necessary, they are not to be effected without this security. Where they are granted, the covenants between the landlord and the tenant are, that the landlord shall find materials for all repairs, and different buildings, as posts, rails, gates, &c.; that the tenant within the distance of four or five miles, shall be at the expense of conveying those materials to his farm, and shall pay all costs of labour, except occasioned by fire, tempest, or extraordinary high winds. The landlord to be at the expense of materials in their rough state; and all other charges to be defrayed by the tenant. Where hops are cultivated, the covenants agreed upon are, that the tenant is to sow one crop of corn between the new and old crop of hops, when they are grubbed up; that one-third of his farm shall be under tillage, and two-thirds in meadow, pasture, and hops; that no grass shall be ploughed up, but for hops: and in old leases, that all manure arising from the farm, shall be g ven to the meadow and hop-ground.

All close fences, yards, stables, barns, and out-houses in general, to be repaired by the landlord.

In some parts, the covenants are, that no grass be ploughed up, under 10*l.* penalty per acre; that the farm shall be sown in four regular *laires*, or divisions,

sions, to prevent the ground from being too much exhausted ; and at the close of leases, that one *laire*
shall be left fallow for the succeeding tenant ; that
no coppice shall be cut under twelve years growth ;
that no trees shall be lopped : rough timber on the
stem, and in some cases brick and mortar, are allowed,
with materials in general ; but all workmanship to be
at the tenant's expense.

SECT. VII.——EXPENSE AND PROFIT.

To draw up any detail of the expenses and profit
of farming with accuracy and precision, such as may
be relied upon as a medium standard for the whole
county, is, I fear, a task so difficult of execution,
that it may be thought to border upon impossibility.
No farmer, for obvious reasons, will lay open to the
view of others a detail of his business; and observation
alone is absolutely insufficient, and never to be depended upon.   It must be founded on documents, and
collected from registered accounts.

In the clayey soils of Sussex, which embrace the
major part of the county, the expenses attendant on
cultivation are high.   According to the common system of husbandry here, fallow, wheat, oats, and clover, the expense and profit of an average acre, may
thus be estimated :

Expense.

	Expense.	Produce.	Profit.
	£. s. d.	£. s. d.	£. s. d.
1. Wheat,	7 7 0	— 8 8 0	— 1 1 0
2. Oats,	4 13 8	— 4 18 0	— 0 4 4
3. Clover,	1 17 9	— 2 11 7	— 0 13 8
	£. 13 18 5	— 15 17 7	— 1 19 0
		13 18 5	

£. 1 19 2; leaving a profit of ten per cent. on a capital of 5*l.* per acre on arable land. If pasture be added, the account will stand higher.

In the very fertile maritime district, the general profit very much indeed exceeds the above calculation.

Expenses and Profit, according to Mr. Woods, at Chidham.

1. *Wheat Expense.*

	£. s. d.
Ploughing,	0 8 0
Harrowing,	0 4 0
Seed and sowing,	1 0 0
Rent,	0 18 0
Tithe,	0 5 0
Rates,	0 3 0
Manure,	0 16 0
Harvesting,	0 10 0
Thrashing,	0 6 0
Total expense,	£. 4 10 0

Produce.

Produce.

	£.	s.	d.
3 qrs. at 44s.	6	12	0
Profit,	£.2	2	0

Straw not calculated, as that goes for dung; the 16s. is for labour.

2. Turnips Expense.

	£.	s.	d.
Ploughing, harrowing, seed, &c.	£.1	4	0
Hoeing once,	0	6	0
Rent,	0	18	0
Tithe and rates,	0	8	0
Total expense,	£.2	16	0

Produce.

	£.	s.	d.
200 fat sheep, at 4d. per week,	3	6	8
Profit,	£.0	10	8

3. Barley Expense.

	£.	s.	d.
Twice ploughing and harrowing,	0	18	0
Seed and sowing,	0	12	0
Rent,	0	18	0
Tithe and rates,	0	8	0
Harvesting,	0	5	0
Thrashing 5 qrs.	0	6	0
	£.3	7	0

Produce.

Produce.

	£.	s.	d.
5 qrs. at 24s.	6	12	0
Profit, ...	£.3	5	0

4. Clover Expense.

	£.	s.	d.
Seed, 12 lb. red, 12 lb. white, ...	0	12	0
Rent, ...	0	18	0
Tithe and rates, ...	0	8	0
Expense, ...	£.1	18	0

Produce.

	£.	s.	d.
Three ton, at 1l. 10s. per ton, ...	4	10	0
Mowing and making, equal to the second crop, ...	1	10	0
	3	8	0
Profit, ...	£.1	2	0

	£.	s.	d.
5. Wheat Expense, ...	4	10	0
Produce—3 qrs. at 44s.	6	12	0
Profit, ...	£.2	2	0
6. Turnips Expense, ...	2	16	0
Produce—200 fat sheep, at 4d.	3	6	8
Profit,	£.0	10	8

7. Pota-

7. *Potatoes Expense.*

	£.	s.	d.
Twice ploughing,	0	12	0
Harrowing and rolling,	0	2	0
Seed, planting, cutting,	2	4	0
Earthing up,	0	2	0
Rent,	0	18	0
Tithe and rates,	0	8	0
Expense,	4	6	0
Produce—80 sacks, at 3s.	12	0	0
Profit,	£.7	14	0

8. *Wheat.*

Expense,	4	10	0
Produce—3 qrs. at 44s.	6	12	0
Profit,	£.2	2	0

9. *Barley.*

Expense,	3	13	0
Produce, 5¼ qrs. at 24s.	6	12	0
Profit,	£.2	19	0

10. *Clover.*

Expense,	3	8	0
Produce,	4	10	0
Profit,	£.1	2	0

Reca·

Recapitulation.

	Expenses.				Produce		
	£.	s.	d.		£.	s.	d.
1. Wheat,	4	10	0	....	6	12	0
2. Turnips,	2	16	0	....	3	6	8
3. Barley,	3	7	0	....	6	12	0
4. Clover,	3	8	0	....	4	10	0
5. Wheat,	4	10	0	....	6	12	0
6. Turnips,	2	16	0	....	3	6	8
7. Potatoes,	4	6	0	....	12	0	0
8. Wheat,	4	10	0	....	6	12	0
9. Barley,	3	13	0	....	6	12	0
10. Clover,	3	18	0	....	4	10	0
	£.37	14	0		£.60	13	4
					37	14	0
					10)22	19	4

Average profit each year, ... £.2 5 11

—————

Calculation of the Expense and Profit of Farming, in the common System of Husbandry, in the strong Lands of the Weald, viz.—1. Fallow; 2. Wheat; 3. Oats; 4. Clover.

1. *Wheat Expense.*

	£.	s.	d.
A fallow, with the crop of wheat upon the land, takes up near two years; say one year and a half, which, at 14s. per acre rent, ...	1	1	0

Carry forward, ... £.1 1 0

Poor,

	£.	s.	d.
Brought forward, ...	1	1	0
Poor, church, and hundred tax, at 5s. in the pound, ...	0	7	6
Road-tax, with labour, at 9d. in the pound, ...	0	1	1½
Tithe, ...	0	4	0
First fallowing per acre, eight oxen; three-fourths of an acre daily, ...	0	10	0
Two stirrings, with six oxen, a man, and a boy, whose wages are 2s. 6d.; the oxen to plough one acre per day; the labour of which set at 1s. 6d. a pair, each day, that is, 4s. 6d. added to the men's work, makes the two stirrings	0	14	0
Lime, 100 bushels, at 6d. per bushel,	2	10	0
Carriage and spreading, ...	1	0	0
Incidental harrowing, and rolling the fallow to occasion a season, set one year with another, at per acre, ...	0	1	3
Three bushels of seed per acre, at 6s.	0	18	0
Sowing, harrowing, clodding water-furrows, crow-keeping, ...	0	3	6
Weeding, ...	0	1	6

Carriage of the wheat, seven men and
   boys, three waggons, twelve oxen,
   four horses, two carters, at 1s. 8d.;
   two boys, at 9d.; the rest at 1s. 6d.

Men and boys, ... £.0	9	4
Twelve oxen, at 9d. 0	9	0
Four horses, at 1s. 0	4	0
Use of waggons, &c. 0	1	0

£.1 3 4 per diem; or, _____

Carry forward, ... £.7 11 10½

One

	£.	s.	d.
Brought forward, ...	7	11	10½
One acre with another, ...	0	3	0
Thrashing 3 qrs. at 2s. 9d.	0	8	3
Winnowing, &c. at 6d.	0	1	6
Turnpikes, 1s. 2d. per quarter, ...	0	3	7½
Expenses, ... £.	8	8	3
Produce, 3 qrs. at 46s.	6	18	0
Straw, stubble, chaff, &c.	1	10	0
Produce, ...,.. £.	8	8	0

Many farmers look upon the wheat as a losing crop. It appears very clear, that a crop of wheat three times ploughed and manured with lime, as is usual in this county, will not more, if so much as pay, the expense of raising it : that all the profit arising, must be from the oats and clover in the two succeeding years.

2. Oats.

	£.	s.	d.
Winter ploughing, &c.	0	10	0
Seed, 7 bushels, at 3s.	1	1	0
Sowing, harrowing with four horses, and cross-harrowing again repeatedly,	0	4	0
Weeding, ...	0	1	9
Mowing, ...	0	1	9
Shoving and turning, ...	0	1	3
Carrying and unloading four acres per day, half a mile, with one waggon,	0	3	0
Thrashing, &c.	0	4	8
Carriage to market, 1s. per quarter, three or four turnpikes, ...	0	4	0
Carry forward, ... £.	2	11	5
		Rent,	

	£.	s.	d.
Brought forward, ...	2	11	5

	£.	s.	d.		£.	s.	d.
Rent, ...	0	14	0				
Rates, ...	0	2	9				
Tithe, ...	0	2	0				
Interest of capital, ...	0	3	0	....	1	2	3
Wear and tear, ...	0	0	6				
£.	1	2	3				

Expenses, ... £.3 13 8

Produce.

		£	s	d
Four quarters, at 21s.		4	4	0
Straw, &c.		0	14	0
Produce, ...		4	18	0
Profit by oats, ...	£.	1	4	4

3. Clover.

		£	s	d
Seed and sowing, ...		0	8	3
Tithe, ...		0	1	0
Rates, ...		0	3	9
Rent, ...		0	14	0
Mowing first crop 20d. making 1s.				
Carriage, 2s.; wear and tear, ...		0	5	9
Second ditto, ...		0	5	0
Expense, ...	£.	1	17	9

	£	s	d
Produce—First crop, two loads, at 25s. }			
Second do. one do. at do. }	3	15	0
Profit by clover, £.	1	17	3

Reca-

Recapitulation.

	Expense. £. s. d.	Produce. £. s. d.	Profit. £. s. d.
1. Wheat, ...	8 8 0	8 8 0	0 0 0
2. Oats, ...	3 13 8	4 18 0	1 4 4
3. Clover, ...	1 17 9	3 15 0	1 17 3
	£.13 19 5	£.17 1 1	£.3 1 7

4)3 1 7

On 5*l.* capital, ... £.0 15 0 per c*t.* profit.

Grass Land.

Expenses.

	£. s. d.
Rent and taxes, ...	1 10 0

Labour—Mowing, ...	2*s.* 6*d.*
Making, ...	2 0
Carrying, ...	2 0
Wear and tear,	0 6
Tithe, ...	0 8
	7*s.* 8*d.*

	0 7 8
Expense, ... £.	1 17 8

	£. s. d.
From Jan. to May, feeding the rouen, 6 weeks,	0 10 0
1½ ton of hay, at 2*s.* per cwt.	3 0 0
Rouen, ...	0 10 0
Produce, ...	4 0 0
Expense, ...	1 17 8
Profit per acre, ... £.	2 2 4

Calculation at the rate of stocking with sheep.

1*st,* Thir-

		£.	s.	d.
1st, Thirteen weeks, three sheep, at 4d. each,		0	13	0
2d, Ditto, ditto, six ditto, at 4d. each,		1	6	0
3d, Ditto, ditto, six ditto, at 4d. each,		1	6	0
4th, Ditto, ditto, three ditto, at 4d. each,		0	13	0
Wool, 3lb. per fleece, from six sheep, at 1s. 3d. per lb.		1	2	8

Produce,		5	0	8
Expenses as above, to which add losses, &c.		2	2	0
Profit,	£.	2	18	8

CHAP.

CHAP. V.

IMPLEMENTS.

IN all the operations of husbandry, how essentially necessary it is to the ultimate success of the
undertaker, that his implements of labour be constructed upon mechanical principles. In the conduct
of operations of so much consequence, and so depending
upon the active knowledge and enlightened minds of
individuals, it is indeed surprising that we find so little
progress made in this branch of rural economy, and
that we so frequently see such a display of ignorance
in agricultural tools. A knowledge of mechanics is so
essentially necessary, that every farmer should be acquainted with the principles upon which the practice
of his profession is supported. So great have been
the improvements brought forward in almost every
other branch of the farmers' art, that it is unaccountable to observe the clumsiness of the ploughs in general. The wheel-plough most common, is the Kentish turn-wrest. It breaks up land from five to seven
inches deep, perhaps better in some instances than
the ploughs of Suffolk and Essex, especially when the
ground is dry and hard; it will then work steadily at
a time when the best ploughman is unable to keep
the other in the earth. There is an advantage which
arises from its use for spring crops on the Downs sown
upon a single earth, for it turns the furrow perfectly,

yet

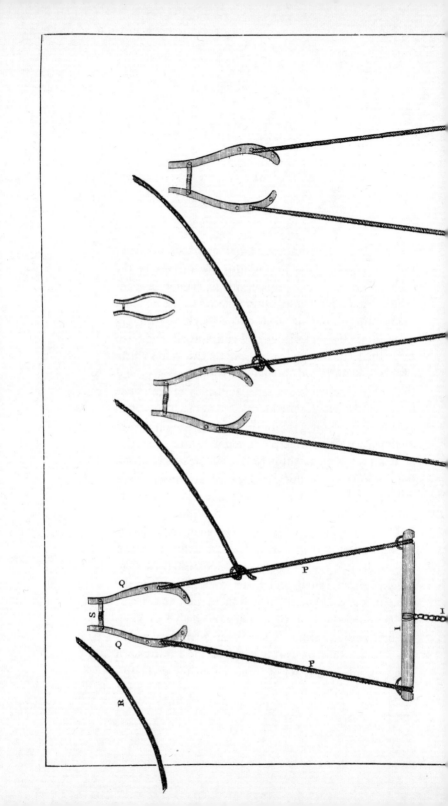

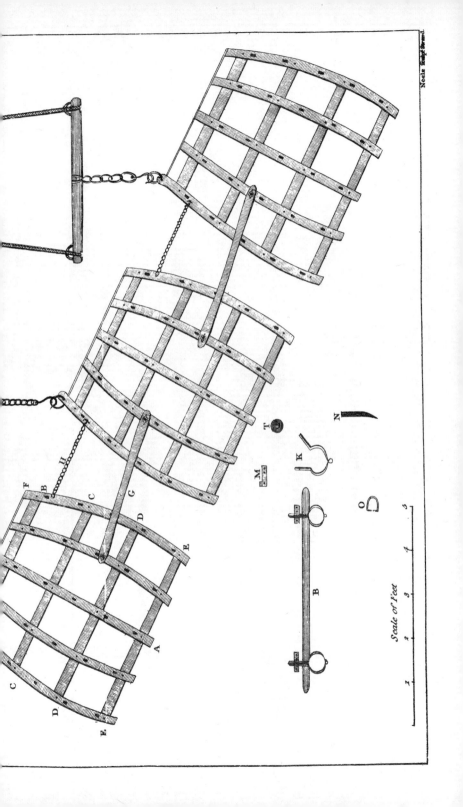

Scale of Feet

1   2   3   4   5

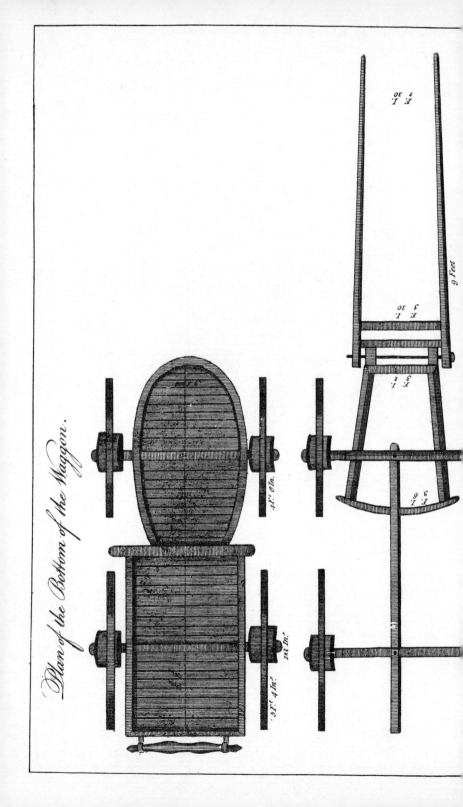

Plan of the Bottom of the Waggon.

Plan of the Frame or Timbers under the bottom of the Waggon.

ELEVATION

Lade 28 F¹ 10 In. in length

12 feet open

6 Feet

Nich. sculp¹ Surrand.

yet leaves the ground in a more crumbly state than most other ploughs (though certainly effected at the expense of a more extraordinary draught). From the weight of this plough, it is absurd to use it in any work where the soil is in a friable or loose state. In all other respects, it is a clumsy and unmechanical plough, and its defects outweigh the advantages. It throws out and drives along almost as much earth on the land-side, as it does on the furrow-side, and the fixed sticks which act in union with the moveable one, as a mould-board, are in so awkward a position, that with deep ploughing they ride on the land on both sides, and keep the plough from going close at heel; to remedy which they sometimes hook on great weights at the tail of it; two half hundred weights are not unfrequently tied to it; and this alone is sufficient to prove the unmechanical construction of this tool; a weight in a plough never acting beneficially, but by correcting some error in its construction. But this tool, which is not very well adapted for any thing except always throwing land the same way, and consequently doing well on steep hills, or for laying land to grass without a furrow, is in this county a great favourite. This is universal. Whatever plough we find in any county, is sure to be called the best in the world.

In the maritime division of this county, a one wheel-plough is much esteemed; it is generally drawn by three horses in a line. This is a much better constructed implement than the other; but the method of harnessing the horses remains for improvement, by substituting two only, and these a-breast. The light plough of the Suffolk kind, introduced by the Earl of Egremont about Petworth, would be a very capital improvement;

Wheel Turnrit Plough usd in Sussex.

Ja.ʳ Lambert Lewes

about 10 Inches   4 in.ˢ

Nose sculp: 55 n. Strand

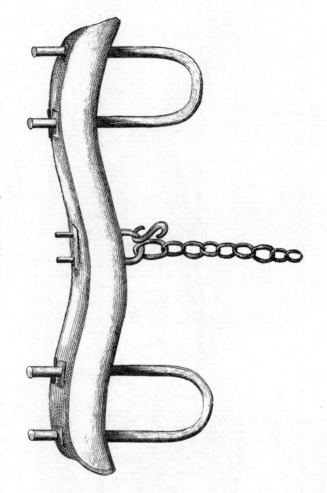

YOKE for OXEN used in Sussex
about 5 feet long.

ment; and it performs its work (upon soils of a light texture) in a more perfect manner than any of the ploughs of the county. The new-invented wheel-plough of Mr. Woods, of Chidham, has gained him much credit. It is drawn by two horses abreast, and without any driver; moves well in stiff land, and ploughs three-fourths of an acre of land in the same space of time that a full acre is ploughed after the common method. A driver and a horse, and some-times two, are thus saved.

Mr. Seaton introduced from Yorkshire the Rother-ham-plough, which the Rev. N. Turner carried into another part of the county, where it was adopted by many, and proved a real improvement.

In respect to the harrows of Sussex, for all strong soils (not kept in small ridges), they are well executed; and at Chidham, the common custom of the driver walking close to the horse's heads, has been improved upon by his holding the reins from behind.

The waggons, taken altogether, are better fitted to a farmer's use, in a country which is far from being level, than any other known in the neighbouring coun-ties. The carts have nothing particularly deserving either praise or censure, but are in general made for the carriage of small loads, from sixteen to twenty-four bushels.

Broad-share.—Whether this admirable tool be-longs to Kent, or is the invention of Sussex, remains a question. The great use of it, of which I have seen many instances near Lewes, is for cutting pea and bean-stubbles, or fallows weedy, that do not require ploughing. It consists of an oblong share two feet long, and four or five inches wide, fixed to the sock

or

or front of the ground-rist, by an iron shank in the middle, and sometimes bolted to the side of the ground-rist of a wheel-plough. It is pitched with an inclination into the ground, raised or sunk at pleasure, by the elevation or depression of the beam on the gallows, and answers the purpose of the great Isle of Thanet shim, for which see my Father's Eastern Tour. After the stubbles are cut with this machine, they are harrowed, raked, and burnt, and the land is left in excellent order for wheat.

The great attention which the Earl of Egremont has paid in improving the farming implements of Sussex, has already had a considerable effect in the neighbourhood of Petworth, and induced some farmers to adopt the use of those which promise the greatest advantage. His Lordship has been at no inconsiderable expense to introduce carts, ploughs, harness, and men, from Suffolk : and the success of the new plough, in the prize-ploughing at Petworth, has sufficiently testified the merit of it. Too much commendation it is impossible to bestow upon his Lordship's unwearied perseverance, so constantly exerted for the benefit of his country.

Amongst a great variety of other implements which his Lordship has succeeded in introducing into Sussex, the following may be mentioned :

1. *The Suffolk Farmer's Cart.*—This farming carriage was introduced by his Lordship for the purpose of removing those errors inseparable from the use of waggons ; and when trial of this cart was made at Petworth, it was immediately found how much superior was the work of a horse or ox, when single, than when he is harnessed with others in a team. These

carts

carts have been found capable of doing every part of
the work of a farm with more expedition than in any
other way; but it is to be observed, that this result
would have been much more striking, had not the
Suffolk wheel-wright made the common blunder of
building it too heavy.

2. My Father's improvement on the Suffolk plough,
from the hints of Mr. Arbuthnot. This plough does
its work with two horses a-breast; and whether the
nature of the soil be a strong clay or a sandy loam,
whether it go six or only three inches deep, it has experimentally done its work in a way superior to all
the tools which have as yet been brought against it.
It bore away the prize at the Petworth ploughing-
match in 1797.

3. *The Mole-Plough.*—This tool was also introduced by Lord Egremont, and at first it seemed to promise great success in laying dry springy and wet pastures. It has been repeatedly tried in the Stag-park, and
it always worked well, forming a circular drain three or
four inches in diameter, by means of a round piece of
iron two feet in length, and tapering from the heel to a
point: it is connected to the beam of the plough by a
strong bar of iron, which either elevates or depresses the
work at pleasure, according as it is found necessary to
plough shallow, or deep; and before it a coulter is
fixed, to cut the sod to the depth of the drain. This instrument will no doubt be found useful in many respects;
but the drains which it formed in the Stag-park, were
so soon filled up after they had been made, that the work
was rendered useless after one year, and other drains
were made in their place.

4. *Horse-*

4. *Horse-Hoes for Beans.*—Various skims applicable to the same beam, and so contrived as to clean intervals of any breadth.

5. *Iron Dibbles*—Invented by John Wynn Baker, in Ireland, and found much superior, in planting beans and cabbages, to wooden ones.

6. *Scuffler.*—Various sorts of this tool have been introduced at Petworth, and with very great success.

7. *Mr. Ducket's Skim-Coulter*—Was introduced by his Lordship, and with such success, that it was adopted by a great number of farmers.

8. *The Rotherham Plough*—Which, the year after the comparative trials alluded to above, beat all the Sussex ploughs, and has since spread much in the neighbourhood.

In the plough described by Mr. Young, the draught applies nearly to the centre of the implement as it stands for use, and goes very near to completely execute in that respect the idea I delivered to the Board upon this subject, in my first notes on the Report for Gloucestershire.—*Mr. Fox.*

What Mr. Young observes, is certainly the case in some soils, but among others, is an instance of the impropriety of adopting one mode for all kinds of land. It is true, a well-constructed plough will work any soil with two horses. But some soils require a deeper furrow than others. Upon such soils, then,

then, it is impossible that two horses can draw as
deep, and plough as fast, as four, or three. The
deeper a fine strong loam is ploughed. we certainly
obtain a finer range for the *food of plants*. It retains
the moisture longer, and consequently defends the
plants the better from the drought. We should not
try an old-constructed plough and four, with a new
one and two horses. It is altogether astonishing what
a saving of friction is obtained by the new-constructed
ploughs in general.

As to the propriety of ploughing good land deep,
one argument seems against it. It requires more manure to impregnate a larger mass of soil, than it would
do for a lesser. Hence the deep furrow constantly
turned up by strong ploughing, must require more
than one less deep. But (if we might venture on a
philosophical argument) as the food of plants is contained in the soil as well as in the atmosphere; and as
much more of that large mass is occupied by the fibres
or roots in quest of their food at one time than that of
the lesser; may we not reasonably conclude, that it
must be less easily exhausted, and consequently need
less artificial manuring? Where a great deal of manure abounds, such land, well ploughed, might be
kept almost in constant tillage, and bear an occasional
application of a stimulus.

The above note is not so much a correction of the
text, as an appendix to it.—*J. T.*

CHAP. VI.

ENCLOSING—FENCES—GATES.

THE very extensive and predominating range of timber, so very congenial to the soil of this county, and the singular custom of their shaws, render Sussex one of the most thickly-enclosed of any in the whole island; and if an exception is made of the wastes that border upon Surrey and Kent, together with the major part of the South Downs, the remainder may be considered as entirely enclosed. And to such a degree is this carried, that if the county is viewed from the high lands, it appears an uninterrupted woodland. No parliamentary enclosures of any consequence have been made, the county having been enclosed from the earliest antiquity.

The custom of shaws cannot be too strenuously condemned, since, wherever it prevails, it has the most pernicious influence on the contiguous land. How glaringly striking is this, by traversing the country with any attention, and marking the state of husbandry wherever these hedge-rows, two, three, and even four rods wide, abound. When corn is enveloped in such fences, impervious to the rays of the sun, it must necessarily experience very great and essential damage. No doubt, the conditions upon which tenants receive their farms, are made compatible with this; but it is nevertheless a loss, and a heavy one, as well to the farmer as to his landlord. I have seen fields of

corn

corn which (excepting in the centre) would never be
ripe.   Perhaps the Sussex farmers may be contented on
this score ; and landlords may think that these hedge-
rows may pay better than corn.   The present condition
of the tenantry in the Weald, is an unanswerable refutation to these ideas.

The history of this custom of the broad belts of underwood is evident.   The country was originally a
forest, and cleared probably among the latest in the
kingdom : fields of tillage and grass were gradually
opened among the woods ; and whilst land was plentiful, no accurate attention was paid to surrounding
them with fences, the forest making a sort of fence.
Carelessness and ill husbandry continued the practice ;
till at last the landlord, finding the sweets of great
falls of timber from these shaws, made it an article in
the lease, to preserve them against those encroachments
which improved husbandry would necessarily make.
A system, however, of greater barbarity can hardly
be imagined : the country being generally so wet, the
means to air and dry it here used are, to exclude the sun
and wind by the tall screens of underwood and forest
around every field ; and these being so small, a great
number are so wood-locked, that it is a little surprising how the corn can ever be ripened.   At the
same time that this mischief is done, the wood itself is
(timber excepted) but of a miserable account, as any
one may suppose, when he is informed. that these
shaws have a fence only on one side, and consequently
are exposed to be eaten by the cattle that graze in
the fields : hence there is an imperfect system of wood,
an injured one of corn, and wretched fences: by aiming at too much, nothing arrives at perfection.

Fences.

Fences.—Under this article, it would be a neglect not to describe the quickset-hedges at Goodwood, which are very capital, and trained in a most masterly manner. The Duke of Richmond planted them about eighteen or twenty years ago: they surround a very considerable farm, and are in a wonderful state of preservation. They form an excellent fence, without the assistance of any ditch, bank, rail, or pale; consist of three rows of white thorn, which spread three or four feet at bottom, but are clipped regularly and gradually to a thin edge at top: the shoots are so numerous, and trained with such care, that even in winter, without a leaf, the thickness is uncommon. By the young hedges in training, it appears that one method pursued has been to plant the centre row first, and when that is well established, to add another on each side of it; at least this is done in these new hedges. They are kept in a state of garden cleanness: the branches are drawn into the line desired, by being tied with mat, or other lines, and the clipping done with the exactest attention: the union of the hedges with the gate-posts, is close and perfect; and as to gap, &c. there is no such thing. How they have been preserved from cattle, but especially from sheep, is marvellous, if either are ever allowed to enter these closes: an attention, never ceasing, and a boundless expense (as far, I mean, as necessary), must have been exerted. They cannot be recommended to the imitation of *farmers*, but as an object beautiful to the farming eye, for its perfection, they merit all that can be said of them.

On Walburton farm are some very good fences, planted about 23 years ago: the quick was set about two inches asunder, and single: they are cut twice in

in a year ; are four and a half feet high, and two feet
thick.   Very little ground lost by the hedge, as it occupies only four feet.   Upon a very extensive scale,
the same excellent sort of fences have been made in
Lord Egremont's new enclosures in the Stag-park farm,
and most neatly kept.

All that remains to be observed under this head is,
that fences are usually, in the new enclosures, two
rows of white thorn on the bank of the ditch.   But
care should be taken that the ditch be not too near the
quick, as it acts as a drain, preventing it from receiving that nourishment so necessary to the growth of
a strong and durable fence.

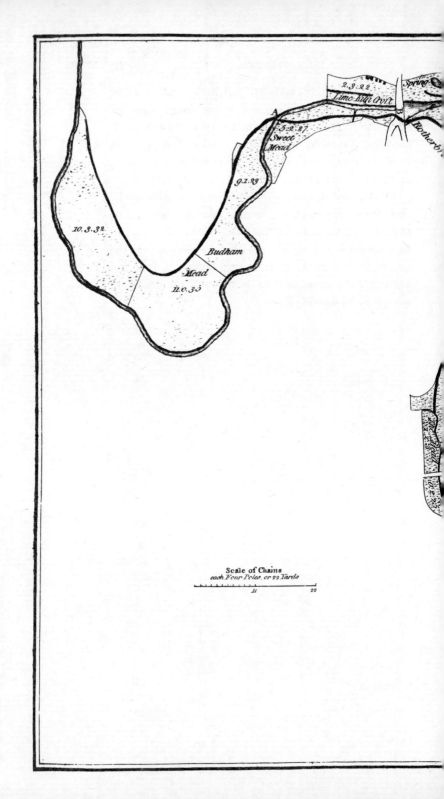

2.3.22

Lime Kiln Croft

Spring

Rotherbi

5.2.87

Sweet
Mead

9.1.28

10.3.32

Budham
Mead

11.0.35

Scale of Chains
each Four Poles, or 22 Yards

11   22

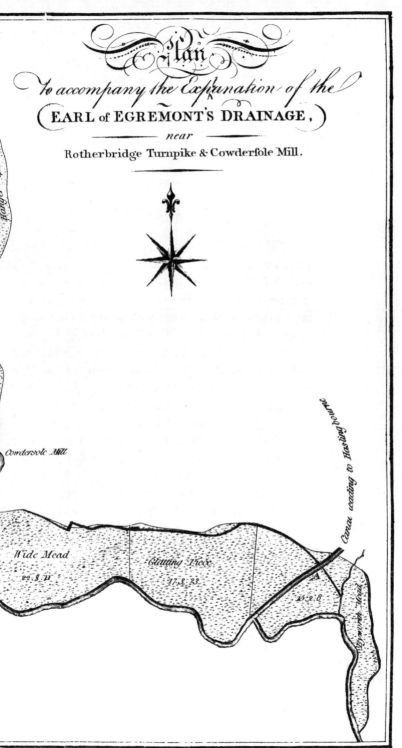

Plan

to accompany the Explanation of the

(EARL of EGREMONT'S DRAINAGE,)

near

Rotherbridge Turnpike & Cowderfole Mill.

Cowderſole Mill

Wide Mead

22.3.11

Glutting Piece

21.3.23.

15.2.8

Canal leading to Haſtingbourne

Neele sc. Strand.

CHAP. VII.

ARABLE LAND.

SECT. I.—TILLAGE.

THE more improved tillage of land, as at present practised, is confined to individuals. Little is there that deserves commendation. The ploughmen are not remarkably adroit in handling their implements, which are for the most part clumsily constructed; but on the light sandy loams about Petworth, the tillage is more perfect. The operations are executed by horses and oxen. Eight of the latter form a plough-team; but ten, and even more, are sometimes in use. They are universally worked, except by a few intelligent individuals, in common yokes and bows, going double; half of the cattle walking in the furrow, and the other half on the unploughed land. At first sight it appears, that in attaching such numbers to a plough, the expenses of farming must be immense: and unquestionably, if these draught cattle were kept for the purposes of their work only, such would be the case: but this must be consistent with the progression in their value, or they consider, and with justice, that the system would lose its principal merit; consequently the work is at all times gentle, and such as will not affect their growth.

With respect to the working of horses, the common management of the county is, to use three or four

with

with a driver, to one plough. Some few persons of intelligence discarding this useless incumbrance, have greatly improved upon this system, by cutting off two of the horses and the driver, and thus ploughing with only two ; and it has been found, upon repeated trials, that four horses and a driver, with the heavy-wheel plough of the county, execute but little more than one-half of the breadth of ground which the same four horses in two teams will perform.

In the neighbourhood of Chichester, Mr. Woods has brought the art of ploughing nearer to perfection, than it had ever attained before in that neighbourhood, by aiming at that standard which prevails in the east of England. His system of tillage is among the best in the county : neighbouring farmers are opening their eyes to the improvements in this line, and are sensible of the beneficial effects that flow from superior tillage.

SECT. II.—FALLOWING.

FALLOWING very generally prevails in the stiff soils of Sussex, where, it is thought, no corn could be had without this necessary preparation. But there is a very rich soil at the foot of the South Downs, all of which is either pure clay, or calcareous earth, and so excessively tenacious, that it adheres to the share like pitch : it is upon this land that the best of farmers never fallow. It has been managed in the fallow system ; and practice has experimentally convinced them, that it is neither necessary nor profitable : others adhere to this system of fallowing every third or fourth year; but in general it declines; is not pursued

sued by the best agriculturists; and most practised by the worst.

The inference to be drawn is obvious: if the most harsh, untractable, adhesive, and strong soils, such as require ten or twelve stout oxen to plough half an acre per day, can be managed without a fallow, to great profit, as is certainly the case; then what becomes of the pretended necessity, or propriety of this practice, on the thousand gradations of soils between common loams and these really stiff clays? To introduce new practices in husbandry, that are full in the teeth of old and inveterate prejudices, is indeed sufficiently difficult. The greatest improvements in the husbandry of this island, have not been established much above a century; but real improvements will work their way in time; and this error, of the necessity of fallowing, will, by degrees, give way to a better system. Those who contend for the necessity of it, from the effect of a particular experiment, or the practice of this or that individual, are not sensible of the mischief they would do the kingdom, if their ideas were universally to prevail. Unfortunately they do prevail too much. Without having recourse to particular instances, Sussex affords general ones, which speak powerfully. On the South Down farms, turnips, potatoes, clover, tares, rape, &c. expel fallows, and rents have advanced from thirty to fifty per cent: more stock is kept, and better kept. But in the Weald, the fallow system is adhered to; and here it is that rents have been far from being in proportion to the advance in the other instance.

Paticular spots and instances, owing to other causes than the modes of husbandry, merely local, do not
affect

affect the general fact. The produce in the Weald is the same at present, or very nearly so, that it was a century back; and consequently improvement stationary. But rents are not to be raised whilst products remain as they were. And how are products to be improved but by the conversion of fallows to turnips, cabbages, coleseed, tares, clover, &c. &c. thereby adding greatly to the live-stock ; consequently to the dung, and saving the expense, nearly a useless one, of time? I will venture to assert, that there is in the line of a great view, and not descending to minutiæ, no other method of doing it. Lord Sheffield has made great advances in proving this material fact, " turn your fallows to crops that shall feed cattle ; do not depend so much upon hay ; mow less, and feed more ; and do this upon an enlarged scale; and never fear but you will grow corn, if you can keep cattle and sheep"—and this doctrine comes from the heart of the Weald.

SECT. III.—ROTATION OF CROPS.

THE rotation adopted by the Sussex farmers, is in a great measure regulated by the nature and properties of the soil under cultivation. The judicious arrangement of a farm, in respect to the succession of its crops, is one of those leading features which so clearly mark the skilful from the inattentive cultivator; and no where seen in a more striking light than upon the different soils of this county. It is, without exception, the most prominent stay of good husbandry, to adopt those crops which are congenial to the qualities of the soil, in such order of succession as will yield the greatest produce ;

produce; whilst, at the same time, the soil shall be kept in the best cultivation.

From not paying due attention to this circumstance, we daily find the old system pursued upon cold wet land, from one extremity of the kingdom to the other; such as was familiar to the Roman husbandmen, and practised in Virgil's time—a fallow succeeded by two crops of corn. Thus it is in Sussex, and the practice is the reason why the clay farmers are so much distanced by the rest of England, and in their practices left so far behind.

The new face which improvement has given to the husbandry of this kingdom, took place on sands, and chalk, and soils of a similar description.

The most general system pursued on the stiffer or strong loamy clays, is in the following order, and may be considered as the standard for the Weald.

1. Fallow,
2. Wheat,
3. Oats,
4. Clover and ray-grass, two or three years;
5. Oats, pease, or wheat.

Upon the very tenacious clay under the northern range of chalk, the clover leys, after having been down some years, are then broken up, and sown with oats; and then summer-fallowed for a crop of wheat; a second crop of oats succeeds that of wheat; it is then laid down with clover and ray, or trefoil. These fail in two or three years, when the land is covered with weeds and grasses indigenous to the soil. A soil which discovers such a tendency to run to grass, should not be suffered to remain in bad tillage.

Upon

Upon those of a lighter texture than the foregoing, an arrangement is practised, which cannot be too much recommended to a more extended cultivation.

1. Turnips,
2. Barley,
3. Clover,
4. Wheat.

Farmers in the neighbourhood of Battel, Eastbourne, &c. arrange part of their land under a system of tillage different from any of the preceding, and bring potatoes to their aid.

1. Potatoes,
2. Barley,
3. Clover,
4. Wheat.

Sometimes, as at Battel,

1. Potatoes,
2. Wheat in succession.

That the potatoes do not decline from repetition, appears by the last potatoe crop turning out better than any of the preceding, and the wheat good. This course is singular, and has been practised with uniform success for more than twenty years, by Mr. Mayo, of Battel.

That fallowing for a crop, even upon the stiff land, is by no means necessary, is proved, by much of this soil being managed without any fallow.

1. Tares,
2. Wheat,
3. Oats.

The arrangement upon the chalk farms is otherwise.

1. Fallow,	1. Wheat,
2. Wheat,	2. Barley,
3. Barley,	3. Clover,
4. Clover,	4. Turnips.
5. Wheat.	

1. Wheat,	1. Wheat,
2. Barley,	2. Pease,
3. Tares or pease,	3. Barley,
4. Oats,	4. Turnips,
5. Clover.	5. Tares.

The following miserable course of cropping is usual on the tenantry *laires* in the neighbourhood of Lewes.

1. Wheat,
2. Barley,
3. Oats, pease, or tares ;
4. Clover, or turnips.

But the more general practice is that of having five crops of white corn in six years. But besides this open-field management, we find, 1. wheat; 2. rye and tares, sown in August and September, fed in May and June for turnips, barley, and clover. Another arrangement put into practice by intelligent farmers on light land is, to manure for wheat after tares or clover, then turnips ; or, clover sown in the spring amongst the wheat: the turnips succeeded by barley or oats, and clover ; which, after remaining one year, the ley is broken up, sown with pease or tares, followed by wheat. But on the stiffer soils, wheat is sown after clover or tares, and seeds with it in the spring ; the clover is either twice mown, or fed in the summer, manured; and sown with wheat upon one earth.

A very

A very common practice prevails in this county, of sowing wheat upon turnip-land*. Those who follow it are compelled to turnip-feed their flocks at that season when the turnips are reputed of the least value, and when a plentiful supply is in existence of all other food upon which the sheep might be supported equally well.

In the maritime district the accustomed mode of cropping the land is in the order of,

1. Tares, or pease,
2. Wheat,
3. Clover,
4. Clover,
5. Wheat,
6. Oats.

This system is only adapted to very rich land. Particular instances occur, of wheat having been sown four or five years in succession, and the produce amounted to four or five quarters per acre. The course of wheat, pease, wheat, barley, pursued in this vale, will afford any person a tolerable insight into the general properties of the land in question.

Throughout the gravelly soils between Chichester and the South Downs, we find,

1. Pease,
2. Wheat,
3. Barley,
4. Clover,
5. Wheat,
6. Pease ;

---

* A very bad practice indeed.—*Annot.*

Or,

Or,
    Turnips,
    Barley,
    Clover,
    Wheat.

A method very commonly pursued is, that part of their bringing land round in six laires, when it is all kept in tillage, in the following manner: turnips, wheat, barley, seeds, oats; all which methods cannot be approved, inasmuch as it is not only a bad plan for keeping stock, but it is farming also at a very considerable expense, as that course will occupy at least five four-horse teams in the management of 700 acres, and we may reckon the expense of each team at 100*l*.; but by laying 200 acres to sainfoin, and as much to pasture, after two years turnipping, the annual expense of ten horses, equivalent to 250*l*. will then be saved. This will enable the farmers to keep the remainder in exceeding good condition, by having so much sainfoin-hay upon which to winter their sheep, besides two other great advantages; for, by having so much sainfoin-hay, the experienced husbandman will always be enabled to feed his seeds, and by that means, will bring his land round in four laires instead of six, and in much better heart.

Farmers are unsettled in their mode of management; many who followed the six-laire course, and others nearly in the same system, are now changing it to four laires on the chalky and gravelly land. Upon the South Downs they substitute a double crop of tares instead of a fallow for wheat, sowing early winter-tares, to be fed late in the spring; then summer tares and rape, fed off in time for wheat; which is altogether very capital husbandry.

In

In place of an unproductive fallow, the skilful and intelligent farmer raises two crops of tares, to answer the great purpose of fallowing (clearing and meliorating) equally well. The ploughing is at a season of the year when the ground can be easily worked; and in the western part of Sussex, with a light plough, two horses and one man, who both holds and guides the plough, which, upon calculation, is a great saving of labour, whilst at the same time, he secures to himself food for his stock at the most critical period of the year, and enriches the ground with the manure arising from the fold, or stock fed on it.

The

The following is the Course of Crops, and the Product of Corn, upon an extensive and highly cultivated Farm in the neighbourhood of Lewes.

Year	A. 35 Acres, Soil, a fine chalky Loam upon a hard chalky rock. (Product Average. Q. B.)	B. 35 Acres, Soil similar to A. (Product. Q. B.)	C. 14 Acres, Soil similar to the preceding. (Product. Q. B.)	D, 21 Acres, same as A. (Product. Q. B.)	E. 18 Acres, Better Soil; greater depth of Earth. (Product. Q. B.)	F. 13 Acres, Similar to A. (Product. Q. B.)
1773	Oats, 7 5	.....	.....	Barley, 6 1	Wheat, 3 5	Barley, 6 1
1774	Tares, fed	.....	Oats, 8 0	Oats, 7 5	Pease, 4 0	Clover, 0 0
1775	Barley, 3 1	.....	Tares, fed	Wheat, 3	Wheat, 3 5	Wheat, 3 4½
1776	Clover, mown	.....	Turnips, folded	Barley, 6 7	Barley, 6 7½	{ Half fallow, Half turnips, } folded
1777	Wheat, 3 1	.....	Oats, 5 2	Clover, fed	Fallow	Wheat, 4 3
1778	Oats, 3 7	.....	Barley, 7 0	Wheat, 3	Wheat, 4 3	Barley, 5 3
1779	Oats, 7 0	Wheat, 3	Rape, fed	Barley, 4 4½	{ Pease, 1 0 / Beans, 3 4 }	Oats, 6 1
1780	Clover, mown	Barley, 4 4	Wheat, 3 4½	Rape & tares, fed	Barley, 4 4	Clover, fed
1781	Wheat, 3 4	Oats, 5 0	Barley, 6 1	Wheat, 3 6	Wheat, —	Wheat, 3 6
1782	Oats, 5 2	Fallow	Clover, fed.	Barley, 4 7	{ Beans, 3 / Tares, fed }	Turnips, folded
1783	Clover, fed	Wheat, 3	Wheat, 3 4½	Oats, 6 0	Wheat, 3	Turnips ditto
1784	Wheat, 4 0½	Tares, fed	Turnips, folded	Wheat, 3 1	Barley, 5	Wheat, 4 0½
1785	Turnips, folded	Wheat, 4 0½	Barley, 6 0	Turnips, folded	Tares, fed	Fallow
1786	Barley, 5 5½	Turnips, folded	Clover, fed	Barley, 4 5	Wheat, 3	Wheat, 4 0
1787	Pease, 5 0	Barley, 6 1	Wheat, 4 0	Clov. mown & fed	Oats, 6	Turnips, folded
1788	Wheat, 2 5	Clover, mown	Turnips, folded	Wheat, 3 5½	Tares, fed	Barley, 5 5
1789	Turnips, folded	Wheat, 3 7	Tares, fed	Turnips, folded	Wheat, 3	Sainfoin, mown
1790	Barley, 5 7	Turnips, folded	Wheat, 3	Barley, 5 5	Clover, mown	Ditto, ditto
1791	Clover, mown and fed	Barley, 7 4	Wheat, 3		Wheat, 3 6½	Ditto, ditto
1792	Wheat, 3 7	Clover, mown	Clover, fed	Barley, 5	Rape & cabbage	

	G. 22 Acres. Very stiff Chalk Loam, over a soil chalky Marl and clayey Bottom.		H. 6 Acres. Soil like G.		I. 21 Acres. Very stiff black and rich Loam upon Clay.		I. 14 Acres. Soil like L.		M. 10½ Acres. Similar to A.	
	Crop	Product (Q. B.)	Crop	Product (Q. B.)	Crop	Product (Q. B.)	Crop	Product (Q. B.)	Crop	Product (Q. B.)
1773	Beans	3 1	Wheat	3 5	Wheat	3 4	Wheat	3 5	Wheat	3 5
1774	Wheat	3 4	Oats	8 0	Fallow		Fallow		Oats	8 0
1775	Wheat	3 5	Clover, mown and fed		Wheat	3 4	Wheat	3 5	Pease	3 2
1776	Pease	2 4½	Wheat	3 5½	Wheat	3 4	Wheat	3 4	Wheat	3 4½
1777	Wheat	3 4	Oats	5 2	Pease	3 7	Pease	2 5	Oats	5 2
1778	Beans	2 7	Rape, fed		Oats	3 0	Wheat	3 0	Fallow	
1779	Wheat	3 1	Wheat	3 1	Wheat	3 0	Oats	7 0	Wheat	3 0
1780	Barley	4 1	Oats	6 1	Beans	3 4	Wheat	3 4½	Wheat	3 4½
1781	Beans	4 3	Wheat	4 3	Wheat	3 4½	Oats	5 0	Barley	6 1
1782	Wheat	3 6	Fallow		Oats	5 2	Tares, fed		Clover, fed	
1783	{ Half barley / Half oats	4 7 / 6 7 }	{ Wheat / Pease	3 0 / 2 2 }	Tares, fed		Pease	3 7	Wheat	3 4½
1784	{ Half oats / Half pease	6 0 / 2 0 }	Wheat	4 0	Wheat	4 2	Wheat	4 0½	Barley	5 1½
1785	Wheat	4 0	Oats	6 0	Oats	5 0	Barley	6 0	Clover, mown	
1786	Oats	6 0	Tares, mown		Tares, fed		Tares, fed		Wheat	3 1
1787	Half tares, half clover, mown		Wheat	2 5	Wheat	4 0	Wheat	4 0	Clover, mown & fed	
1788	Wheat	2 5	Oats	8 7	Wheat	2 5	Pease	2 6	Wheat	2 5
1789	Oats	8 7	Wheat	5½	Barley	5 5	Wheat	8 7	Clover, mown & fed	
1790	Half tares, half rape, fed		{ Tares, seeded / Oats	6½ / 2½ }	Tares, fed		Oats	8 6	Wheat	3 5½
1791	Wheat	3 6½	Wheat	3 5½	Wheat	3 5½	Tares, fed		Turnips, folded	
1792	{ Half pease / Half tares, folded	2 2½ }	{ Half pease / Half tares, folded	}	Oats	6 2	Wheat	3 7	Wheat	3 7

I shall

I shall close this article with noting two arrangements practised by the Earl of Egremont at Petworth. Upon a cold springy land, which in frosty weather works well, and becomes mollified, but if dry succeeds, it binds like stone, his Lordship sows,

1st—Tares and rye; or, if the land is foul, he ploughs it four times, and three or four inches deep, according to the depth of the clay beneath the surface earth; but never with a view to bring up the clay. When tillage has brought it into order, the next crop put in is,

2d—Turnips. This land having been very judiciously drained, answers well for this root: these are folded, the largest having been previously drawn for fattening cattle; the manure for the turnips, is either dung by itself, or compost of earth and lime; lime without mixture, not answering for turnips.

3d—Oats, one ploughing; six bushels of seed.

4th—Clover, one gallon; trefoil, one gallon and a half; rye-grass, two gallons: but when wheat succeeds upon the layer (one year's duration), his Lordship sows only clover and trefoil, and no ray-grass, as it is a plant unkindly for wheat, which is the last crop in the course.

Where the staple of the soil is fleeter, and the clay rises nearer to the surface, the same system is pursued, as far as the seeds, which instead of one, are kept two years; the reason of which is, the soil is so poor, that the layer of one year's duration is insufficient.

He has introduced, since this report was written, beans as a fallow, thus:

1. Break up a layer for beans,
2. Wheat,
3. Manure for beans,

4. Wheat,

4. Wheat,

5. Turnips,

6. Oats and grass-seeds: and better husbandry can no where be found*.

SECT. IV.—CROPS COMMONLY CULTIVATED.

I. WHEAT.

1. *Preparation.*

THE tillage for wheat depends upon the crop it succeeds. It is, 1*st*, a fallow three times ploughed, the first earth fallowing up; the second, a stirring, and the third landing up; but the number of earths is regulated by the condition of the fallow : if foul, another earth is given, or more. 2*dly*, If it succeeds clover, the practice of bastard-fallowing is in many cases adopted. This method of breaking up a cloverley, as a preparation for wheat, is supposed to be

---

* It never can answer the purpose of fallowing. Clay and strong loams can never be kept clean, without being ploughed in summer, as working them at any other period, will never kill either couch-grass or thistles. The condition of land in the vicinity of every large town, is a pregnant proof of the truth of the above, as from the great rents paid in these situations, the farmer is too often tempted to neglect the beneficial practice of giving his ground a complete summer fallow.— *Mr. R. Braun.*

The instance that I have cited above, and the facts that we quoted in illustration of the anti-fallowing system, so opposite to this gentleman's opinions, are sufficient to speak for themselves, where we see that clay and strong loams, by the attentive management of good husbandmen, can be kept perfectly clean without recurring to fallows. But the instance given of this tare-husbandry has nothing to do with clay ; and therefore the observation is wide of the mark.—*A. Y.*

caused

caused by the ravages of the worm and slugs, if it
were sown upon a single ploughing; for just as the
wheat appears in the blade, these insects eat the root.
It is effected about Midsummer, by ploughing and
harrowing till the fibres and roots of the clover, by
the operation of the harrows, are separated from the
earth, and die away, from being exposed to the atmosphere and the effects of the sun. It is imagined
that the wheat upon clover-leys of this nature, cannot be trodden too much. It is therefore harrowed
with double implements, and six horses are used for
this purpose. The business of harrowing is deemed
so necessary, that land has undergone this operation
not less than a dozen times. *3dly*, If the crop of
wheat is preceded by pease, a single ploughing is by
some farmers thought sufficient, provided the land is
in tolerable order. Some plough the pea-stubble in
harvest, harrowing from four to six times, and again
stir it between harvest and Michaelmas; and in October they ridge it up in the usual manner (nine bouts
to a land of thirteen feet and a half), and sow immediately after the plough. *4thly*, If the wheat is sown
upon turnip-land, one earth about Christmas is the
general method. *5thly*, After tares, one earth also.

2. *Manure.*

Stable-dung, Mr. Woods, of Chidham, observes,
should be laid upon a clover-ley, or other land, just
before sowing, at the rate of sixteen to twenty loads
per acre, spread and ploughed in immediately.

There is about Eastbourne, Jevington, &c. a bad
custom on the arable lands of that neighbourhood,
spreading in July forty large loads of dung per acre,
to be sown with wheat at Michaelmas; and they
leave

leave it till then on the surface, exposed to the sun
and wind.   Upon what system they can follow this
custom, it is difficult to conjecture.   If they would
reflect on the fact of the volatile alkali being the food
of plants, and that one of the principal causes of the
fertility resulting from dung, is its containing that
evaporative salt, surely they would think that some experiments on this point would not be undeserving their
notice.   If they will try the effects of spirits of harts-
horn applied to common-field earth in a garden-pot,
they will presently be convinced of one fact ; and if
they then expose some of the same spirit to the atmosphere in a plate, they will soon understand another
fact not less important : these two trials are very easily
made ; and he who tries them will not be ready afterwards to expose his dung-hills one moment longer than
necessary.

3. *Sort.*

Of the several sorts of wheat in cultivation in Sussex, the velvet-eared is preferred in the Weald, having by much the thinnest skin : they call it *fluffed.*
It weighs upon an average 59 to 60 lb. per bushel.   It
is an observation of Mr. Gell, of Applesham, one of
the most spirited and intelligent farmers in the county,
that the white fluff on good land answers best, as being the most saleable ; but on poor land, subject to
poppies : the strong-strawed sort that overpowers this
weed, should certainly be sown.

A sort of wheat, obtaining much on the Downs,
is what they call *Clark* wheat.   It is not bearded ;
red blossom, red chaff, and red straw ; white grain ;
the sample coarse, being in price under the finest sorts.
It is a great yielder, and requires to be cut forward.

SUSSEX.]   Mr

Mr. Woods, of Chidham, a very excellent and spirited farmer, has found by long and attentive experience, that a change of seed-wheat is of essential importance to the farmer, as that seed which has been repeatedly sown over the same ground, at length degenerates, and the produce becomes each succeeding year inferior in quality ; for which reason he sows wheat that is apt to run to straw upon ley-land, and the Hertfordshire white upon pea-stubbles.

The farming world is certainly indebted to Mr. Woods for a valuable acquisition, in bringing into cultivation, what with justice has been called, and is a new sort of wheat, the *Chidham white*, or *hedge-wheat*. The origin of it was this: as Mr. Woods was occasionlly walking over his fields, he met with a single plant of wheat growing in a hedge. This plant contained thirty fair ears, in which were found fourteen hundred corns. These Mr. Woods planted the ensuing year, with the greatest attention, in a wheat-field: the crop from these fourteen hundred corns produced eight pounds and a half of seed, which he planted the same year ; and the produce amounted to forty-eight gallons: this he drilled, and it yielded fifteen quarters and a half, nine-gallon measure. Having now raised a large quantity of seed, he partly drilled, and in part sowed, the last produce broad-cast, over rather more than fifty acres of land, and he gained 38½ loads. Twenty loads of this quantity was sold for seed, at 15*l*. 15*s*. per load. The wheat, upon trial, was discovered to be so fine, that Mr. Woods had an immediate demand for a far greater quantity than he could spare for sale. 1792 turned out a bad yielding year, otherwise the last produce would have fully equalled forty-five load. With respect

spect to the sample of the *Chidham wheat,* it is white, of a very fine berry, and remarkably long in the straw, so as to stand, in a wet summer, full six feet in height. The seed is now dispersed over Hampshire, Surrey, and other counties, and much cultivated about Guildford.

4. *Steeping.*

The method of using lime in preparing the seed-wheat, practised by Mr. Ellman, of Glynd, is, to have a sieve made about ten inches in depth, containing three pecks of wheat, which is dipt into a tub of sea-water or brine; this causes the lime to take effect, and thereby to destroy the seed of the smut; it leaves a coat of lime upon the wheat. By making the brine sufficiently strong to swim an egg, where no sea-water is to be had, all the light corn floating on the surface is skimmed off, and the good wheat remains at the bottom.—*Mr. Ellman's Experiment.*

The common method of preparing seed-corn is, to soak it in briny or sea-water twelve hours; after this the water is let off, and the lime sifted on the corn, mixing the whole together. This operation is performed at five or six o'clock in the morning, and the seed is carried into the field at seven; consequently the lime, remaining so short a time on the grain before sowing, has no time to penetrate into the corn; whereas, by wetting the wheat, and leaving it until the succeeding morning well limed, the lime has a greater power in destroying the smut-powder than when it remains on it for half an hour only, when most of the lime is rubbed off the corn. Sometimes the brine has been heated, and then poured out of a pot upon the seed.

Another

Another process, practised by Mr. Woods, of Chidham, is this: about two sacks of seed, at each time, are shot into a leaden cistern constructed for that purpose, filled with salt-water, in such a manner, that the water is made to flow over the seed, and to float all the light and hood-corn, which is then skimmed off. It remains in this manner about the space of six hours, when they are in haste for sowing ; but at any time twelve hours are sufficient ; the wheat is then thrown out of the cistern upon a brick-floor, the water being first drained off through a tap-hole, and fresh lime, which is newly slaked for the purpose, is then sifted over it, to the amount of half a bushel of the strongest grey lime, to one quarter of wheat : it is then turned over and mixed into an heap, where it remains in that situation till the following morning, when it is taken for use. Every morning, previous to the sowing next day, the wheat should be steeped till evening, and then limed as above, and left till the morning.

Seed-wheat, prepared by steeping it twelve hours in sea-water, dried with lime, has been known to be effective ; a headland sown dry has been smutty, when the rest of the field, steeped, has escaped. The smut in the corn is an evil which the South Downs are little subjected to. It has been attributed to the practice amongst the farmers, of sowing the same sort of seed for a length of years, without giving the land the least change ; or it may be owing to a negligence in improperly preparing the seed. Lime is the best preventative*.

5. *Seed.*

* " So great is the variety of cases on this head, that it cannot easily be ascertained to what cause to ascribe the wonderful effect of smut in wheat

5. Seed.

The quantity sown depends upon circumstances; the crop it succeeds, &c. In general, it may be estimated from two to three, and up to four bushels, per acre. Mr. Gell sows at Applesham four bushels upon ley land, and three upon tilth. Mr. Woods three and a half; and when wheat succeeds pease, he sows only three bushels, provided it is early in the spring; but if late, more; since the vegetation is not then equally flou-

---

wheat. If to an insect in the grain, why is one part of a field affected, and not another at the same time, and under the same circumstances? But this now and then actually occurs. The same objection lies against an insect in the soil, the season, or any general circumstance in which all the field equally participates. We are equally uncertain to which of the two, lime or pickle, we can attribute the cure. Accordingly, some only wet it, on purpose to make the lime adhere; others are extremely attentive to the pickle, and only use the lime to dry it for sowing.

" Some farmers sow without using either, and sometimes successfully for a number of years together, but always suffer more in one unlucky year than would pay the expense, and reward the trouble for a long lease.

" A report prevails in Scotland, that the practice was accidentally introduced by some seed-wheat having been sown from a wreck after having been steeped in the sea a night or two; that the wheat sown dry on the same field was blackened much, whilst the other was clean and healthy. Both pickling and liming are practised generally in Scotland, and the pickle most in use is urine. When the grain is pretty equal, there is no need of floating; and blacked wheat ought to be avoided for seed."

The observation is ill founded. The smut is perfectly well understood. Every scientific farmer knows that it is occasioned solely by smutty powder adhering to the grain, which at once accounts for all the cases this gentleman starts. Any operation that completely freed the grain by washing, or destroys it by acrid, corrosive, or poisonous application, will have the effect infallibly of securing a clean crop.—
R. Braun.

rishing.

rishing. The medium quantity may be estimated from two and a half to three bushels*.

6. Time of Sowing.

This too depends upon contingencies; seasons, preceding crops, &c. Respecting the fittest time for sowing wheat, experience only can determine. Early sowing seasons are well adapted to some soils, but pernicious to others. Mr. Gell is of opinion, that the earlier the seed is put in, the better will the crop turn out; he therefore sows as early as possible, to allow it a sufficient time to take root, and be enabled to stand against the frosts. The wheat sowing season commences about the beginning of October. Mr. Woods prefers early sowing also, as a less quantity of seed is demanded, and vermin destroy more of it in cold weather; much too is apt to rot in late sowings. By sowing after the month of November, the corn remains in the ground so long before vegetation arises, that much of it is destroyed; and if hard frosts come in a late seed-time, the grain is cut off before the nourishment takes place. However, notwithstanding, farmers will regulate their sowing by the nature of the soil they cultivate. In many places it is seldom finished before new Christmas.

7. Culture whilst growing.

The culture which the wheat receives, is in proportion to the active industry of the farmer, and the

---

* The reason assigned for sowing a large quantity of seed is, that it choaks the rubbish, such as charlock, poppy, &c. if the ground be not well covered with corn,—*John Ellman.*

means

means he exerts to keep his crops in the most per-fect state of cultivation. The practice which is most generally adopted is, to hand-hoe in the spring : this operation is effected sometimes only once, but fre-quently twice, as it depends on the preceding crop. Women are usually employed, at 8*d.* per day.

This operation of hand-hoeing wheat is disapproved by Mr. Ellman, who never hoes his white corn, hav-ing given it up from a conviction that his crops were never benefited by the practice ; but on the contrary, that it always did mischief. Should the practice sometimes be right, and sometimes wrong ; or right on some soils, and wrong on others, these contrary facts may probably depend on the spring roots, which are said to strike into the air, and enter the ground at some small distance from the stem. If a hand-hoeing is given just before the appearance of those roots, it may, on a bound surface, prepare for their easy en-trance ; but if given afterwards, it should seem pro-bable, that the effect would be mischievous, would re-tard the progress of the plant, and force it to do its work over again, perhaps at a worse season. If this is the case, the benefit which results from hitting the moment exactly, may by no means equal the probabi-lity of mischief upon a scale of any extent ; in which the right time can scarcely be taken for the whole of a crop.

I have heard excellent farmers declare, that if a man would pay for the hoeing their wheat, they would not permit the operation, being convinced that it did more harm than good.

Feeding.—The custom of feeding the young wheat is practised in various parts of Sussex. Upon the rich arable vale upon the coast, sheep are turned into the

wheat

wheat from Christmas to March.   Many farmers have
thought that the wheats are the stiffer, and rise more
abundant for this practice.   The truth appears to be,
that this is done not so much to benefit the wheat, as
through mere necessity ; since it is allowed, that in
proportion to the searcity of turnips, and other artificial food, will this practice be in vogue.   Εν δε ταῖς
αγαθαις χωραις προς το μη φολλομανεῖν επινέμϗσι και
επιχείρϗσι τον σιτον.

Meadow and pasture, in various parts of Sussex, is
so small in quantity, that it requires a greater abundance
of artificial provision, and brought into cultivation in
a different rotation, to keep sheep in much greater
numbers ; and we see, that to feed the present stock, is
sometimes found to be a matter of no small difficulty,
and highly hazardous ; so that the resort of the farmer
is to turn his sheep upon his wheat, which at best is a
measure of questionable policy. Sheep are often turned
into the wheat to tread and bind it, and give the soil
a cohesion grateful to that plant.

Treading.—Mr. Kenward, of Fletching, uses six
and eight oxen in drawing a light pair of harrows ;
and he remarked, that they were not, on such occasions, used either at harrows or ploughing for the
draught; but for the treading on such of the Weald
lands as tend pretty much to sand, or rather a sort of
soft abraded stone.   He named a farmer who could get
no wheat, until he drove all his oxen, cows, and
sheep, repeatedly over his land, directly after sowing.

Without these precautions, the plant is root-fallen,
and eaten by the cockchafer-grub all winter, and by
the red wire-worm in the spring.   The best wheat
                                                    on

on these lands is, when the seed is, from a wet season, poached in at sowing.

The county, it is to be noted, is in general more inclined to wet loam and clay, than to sand; but the sides of hills have a soft friable stone, which moulders into sand for an under-stratum; and in proportion as this rises to, and mixes with the surface, the evil here complained of, takes place. To avoid such expensive remedies at a busy season, rolling appears practicable: but it has been tried; it makes the sand blow more: this is known in Norfolk turnip-fallows. Fallowing is the bane of such a soil. The farmers admit the best crops to be on clover. But care should be taken that the clover be sown really clean, and let it remain two years, never mown; and in a moist season, before sowing the wheat, it should be rolled with a heavy roller. It would be the best and cheapest culture.

Upon dry soils subject to poppy, Mr. Ellman, of Shoreham, ploughs his tare and rape land for wheat, the beginning or middle of September, to sow the wheat the middle of October. The harrowing kills the poppy; and in putting in the seed, he likes to tread much with oxen, or with sheep. A neighbour treads his with oxen in March, which he thinks better against poppy, than doing it at sowing.

8. *Harvest.*

The wheat-harvest commences, in forward seasons, about the latter end of July; in late seasons, about ten days or a fortnight after. The operation of gathering in the harvest, is performed by a contract between the farmer and his men, sometime previous to the harvest, when the wages are agreed upon, and the proportion of corn allotted respectively to each man. The

reaping.

reaping-hook and sickle, both the jagged and the smooth edge, are the instruments made use of. The duration of harvest depends upon the season; but generally varies from four to six weeks. Wages (1793) were usually 3*l.* for a month, and board: but since that time, this rate has advanced. Mr. Woods begins his harvest about new Lammas-day, and observes, that if wheat be fit to reap before that time, a greater crop is expected in proportion to the number of days preceding August 1st, and so *vice versa.*

After harvest is finished, it is every where, I believe, customary for the farmers to give a harvest-home, or supper, to their harvest-men.

At Mr. Ellman's, above eighty men, women, and children, generally sit down to his harvest-supper. The supply of provision for this numerous company is abundant: beef, 16 stone; mutton, 8 stone; plumb-pudding, 1 cwt.; beer, 50 gallons; bread and cheese, &c. &c.; what remains is distributed to the poor. The origin of this custom is thought by Mr. Ellman to be this: that when labour was scarce, the neighbouring artisans assisted the farmers in their harvest for two or three days, gratis; and the harvest-home was a recompense for it.

Reaping wheat is done by the acre: it varies from nine to eleven and twelve shillings. A good labourer reaps an acre in three days.

9. *Thrashing.*

Thrashing the wheat is every where performed by flail-work, and cleaned either with a shovel and broom, or by winnowing-machines. Three instances occur of thrashing-machines having been erected, namely, that of Sir Richard Hotham's, at Bognor, which

which has been out of repair; Mr. Pennington's, at Ashburnham : and the Earl of Egremont's, at Petworth; of which more will be said hereafter. The prodigious saving that might be made in the expenses of labour, in the article of thrashing only, by substituting machinery in lieu of the common system of thrashing, ought to induce gentlemen, and large farmers in this county, to improve this branch of rural economy. These machines, where they have been erected upon proper plans, have facilitated the common operations of thrashing, lessening expenses considerably.

It has many advantages to recommend it : the straw for fodder is better by passing through a mill ; and what is a point of much greater importance, it has been discovered, that when the work has been executed after the usual manner, by the flail, one pint and a half of wheat usually remains in each truss of straw; since straw that has passed through good thrashing-mills, has been found to yield that quantity. The vast utility of them, upon large corn farms, is at once obvious : the saving of labour is considerable; and when they come to a higher state of improvement, will unquestionably be adopted over the great corn farms of this kingdom. The plea, that the poor would be dreadfully injured, is more visionary than substantial; and will, in many cases, hold to be equally as fallacious in agriculture as it once did in manufactures. Clear enough it is, that the great object in farming is to cultivate land in the best possible manner, at the least expense. By the means of machinery, which enables him to thrash and to dress at the same time, he spends more money in improving, and raising a greater produce for the market.

10. *Pro-*

10. *Produce.*

Respecting the produce of wheat, this depends upon so many circumstances, that all that can be said upon it is, to draw an average of the produce of several parishes chiefly in the district of clay, and scattered over a very considerable tract of land, which will enable us to form an idea of the corn products of Sussex.

Worth, Slaugham, 12 bushels.

Horsham, Rusper, Balcomb, 14 bushels.

Lower Beeding, Crawley, Nuthurst, 16 bushels.

Rudgwick, Billinghurst, Kindford, Green, Hitchingfield, 16 bushels.

Bolney, Cuckfield, 20 bushels.

Luggershall, Warneham, Slingfold, Cowfold, Shermanbury, Henfold, 22 bushels.

Salehurst, West Grinstead, Ashurst, Pulborough, Chiltington, Shipley, 24 bushels.

Hursterpoint, Albourne, Ditchling, Haylsham, Ashburnham, Winchilsea, Westham, 32 bushels.

Average, 21 bushels 2 pecks.

Individual instances of high corn products do not affect the general average. Upon the very fertile land which borders upon the coast, products of wheat much greater than the above, are frequently met with. A fair crop and an average one, vibrates from 34 to 40 bushels, statute measure, upon the same land. At Felpham, adjoining Bognor, 52 bushels have been raised over an eighteen-acre field of Sir Richard Hotham's; and land at Winchilsea has yielded up to 48.

In 1794, over 54 acres of very strong clay loam, but all of it drained in a very masterly manner, the Earl
of

of Egremont gained three quarters and a half per acre;
and in the following year, upon land of a similar description, his Lordship raised four quarters and a half
per acre. These are extraordinary products, when the
nature of the soil is considered. But the merit, in this
instance, arises from the corn having been produced
upon land which has lately been converted by his
Lordship from a forest into a capital farm.

One of the most extraordinary experiments that was
made in this county, was by the father of the present
Mr. Car, of Bedingham, who upon a piece of land
that had been left by the sea at Bishopstone, tried how
often in immediate succession it might profitably be
sown with wheat; not so much from an experimental
intention, as from the circumstances arising in the
trial. The first crop was seven quarters; the second,
the same; the third six; the fourth, fifth, and sixth,
each five quarters, upon an average. This is perhaps
the most extraordinary instance of fertility upon record.
(See *Annals*, vol. xii.) The same piece of land (31
acres), in 1795, yielded an extraordinary crop, as
the following letter, communicated by Lord Sheffield,
sufficiently evinces.

"*Bishopstone, Dec.* 1, 1795.

"DEAR SIR,

"It is but just you should be apprised, that this
fruitful spot is part of my Lord Pelham's farm at
Bishopstone. It is indeed an undeniable fact, that
this single piece of ground, containing 31 acres, produced more than 40 loads of wheat in the last year.
My authority is the miller who has lately purchased
Bishopstone-mill, to which the land in question is
conti-

contiguous, and who puchased the whole of the produce. He told me, he believed there were 42 loads.
He gave different prices for the crop, as it was brought
in, from 20*l.* to 24*l.* per load. Forty-two loads at only
20*l.* amounts to 840*l.* The tenant's rent to my Lord
Pelham, the miller told me, is 50*s.* per acre. He has
then the whole piece at 77*l.* 10*s.* This satisfies me
that the miller told me truth, when he assured me,
that he knew the tenant cleared more than 700*l.* in the
last year by this single piece of land.

" You may remember, Sir, that I once pointed
out to you this rich part of Bishopstone-farm; and
that I informed you it yielded Car, when tenant, six
fine crops of wheat in as many years. In the seventh
year he sowed it with pease, that it might be cleared
of weeds by hoeing; and in the eighth year sowed
wheat again. There are two pieces of the land which
have this fertile property, which are separated only by
the road leading to the mill. The one is the piece we are
speaking of, which contains 31 acres, the other contains 17 acres.

" I do not remember to have ever seen manure laid
on any part of the whole 48 acres; and were manure
never to be laid, I think it would not be impossible
soon to make a small fortune out of the land, though
wheat were at 10*l.* the load.

" Adjoining the above two pieces of land, is a large
piece of grass-land on the right, and several smaller
pieces on the left, which I make no doubt, were they
to be broken up, would be found equally fruitful, being
on the same level, and having, no doubt, been re-
scued from the sea at the same time. They cannot
contain much less in quantity than the other two pieces.

" If the above intelligence proves of service to you,
it

it will add greatly to the pleasure of your obedient humble servant and friend,

" C. Hurdis."

11. *Manufacture of Bread.*

The common preparation in the manufacture of wheaten bread, is too universally understood to need any recital in this place; but the late very high price of bread corn, which has been so affecting in its nature, and so alarming in its tendency, induced several gentlemen and others, who were friends to their country, to set about trying experiments in order to ascertain what other substitutes could be devised in the hour of scarcity, equally nutritious as wheaten bread, and sufficiently palatable so as to ward off those evil consequences which threatened so speedy an approach. Among other substitutes for this end, none appeared more efficacious, none that bore a stronger analogy to wheat with respect to the above essential requisites, than a mixture of potatoes. In order to obtain this end, the Earl of Egremont, with that regard for his country, to which it is out of my power to do justice, undertook various experiments in the composition of bread with the meal of potatoes, and wheaten flour, with a view to determine the true quantum of potatoes which should be a standard for making a sufficiently agreeable, and nourishing substitute instead of wheaten bread. By his Lordship's directions the loaf was composed of

1st,   1 lb. of potatoes .... 1 lb. of flour
2d,   1 ditto .... 2 ditto
3d,   2 ditto .... 1 ditto.

These several mixtures were baked on the 18th of
December,

December, 1795, and were not proved till six days afterwards; for it was affirmed, and with great jus› tice, as the result demonstrated, that potatoe bread, when new, is certainly apt to pass off too quickly, without affording that sustentation so necessary to the labouring class of the community. Now this defect in the potatoe bread, by being kept for some days previous to its being eaten, is taken away, and the composition of the materials admit of longer keeping than other bread, without being deprived of any of its good qualities. When the bread made of the above materials was eaten, the result was, that between the bread which was one half potatoes and one half flour, and that which was two thirds flour, and one third potatoes, the difference was so trifling as to admit of little observation or remark. Both were equally pleasant, and the latter, of two thirds potatoes and one third flour, though not equal to the foregoing, in point of flavour, as it yielded rather a bitter taste, from the preponderancy of the potatoes in the mixture, yet it turned out far from unpleasant or disagreeable, and indeed highly superior to that which is the ordinary bread in many of our Northern and other counties.

Relative however to the general result of these trials; great doubt was entertained whether the bread answered in point of nourishment: it is good enough for those who have plenty of other food; but deficient for others who depend altogether, or very much on the staff of life: the general opinion was *against* the practice.

His Lordship has also tried rice in the manufacture of bread, and none could possibly be finer.

Another valuable and interesting piece of intelligence at the present crisis is, the discovery of a substitute for
yeast,

yeast, or for lessening the quantity commonly used,
which is made at Petworth from the fermentation of
potatoes.   The bread from the result of the foregoing experiment was kneaded with the yeast prepared from potatoes, and a small quantity of common
yeast.—Three pound of potatoes put into three pints
of water, boiled till it becomes a mash, then taken off
the fire, and the liquor and potatoes strained through a
cullender : one pint, or rather more, of milk is then
mixed with it, and left to ferment, and this quantity
is sufficient for a bushel of flour.

II. BARLEY.

1. *Preparation.*

The preparation for barley, if preceded by wheat,
is two or three earths : as soon as the harvest is over
the wheat stubble is, by intelligent farmers, fallowed
up, and whatever other tilth is requisite, is given in
the spring : the winter frosts ameliorate and pulverize
the clods, and render it better prepared for the reception of the seed, and land can hardly be too mouldy for
barley.   After turnips, three earths are usually given,
ploughed cross ways ; or if it follows pease, the tillage
is much the same.   When the pease are off the ground,
the stubble is fallowed up, and the remaining earths
given in the spring : other variations in the rotation
and number of earths are certainly found, which depend upon the degree of intelligence and skill, or to
the want of it, of which the farmer is possessed.

After the wheat has been carried, Mr. Woods turns
his sheep into the stubble, which is soon after begun to

SUSSEX.]   be

be ploughed for barley, that is, when wheat seed time is past. He fallows up the stubble about seven inches deep, and ploughs three times.   In the tillage of his land he has a practice which is peculiar to himself : that of opening his furrows in his wheat stubbles, where barley is intended to be sown in the spring.   First, the furrow is opened by ploughing and throwing the soil upon the stitch, leaving a small space of about six inches in width, which is ploughed the third time.   This operation is performed through the field, in dry weather, early in October, and remains in that state until the wheat sowing season is over, when the three furrows are ploughed together to form one ridge.   This work should be executed in dry weather, or at least in the driest season that occurs in November or December ; and afterwards the remainder of the land to be ploughed as a fallow for barley.   His reason for this mode of ploughing is this ; that when the land is ploughed a second time in March, the ridges open the more freely, and the furrows will not be found to be stubborn and difficult to work, nor the ground to be rough and cloddy in the furrow after that practice, so as to produce an unkind ridge in the third ploughing, which is often the case when such opening is neglected ; insomuch that barley frequently makes no appearance for the space of three feet on each ridge.

It is to be observed, that Mr. Woods is particularly careful not to lay his ridges too round, nor too high at the wheat season, least the lands be too flat when fallowed up for barley, as he is clearly of opinion that the land sustains a greater injury by the retention of the water in the fallow, when the ridges were reversed and lay hollow in the middle, than at any other time ;

and

and in addition to this, greater loss is felt in the crop of barley by such a method, than can possibly be the case in the wheat by laying the land flat.

Potatoes are a good preparation for barley. Mr. Gilbert dunged for wheat, and after the wheat planted potatoes, which gave four hundred bushels per acre; he then took barley, which was a better crop than another piece sown after barley following wheat, which was dunged for equally with the other.

2. *Sort, Seed, Quantity sown, Time; Culture, Harvest, &c.*

The only sort of barley which is in general cultivation, is the common English barley : it is never steeped ; the quantity of seed is various, and depends upon circumstances ; but it vibrates from four to five bushels and a half to the acre. The time of sowing is either the latter end of March or beginning of April.

Mr. Woods, after the wheat has been reaped and carried, fallows the stubble, and in April sows four bushels and a half to the acre ; it is hoed after weeding the wheat is finished, and after wheat harvest is finished he cuts his barley : if foul, it remains three or four days on the ground before it is cocked, which is done in small heaps, as they dry sooner, besides not being so subject to be trod by the pitcher. Mr. Woods obtains the finest crops of barley upon a pea stubble, but upon a cold winter fallow, the barley is not so productive ; this necessarily depends upon the soil, culture, situation, &c. The produce may be estimated to vibrate from three to six quarters. Perhaps the average is four quarters. The Weald of Sussex is for the most part composed of too heavy a soil for the culture

of

of barley, and the proportion which it bears to the remainder of the county, is too considerable to call Sussex
in general a great barley district.   In some few places
of the Weald, where it is cultivated, the average is thus
estimated :

Slaugham, Worth, 16 bushels.

Cuckfield, Horsham, 24 bushels.

Shipley, West Grinstead, Ashurst, 26 bushels.

But these products bear no comparative proportion
to those which are obtained on the Downs.

III. OATS.

The Weald of Sussex is well adapted to the growth
of this crop.   It generally follows either wheat, barley,
turnips, potatoes, or beans.   Two ploughings are
given, the first in winter, from three and a half to five
inches.   When the crop is on ley ground, the field is
broken up with a single ploughing.   The quantity of
seed is various ; from four to six bushels is the accustomed allowance, which is sown in March and April.
Mr. Gilbert, of East Bourne, had a field of oats, which
at its first appearance out of the ground was very unfavourable, so that he had thoughts of ploughing it
up : however he drove a large heavy roller of 35 cwt.
and twenty-four oxen in it, repeatedly over the field
in the spring, and it turned out a most abundant crop.
Many soils in this county require a similar treatment,
before any produce can be expected.   The crop is various, and depends altogether upon circumstances :
from four up to eight and nine quarters are gained.

On the fertile land about Walburton, Mr. Henry
Murtel has grown 1320 bushels upon ten acres ; which
is

is at the rate of sixteen quarters and a half per acre, upon a very adhesive clay loam. The Earl of Egremont, over a sixteen acred layer, broken up and sown with Dutch blues, has gained one hundred and sixty quarters; ten quarters per acre. It is this land, now indeed bearing such noble crops of corn, that was lately a forest, and absolutely unproductive. Above seven hundred acres have been thus improved! What a noble undertaking !

Slaugham, Rusper, 16 bushels.

Worth, Horsham, 20 bushels.

Rudgwick, Kindford, Wisperer-green, Billingshurst, Hitchingfield, Crawley, Ifield, Balcomb, 24 bushels.

Shipley, West Grinstead, Ashurst, Warneham, Cuckfield, 28 bushels.

Horsham, Slingfold, Pulborough, Chiltington, 30 bushels.

Salehurst, 32 bushels.

IV. RYE.

Rye is much cultivated on the South Downs as food for sheep. It is sown in August and September; the earlier, the better it is. In spring, when other food is scarce, and in the lambing season, ewes and lambs are turned into it : a certain portion is hurdled off for this purpose.

V. PEASE.

Pease are much cultivated in Sussex, especially on the South Downs, and along the maritime district. The common preparation is to sow them after one ploughing, either upon a wheat, barley, or oat-gratten; the land is ploughed from four to five inches; four or five bushels of seed are sown. The produce is very various—from two and a half to four, and even five quarters per acre. They are often drilled; many farmers preferring this method to the common one of broad-cast.

When Mr. Ellman drilled pease, he used the Kentish drill, and found great advantage in shifting the draft by a staple in the axle, and a notch in the pillow; drilling thus at eighteen inches instead of two feet; the wind drove them together so, that they united well, which they will not equally at two feet. His greatest crops, however, have been broad-cast, in which way he has had as high as five quarters and a half per acre.

Mr. Carr approves of drilling by skimming and hand-hoeing; has thus had four quarters of Marlborough greys; but Mr. Davies, pursuing the same system, has not gained two quarters.

Mr. Woods prepares his land at Chidham in the following manner: after harvesting, the barley-stubble is occasionally fed until January, when it receives a single ploughing for pease, which he drills in rows eight inches asunder. It is a rule with him never to plough twice for pease, especially too on cold ground, as he finds by experience, that the soil is put into a much worse condition by this practice; besides, cold land

land is not able to receive the plough early enough
for pease to be sown in that manner. Mr. Woods
ploughs for pease six inches deep, drilling four bushels
to the acre. The wheels of his drill-machine, by moving after the drill, covers the seed, and obviates the
necessity of harrows. He has tried wheat and barley
drilled, but without effect; but for pease, drilling
answers well. When the plant is three or four inches
above ground, Mr. Woods harrows, and frequently
rolls them in March, to loosen and prepare the ground
for hoeing them in April. Two five-inch hoes are fixed
at three inches apart, and between which the drill
passes in such a manner, that a man draws it after
him : of this work a man will hoe an acre in a day.
They are cut about the middle of July, by hacking
them with a long handled hook, and wadded into
small parcels or locks ; as soon as they are harvested,
the stubble is well harrowed, and carried into the
yards for making dung. I shall close this account
of the cultivation of pease with noting the average
product of this crop in several parishes where they
are cultivated in the Weald ; which it may be proper
to remark, is not a soil well adapted for them.

West Grinstead, Slaugham, 10 bushels.

Worth, Rusper, 12 bushels.

Balcomb, Horsham, 14 bushels.

Ifield, Cuckfield, Rudgwick, Kindford, Wisperer-
green, Billinghurst, Hitchingfield, 16 bushels.

Warneham, Horsham, Slingfold, Pulborough, Chiltington, Shipley, 20 bushels.

Hurstperpoint, Albourne, Bolney, 24 bushels.

Haylsham, Ditchling, 30 bushels.

VI. TARES.

VI. TARES.

The cultivation of tares is well understood, and in many parts successfully practised. They are used for cattle, horses, and sheep; and sometimes hogs have been folded upon them. From two to three bushels are sown upon the stubbles in autumn, and in the spring they are wattled off with sheep; one acre, at 4*d.* per week for ewes and lambs, is worth 40*s.* to 60*s.* In summer, horses are soiled with tares; and they are of such infinite importance, that not one half of the stock could be maintained without them: horses, cows, sheep, hogs, all feed upon this valuable plant. Upon one acre, Mr. Davies maintained, at Beding-ham, four horses, in much better condition than with five acres of grass. Eight acres have kept twelve horses and five cows for three months (June, July, and August), without any other food. Spring tares are sown from April to June. Horses thrive upon them surprisingly, and no plant is able to vie with this excellent food.

Mr. Halstead cultivates them at Lavant with great intelligence and success. He has sown three bushels of seed to the acre upon a wheat stubble, 5th of Sep-tember. When these have made their appearance above ground, and are strong, he throws in a second crop, and then, in like manner, a third, about a month intervening between each sowing. By one crop of tares succeeding the other, he ensures a crop for the whole summer, of the best food that can be given to cattle.

They have on the South Downs an admirable practice in their course of crops, which cannot be too much commended;

commended; that of substituting a double crop of tares, instead of a fallow for wheat. Let the intelligent reader give his attention to this practice, for it is worth a journey of 500 miles. They sow forward winter tares, which are fed off late in the spring with ewes and lambs; they then plough and sow summer tares and rape, two bushels and a half of tares, and half a gallon of rape; and this they feed off with their lambs in time to plough once for wheat. A variation is for mowing—that of sowing tares only in succession, even so late as the end of June, for soiling. October 6th, a crop was finishing between Lewes and Brighton, on land which had yielded a full crop of winter-sown ones. The more this husbandry is analyzed, the more excellent it will appear. The land in the fallow year, is made to support the utmost possible quantity of sheep which its destination admits; the two ploughings are given at the best seasons, in autumn, for the frosts to mellow the land, and prepare it for a successive growth of weeds, and late in spring to turn them down; between the times of giving these stirrings, the land is covered with crops; the quantity of live-stock supported, yields amply in manure; the treading the soil receives previous to sowing wheat, gives an adhesion grateful to that plant; in a word, many views are answered, and a new variation from the wretched business of summer-fallowing discovered, which, by a judicious application, would be attended in great tracts of this kingdom with most happy consequences to the farmer's profit.

A practice which Mr. Thomas Ellman adopts at Shoreham, is that of breaking up his layers (clover, ray, and trefoil) for summer-tares and rape. What an an immense improvement is this upon the common slovenly custom in Norfolk, of ribbling, or half, or bastard-ploughing such layers! a miserable practice, yet very general amongst the spirited cultivators of that celebrated county. Preparatory to this practice, Mr. Ellman, in his system of tillage, sows rye-grass with his spring corn, which is laid for two years During this time it is twice folded, when he breaks it up in May and June, and sows rape and tares, fed with sheep in August and September.

The benefit of sowing rape and tares in this manner, Mr. Ellman discovers to be inestimable. The common system of cultivation in this neighbourhood would be, to break up the layer, and fallow it for wheat, at an expense of full 4*l*. per acre ; but this experienced farmer pursues a very different course : instead of an unproductive fallow, he gains a noble crop of rape, with all the expense of raising it paid by the crop; besides thoroughly preparing the ground for the succeeding crop of wheat.

VII. COLESEED.

Cole is deservedly in high repute amongst the flock farmers of the Downs. It is sown either with tares, or by itself, as food for sheep ; not frequently for seed. Ewes and lambs are wattled upon it in spring, and it is generally allowed to be most efficacious and highly nourishing to the young lambs. Mr. Gilbert, of Eastbourne, at the lambing season, seldom allows his ewes any other food but this, as the rape produces a larger supply of milk than turnips; which he thinks has the effect of extending the udder, without affording any
consider-

considerable flow of milk.   This gentleman, some
years ago, lost 80 or 90 of his ewes by slipping their
lambs, which he attributed to feeding them on rape
about Christmas ; yet he had fed them on it before, without being attended with any such effect ; the sheep had
been hard kept.   He has since heard of the same thing
happening amongst other farmers ; but it is remarkable, that a neighbour fed his rape over the hedge at the
same time, without any inconvenience of the kind.   Mr.
Gilbert sows ray-grass with his rape for sheep, on Down
land ; one gallon of rape-seed, and two of ray-grass.
The rape is fed off first; and after that the ray-grass
rises and affords a spring bite.   June and July is the
usual season for putting in this crop, one gallon to the
acre : when folded, a rood and half is a sufficient daily
consumption for 600 sheep.

VIII. TURNIPS.

The cultivation of this very valuable root is thoroughly well understood ; and the high degree of importance which is attached to it in the economy of a
flock farm, renders it an object of the last consideration
among the South Down farmers. Turnips for many years
have been cultivated in this county, and with increasing success.   Indeed, so great is the dependence upon
them, that it is the first object to secure an abundant
crop for the winter and spring provision of their flocks.
The common tillage is to plough three or four times,
or more, to pulverize the soil, and render it as fine as
possible, and to extirpate all weeds ; the preceding is
either a crop of corn, or pease, or tares, &c.   Many
farmers carry the dung rough out of the yards.   In
                                                    this

this manner Mr. Ellman carries all his for his turnips, without giving it any previous stirring, or mixing it with earth or lime, &c. ; for it is clear with him,
that much of the virtue of manure is lost by stirring.

In some parts of this county, liming for turnips is
practised : it was first adopted in the neighbourhood
of Hastings, and the effect has been such, that the
practice has not declined. Mr. Clutton limed nine
acres at Cuckfield, in 1793, the expense, 30 guineas ;
two horses ploughing one acre and a half per day :
six oxen will finish one and a quarter in the same
time. About Midsummer, the seed is put in ; from
one to two pints of seed to the acre. Good farmers hoe twice.

Mr. Ellman observes, that in hoeing with the common Norfolk hoe, more of the weeds are drawn together than are cut up, and if rain come, most of these
weeds shoot again ; but his own hoe, the blade of
which is but an inch wide, effectually cuts up every
thing, whilst the weeds and earth pass freely over it,
at the same time that none of the earth is collected.
This hoe ought by all means to be used on turnip
farms, where the soil is inclined to be light and sandy,
but on those of a heavier tendency, the hoe should be
wider.

The attention which Mr. Ellman has given to eradicate weeds, is another instance of good management.
Kilk or charlock, is the most destructive foe to which
the chalk hills are liable, yet a blade of it is never visible upon his farm ; whilst between Lewes, Eastbourne,
and Brighton, almost every farm is overwhelmed with
this weed. His neighbours have been frequently surprised at seeing his turnip crops upon land similar to
their

their own, and apparently with similar management, whilst they are not able to grow any.   This has been a frequent object of remark; but there are some circum-stances in his management which will explain the rea-son.   Mr. Ellman pays great attention in saving his seed, by transplanting some of the largest and roundest turnips in his garden, and in rejecting all those large ones which indicate any hollowness in the crown of the plant, which forms a cavity for the rain to lodge on it, and thus cause the turnip to rot.   By constantly sow-ing such seed, which he annually saves, he contrives to get fine crops ; and by setting them out very thick, he raises very heavy ones. He begins to sow early, and raises several pieces in succession. His turnips are this year (1797) upon rye grass, which he folds in spring ; he then ploughs in June four or five times for turnips, hoes twice, setting them out very thick, remarking at the same time, that the small crop and thick one will exceed the other considerably.

December 9, 1796, he measured, numbered, and weighed, two perch of turnips.

	Tons.	cwt.	qr.	lb.
One perch of middle sized contained one hundred and ninety four, which weighed 437 lb. which is per acre,	31	4	1	4
One perch of the largest sized con-tained one hundred and forty-five; which weighed 399 lb. which is per acre,	28	10	0	0
In favour of middle sized,	2	14	1	4

One

One hundred and ninety-four turnips to a perch, is allowing a space of sixteen inches and three quarters and a fraction for each turnip : the other is at the rate of twenty-one inches.   Mr. Ellman is clear that fifteen inches is fully sufficient for each turnip, or two hundred and eighteen turnips for every perch.   Some experiments of this sort have been registered in the Annals of Agriculture, which leave no doubt as to the advantage of setting them out thick, and close.

In 1793, Mr. Ellman had thirty-five acres and a half of turnips : he began folding the beginning of October, and fed twenty-seven acres and a half till the beginning of March : six months complete, with five hundred ewes and three hundred and twenty lambs, besides carting off eight acres for cattle.

The generality of farmers pay little attention in cleaning their land of kilk, &c. nor do they dress it fine at each time of ploughing, but lay it rough ; nor sufficiently observe to let the harrows and roll follow the plough as quick as possible, to prevent the earth from drying, &c.   With respect to the manure, Mr. Ellman lays it on after the second ploughing, carried out in small heaps into the field from the farm-yard, as conveniently as possible for running it out : first, it prevents the carts from treading much over the ground when it is dressed very fine, which would cause it to bind and turn up at the succeeding ploughing very close, the carts going with three horses and four oxen.   When the dung is got into high fermentation, he then sets on to plough the third time, and lets the harrows and roll follow the plough immediately, to break the clods occasioned by the carts and cattle.   In this system, the ploughing, dunging, and
                                                sowing,

sowing, is in such succession, that the seed is laid as it were in a hot-bed, which makes it come up rapidly, and vegetate remarkably quick. By this means, he has never failed securing an abundant produce. He sows his light lands every fourth year, which is not the common practice of the county. It should be remarked, that he never sows in wet weather, or whilst it rains ; since after the land is worked in that light, or pulverized state, before the last time of ploughing, or before sowing, the harrows, by going over the land, encrust the surface, which, when it becomes dry, the young plants find a difficulty in penetrating through. After ploughing in his dung, he runs a light roller and a brisk harrow, to raise a little dust on the surface.

Hoeing is done by the acre ; 6s. 6d. the first time, and 3s. the second. In folding his sheep, Mr. Ellman draws them out of the ground two or three days before the sheep are turned into the field ; by this method, which begins to be general, the turnips lose their watery property, and the sheep thrive on them much better.

Similar to this is the practice of Mr. Carr, who, in folding his sheep, draws up all the turnips within the fold, a day or two before the sheep are allowed to enter, in order that the turnips might wither, and evaporate their water. The reason is, that when the sheep ate them without this precaution, many were lost.

With respect to the distribution of stock to ground, Mr. Ellman finds, that twenty acres will fatten one hundred sheep, if turned in in a lean state, and feeding from October 1st to the end of March. In other parts of the county, about Chichester, they calculate,

that

that one acre maintains one hundred ewes with their lambs a week. The tankard turnip is sown when wheat succeeds, as it yields more early food.

In the summer of 1797, the Earl of Egremont, as an experiment, sowed one acre with turnips in the park. As they grew up, part of the acre failed. In September his Lordship filled the vacant spot with plants drawn from the neighbouring crop, and the whole is now (January) one continued mass of green food, and the roots of considerable size and dimensions. These transplanted turnips are very flourishing, although the experiment was undertaken too late in the summer to expect so favourable an issue ; and it answered so well, that his Lordship means to extend this beneficial practice over his whole crop of turnips.

SECT. V.—CROPS NOT COMMONLY CULTIVATED.

I. BEANS.

IN the few places where beans are cultivated, they are generally after wheat, as in the neighbourhood of Shoreham, and some other places. Mr. Breseton, grows the mazagan at Pagham, and horse beans are planted by Mr. Peachey. It has been frequently asserted, that the bean system might be introduced to great advantage in the heavy soils, and most materially tend to ameliorate the present system of husbandry, substituted in lieu of a fallow. This idea struck the Earl of Egremont, who, in March 1795, planted two acres, three feet row from row : the land

was

was covered with a mixture of thirty loads of stable
dung and good mould ; the tops were cut off when in
blossom, and reaped in October, but it was an indifferent crop.

In 1794 his Lordship tried them before, and the
result much the same ; but as that year was very unfavourable to beans, he attributed it to the unkind-
ness of the season. The same land bears excellent
pease, turnips, &c. and other grain in abundance.

That beans may not answer in some years, is certain. The cultivation has been attempted in other
places, but with little success.

Early in the spring of 1797, the Earl of Egremont
made another very capital experiment on this subject :
he ploughed up a grass layer five years old, of seventeen acres, and as fast as the land was ploughed, every
other flag was dibbled with horse beans : iron dibbles
being used, and about two bushels of seed per acre put
in, the moment the rows were visible, they were ef-
fectively hand-hoed, and throughout their growth a
shim of various shares was constantly going through
them. These shares consisted of one cutting plate of
twelve inches, two of three inches, nine apart, and
a central one of nine inches, and a double mould-
board plough expanding at pleasure for earthing up :
by means of these tools skilfully applied, the crop was
kept in garden cleanness, notwithstanding the incessant rains which fell that year ; all weeds which grew
among the plants were carefully extracted by hand.
The crop was viewed by many noblemen, gentlemen,
and farmers, as a beautiful exhibition of perfect husbandry. They were pulled, the crop good, but not
thrashed at the time of writing this account.

The stubble was broad-shared, harrowed, and

SUSSEX.]   ploughed

ploughed once for wheat, which now makes a very good appearance*.

After having viewed the stiff and rich soils of this county, I venture to recommend beans in the following course, as a modification of their own :

1. Tares,
2. Oats or wheat,
3. Clover,
4. Beans,
5. Wheat.

But that beans would not answer upon their land, was the general opinion of all the farmers; that they had been tried, and did not give equal crops with pease. If the trials that have been made were all done with skill and intelligence, and often repeated, this is satisfactory : after clover, beans have not generally, if ever at all, been tried : on this soil they should perhaps be dibbled by hand (provided a tool was not to be had that would drill them) in a straight line along exactly the middle of every other furrow, that is to say, in rows at eighteen inches asunder, on the richest land : on soils not equally rich, the same, with double rows at nine inches, and then one missed, in which way they would come up in rows at nine inches, with intervals for the shim at eighteen. And on still poorer soils, every furrow to be planted three or four inches from bean to bean. If some of these intelligent farmers will make this experiment with care, and keep the beans by horse and hand-hoeing clean, it may possibly be found a valuable acquisition : nor let them forget, that to have winter plough-

---

* The crop turned out greatly, and the husbandry continued with success.

ing in a dry time, on a clover ley, is, on such ticklish soils, being as much at their ease as they can be.

II. POTATOES.

The cultivation of this very valuable root is in high repute, and the management of it ordered with the greatest success. Indeed the culture of it might form so important an article as an ingredient in the food of men and cattle, that it is not a little singular, that it has not spread with greater rapidity, when allowed to be of such infinite utility. The late very high and alarming price of bread-corn ascertained the value of potatoes, and directed the public attention to the production of this root, which, in case of necessity, might prove a substitute for wheat; and the inquiry which the Board of Agriculture instituted with a view of determining the comparative merit and qualities of potatoes as a *succedaneum*, has naturally excited much attention. This root certainly possesses great merit as food for man, and doubtless, when the culture of it is more extended, may be found upon further trials to be as equally beneficial and nutritious as bread-corn.

Preparation.

In the neighbourhood of Battel, Eastbourne, and Chichester, are cultivated the greatest quantity of potatoes. It is upwards of twenty years since the first introduction of them into the Sussex husbandry, for fattening bullocks; and the farmer (Mr. Mayo, of Battel) to whom the county stands so highly indebted, has had the most productive crops of wheat sown upon potatoe land*.

The

---

* Wheat after potatoes may answer in the neighbourhood of Battel,
but

The course in which they are introduced, is various. Mr. Mayo has them in the singular course of, 1. Wheat; 2. Potatoes; confining the culture, for the convenience of vicinity to the potatoe-house and yards, to two fields, which are alternately under those crops, and have been so for twenty years, being manured every potatoe year. That they do not exhaust or decline from being on the same land, appears by the last crop being better than any of the preceding, and the wheat always good. The soil, a loam on a moist bottom. The manure is put into the furrows at three feet asunder, to the amount of sixty loads to the acre, each sixteen bushels. About Eastbourne they are planted upon three ploughings, from three to seven inches deep. About Chichester, the crop is put in after turnips. For the growth of potatoes, the weather, according to Mr. Mayo, cannot be too hot and dry; he finds that potatoes do not draw the land more than clover, and he builds his theory on the fact, that where the ground is overshadowed and covered, fermentation, so favourable to vegetation, is thereby excited.

Sorts.

The sorts in cultivation are various; about Chichester, chiefly the *golden-dun*, and *ox-noble*. Mr. Mayo prefers the *cluster*, before any other sort, as the oxen like them best. Mr. Clutton planted as well he *hog*, or *cluster*, as the finer sorts, and thinks that he former pushes an ox as forward as any other sort. Mr. Fuller, of Heathfield, plants the *golden-globe*.

---

but with me, on a light strong soil, it has generally been unproductive. I have found oats after potatoes answer much better than wheat.— *W. D.*

Quantity,

Quantity, and Method of Planting.

From sixteen to twenty bushels are planted. Mr. Gell plants at Applesham twenty, one foot distant from plant to plant. The method of planting them practised by Mr. Mayo is, to open the furrows at three feet asunder, in which he drops the sets one foot from each other, covering the sets with dung, and then covering by hand, by drawing the earth over them with hoes. He has tried whole potatoes, also small, and cuttings of different sizes; but little or no difference in the crop. One year he made a variation in the distance of the rows, putting them in at four feet, and gained a very fine crop; but he prefers three.

Time of Planting.

From the latter end of April to the middle of May, is considered as the properest season for ensuring a plentiful crop; but the season for planting should be regulated by the sort. Mr. Mayo's season used to be the end of March, or the beginning of April; but he has for some years been steady to the beginning of May, from experience having clearly convinced him that it is the best season.

Culture.

Mr. Mayo both hand and horse-hoes his potatoes, as well as earths them up. Mr. Gell weeds, by hand-hoeing close to the plant; after which he runs a double-breasted plough to earth them up, and they are generally put out to be taken up by the bushel. About Chichester, they use the prong for this purpose.

Produce

Produce.

The produce is various, and depends upon the fertility of the soil, culture, season, &c. A common crop on Mr. Peachey's land near Chichester, is 400 bushels per acre : the soil a hazel mould upon a red brick earth ; a soil which agrees remarkably well with chalk ; of which manure he lays eight bushels to the perch. Mr. Mayo's average crop varies from 350 to 400 bushels ; he has grown 500, and even 600 ; Mr. Gilbert from 300 to 400, without manure, on good land, in rows at two feet and a half. Mr. Calverley, of Broad, on an old hop ground, has raised 700 per acre; Mr. Fuller, of Heathfield, from 400 to 450 ; the late General Murray 400.

All these are great products, and cannot fail impressing us with a high opinion of the culture and soil from which such considerable crops are gained.

Method of Preserving.

Perhaps the greatest objection to this root has been the difficulty in preserving it through severe winters ; and to guard against such hazards, is a point of some importance in the cultivation of them. Mr. Mayo preserves them by digging a hole proportioned to the quantity to be put in, usually two feet deep, and over this to build a house ten or twelve in height, with walls six feet in thickness, made with hay and chopped straw plaistered ; the entrance is filled with haulm, or straw. Sometimes, in very severe weather, a charcoal fire is kept up in an iron kettle.

In the severe winter of 1788-9, General Murray, who was one of the greatest cultivators of potatoes in the county, preserved all his crop during that winter in in the utmost safety and security, notwithstanding the intensity of the weather; and hardly had a rotten potatoe in a hundred bushels.

When this circumstance is well considered, in the pinch of such a season as that was, every one will agree, that the vast experiment made by General Murray in the introduction of this root, as a winter and spring food for sheep, was truly important.

His magazines for preserving the potatoes, are holes cut in the side of a hill, five or six yards wide, ten feet deep, and of an indeterminate length. The carts from the field unload at the top, shooting them at once into the hole, and they are taken out at that end at bottom, which opens to the slope of the hill, where a wall is built to it, with a door, &c. When full, a stack of stubble or straw is built over the whole, wide and large enough for security against all frosts. In this method, it seems, the largest quantities may be kept together; for no earth or other means of keeping the effluvia of the roots being used, it rises through the stubble, and does not occasion their rotting from heat; on the other hand, the stubble is thick enough to exclude frost.

The preservation of potatoes in severe weather is a difficult business, when very large quantities, as in this case, are laid together. Whilst the magazine is full and kept untouched, I have no doubt of the preceding method; but it is doubtful, after it is begun, and there is a vacancy in it: the air in that vacancy, it is apprehended, would rot them. Quere therefore, if stubble or straw must not be supplied to fill it close, as fast as the potatoes are taken for use?

Appl=

Application.

The chief use and object for which they are cultivated in Sussex, is the fattening of bullocks.   Mr.
Mayo has entered largely into this practice, and with
uniform success, for upwards of twenty years, and is
decided in the conviction of the profit of it.   He fattens every year six oxen, two steers, and four cows
or heifers.   They complain at Battel, that they have
no hay good enough to fatten a bullock, but with potatoes, all difficulties vanish.   An ox of 140 stone eats
rather more than a bushel per day, and ten pounds of
hay.   He has had beasts on turnips, that ate each
three bushels a day, and as much hay as if they had
no other food.   Some graziers that feed with oil-cake,
have come to see M.r Mayo's beasts, and have been
of opinion, that they fed as fast as on that expensive
food.   One farmer resisted the practice for many
years; at last he made an experiment, and found it
so beneficial, that he much feared the profit would
turn out little, from every one getting into it; thinking from the great advantage, that it certainly would
become general.

Mr. Fuller has fed many sheep till they were
quite fat, upon potatoes, and has kept to the practice.
Mr. Mayo has fed horses with them, and with success;
and Sir Charles Eversfield fed all his horses upon them
at Horsham.

Mr. Ellman, at Shoreham, has fed his oxen with
potatoes, at the rate of four gallons daily to each
ox : one is given in the morning; another soon after,
and the remainder at different times in the course of
the day.   In other places, three gallons is the usual
allowance.   The common quantity to an ox on Mr.
Gilbert's

Gilbert's farm, is from one and a half to two bushels, unwashed and uncut, except a few of the largest.   Mr. Gilbert finds that an ox of 160 stone, eats from one and a half to two bushels, but consumes little hay: this is a great saving, as he considers potatoes a much cheaper food than hay.   The cattle will rarely eat them for the first two or three days, but like them much afterwards, and they fatten upon them much quicker than on hay alone.

Mr. Clutton, of Cuckfield, has fed his oxen largely with potatoes; but the experiment did not answer. When he fed on this root, the usual allowance to each ox was one bushel and a half, and as much hay as they chose.   The bullocks choaked, and the moisture loosened them, besides being blown and much physicked. Mr. Clutton has fattened twelve oxen at a time: beasts which have arrived at 140 stone, and which fed on hay alone, ate about half a hundred weight a day; have had a bushel of potatoes, and a quarter of a hundred weight of hay. Mr. Clutton has been paid 9*d*. per bushel, but not so much in general: in the above feeding, a bushel is set against a quarter of a hundred weight of hay, which, at 3*s*. per cwt. is just 9*d*.   But hay on the farm cannot be reckoned so high ; 2*s*. per cwt. would make the potatoes 6*d*. per bushel.

Feeding with potatoes well known about Lewes, but opinions do not very well agree.

Mr. Carr feeds oxen with potatoes and hay, and afterwards with oil-cake; but neither answered.   Mr. Davis thinks they answer when well got up, that is, dry : he used 200 bushels for young beasts, and also for hogs : a heifer took well to them; had half a bushel with straw; 60 bushels made her nearly fat.   Their use

use appeared to him so considerable, that he would buy them if he could at 6*d.* a bushel for cattle.

Mr Saxby, of North Ease, bought 120 bushels at 6*d.* carrying them himself eight miles ; gave them to fat his heifers ; and they paid so well, that he desired to have more this year at the same price : hog-potatoes. Nobody washes them, even when dirty.

Mr. Hicks, and Mr. Sharp, of Laughton, gave potatoes to working oxen, and they did well : have had 700 bushels stacked up for the same use. Upon inquiry, if they ever gave them artificially sprouted, or remarked them to be better when naturally so in the spring ; they replied in the negative, but approved the idea ; Mr. Davis remarking, that one bushel of barley malted had, with him, been better than two bushels of oats not malted.

Steaming.

At Petworth is an apparatus belonging to Mr. Fawkner, for steaming potatoes. He had often fed hogs on raw potatoes, but found that they fell off their flesh, and throve badly, which induced him to steam them. His contrivance is nothing but a hogshead cut in halves, the bottom raddled, and mortared half way into a small copper, and coarsely covered with a wooden lid. The potatoes are done so quickly, that six tubs are steamed in a day, which is nearly double the number that could be boiled in the water : there should however be an easier way of clearing the copper from the dirt that will, in spite of washing, gradually collect, without the necessity of breaking the mortar joint.

The

The expense of cultivating an acre, is thus estimated by Mr. Mayo:

	£.	s.	d.
Rent, tithe, and rates, ...	0	14	0
Ploughing, ...	0	19	0
Harrowing, ...	0	2	0
Rolling, ...	0	0	6
Manure, 70 loads, at 16 bushels, three carts, one acre and a half per day,	1	10	0
Filling and laying in the furrow, ...	0	7	6
Cutting 18 bushels, at a halfpenny, ...	0	0	9
Planting in the furrow, ...	0	1	0
Covering with hoes, ...	0	2	0
Hand-hoeing once, ...	0	2	0
Weeding, ...	0	1	0
Earthing up, ...	0	2	0
Horse-hoeing, one horse, one man and boy, five acres daily, twice thrice, at 9d.	0	1	10
Taking up and putting in carts 450 bushels, at three farthings per bushel, ...	1	8	0
Carting home, ...	0	4	3
	£.5	15	11*

The greatest and most important point of all, and which should be ascertained in the clearest manner, is the value which bullocks pay for them in fattening. Mr.

---

* These particulars are from Mr. Mayo himself; but I must confess they seem to me to be near fifty per cent. under what they would be in most situations with which I am acquainted.—*A. Y.*

Mayo

Mayo is clear, and has no shade of doubt, that they pay 4d. per bushel.

Produce.

450 bushels, at 4d. .... £.7 10 0
Expenses, ... 5 15 10

Profit, ... £.1 14 1

Decisive experiments resulting from weighing the bullocks alive to and from the food, would be more satisfactory; but this valuation of 4d. is the lowest we have met with : and we are not entitled to doubt of the accuracy of observation and calculation of a man who has been fattening oxen for sixteen years on this food.

The late General Murray was in the constant habit of feeding a very large flock of sheep on potatoes; they were given in a manger: 710 ewes in winter, ate one-third of a ton of hay, and 22 bushels of potatoes, every day, which is a quart to each. He used potatoes for fattening sheep, as well as for lean stock; 196 fat wethers ate 14 bushels and 1 cwt. of hay daily: it may be reckoned 14 bushels for 200 sheep. As fat sheep are to be supposed to have as much of both as they will eat, it should seem, that if they have as many potatoes as they will eat, they do not require more than half a pound of hay each. The General gave potatoes to his working oxen, and found that half a bushel with oat-straw was equal to 40lb. of hay.

280 lb.

280 lb. of hay for a week, at 2s. 6d. per cwt.	£.0	6	3
Deduct oat-straw for seven days ; suppose it	0	1	0
Remains value of three bushels and a half of potatoes, ...	£.0	5	3

This is 1s. 6d. per bushel ; a higher value than has been found in any other application*. It should be calculated from a bushel a day, and that would make 9d. for the value of a bushel.

A variety of experiments respecting the culture and growth of this valuable plant, have been undertaken by the Earl of Egremont, but more especially by raising them from shoots. The following are well worthy of attention.

May 12, 1795. A potatoe weighing 6¾ oz. was put into a pot full of earth, and plunged into a hot-bed.—May 22, six shoots were taken from it, and the potatoe was replaced in the hot-bed.—June 9, fourteen shoots were taken from it, and planted. The potatoe was then weighed, and found to have increased in weight 2½ oz. weighing 9¼ oz. The potatoe was immediately replaced in the pot, but was not put into the hot-bed.—July 3, twenty-five shoots were taken from it, and it was placed in the pot, and removed to a hot-house.—July 13, fourteen shoots were taken, and the potatoe returned to the pot, and replaced in the hot-house.—July 23, it produced 26 shoots. The potatoe now weighed 9 oz. and was quite firm, and not in the least degree shrivelled. Many

---

* Indeed so high as to appear to me extremely questionable.

more shoots might probably have been taken from it, but it was boiled, and found to be a very good eatable potatoe, although it required much time in boiling : in this process it lost an ounce. The shoots were planted, 95 in number, all from one potatoe, and were thriving and strong plants, according to the different dates of planting. The experiment was begun much too late in the season; but it was only suggested at that time, by some appearances which were observed in some potatoes from which the shoots had been taken. Three potatoe plants from these shoots were uncovered during their growth, and appeared full of young potatoes; at least 20 were counted to each plant, and they were only partially uncovered, and the earth immediately returned.

Planting with shoots appears to be as productive, and more so, than is the case with eyes. In another experiment of his Lordship's, 140 yards were planted with shoots; these produced at the rate of 377 bushels per acre: 180 yards were planted with cuttings; yet the former produced one bushel more than the latter: the whole yielded 21 bushels: they were planted in June, and taken up in October. The early kidney, if planted in June, comes to perfection in October. With respect to the method of breaking off the shoots from the potatoes, there is no reason to be apprehensive how they are taken off; and if the shoots, after they are separated from the potatoe, are put into a basket, and have a little earth thrown over them, they will keep in this state, if not immediately wanted, for months. The kidney, after its second cropping, decreases in the number of its shoots. No sort equals the red cluster. The early kidney comes up before any other; the cluster is next, and the ox-noble last.
The

The cluster throws out more shoots than any other; up to thirty at a time.

Part of another field was planted with shoots, and without manure ; yet these latter turned out a greater crop than any of the former, trenched, mucked, &c. Some of the shoots were planted early in March : a fortnight's severe frost afterwards affected them : the consequence was, that the frost cut them off ; yet they again recovered, and were equally good as the others. All the flowers of these potatoes, for experiment, were broken off : if the leaves and tops are taken off, the root is materially injured ; for the stalks of several were cut close to the ground : the earth was afterwards uncovered, and not a single potatoe was to be found, which appears to be a proof of the bad effect of cutting the tops, which some people so zealously contend for. The shoots are planted promiscuously from one to six or seven inches long : till the third or fourth month after planting, the shoots have but a small apple, not above the size of the end of a person's finger ; but afterwards they wonderfully increase their size : hence it follows, that the eyes having a greater and more substantial root to support the vegetative power of the plant, comes easier to perfection ; and that the shoot, though stationary at first, will in the end more than equal the other in produce ; and if we add the saving of seed, the advantage will be still more considerable.

Hence we may infer, that this method of cultivating potatoes, which is practised with success at Petworth, merits the attention of farmers ; for an early market it is the only method of raising them, and the seed and expenses of cutting are saved. In addition to the
above

above intelligence, some further valuable information has been inserted in the Annals of Agriculture, from the same quarter, where the cultivators of this root will find these experiments more amply detailed.

III. BUCK-WHEAT.

Mr. Davis had one year eight acres of buck-wheat at Bedingham, which his shepherd fed with the flock when in full blossom, for two hours: all were drunk; the glands of three were swelled quite to the eyes; none blown; but were staggering and tumbling. On hogs it had the same effect: bleeding made the sheep worse; however he lost none.

IV. LETTUCES.

The same gentleman made a remarkable experiment on lettuces for hogs. He has practised it often, but not with equal attention. He sowed four ounces of white coss-lettuce seed the beginning of March, very thick over two perches of ground. His crop of potatoes was in rows at three feet. In May, he planted a double row of lettuces between the rows of potatoes. After that, both crops kept clean by hand-hoeing.— June 7, they were begun to be used for three sows with little pigs; they were kept on these lettuces six weeks; then the pigs were weaned a fortnight earlier than usual; and after weaning, the great use of the lettuce is found, for the pigs did admirably well on them, till all were gone.—August 15, they were fed with cabbage, turnip-tops, &c. as usual, but fell off at once for want of lettuce. The sows had wash.

This

This trial deserves attention: weaning pigs without a profusion of milk and some corn, is a difficult business; and if lettuce will do it, a man ought never to be without a rood, or half an acre, for this purpose.

V. HOPS.

In the eastern part of Sussex, they are much cultivated; but the expenses and uncertainty of the crop, tend strongly against the culture. The expenses are indeed very great. At Battel they are thus estimated:

Rent, tithe, and taxes, ...	£.1	8	0
Digging, &c. by contract, ...	3	0	0
Tying, ...	0	7	0
Poles, at 15s. per 100, ... £.7 10 0 ⎱	8	12	0
Interest, ... 1 2 0 ⎰			
Carriage, at 3s. per 100, ...	1	10	0
Picking, at 7s. per cwt. 6 cwt.	2	2	0
Drying and bagging, at 6s.	1	16	0
Tax, ...	2	10	0
Oast, 500l. for 20 acres, interest, ...	1	5	0
Manure; dung 100 loads once in three ⎱ years, 33 loads a year, at 1s. 3d. .... ⎰	2	0	3
	24	10	3
Interest, ...	1	4	0
	£.25	14	3
Produce.—Average crop, 6 cwt. and ⎱ average price, 3l. per cwt. .... ⎰	£.18	0	0
Loss per ann.	£.7	14	3

But

But here comes an observation which must have its weight, and which seems to prove that there must be a fallacy in such calculations. The number of men who are fond of hops, is great; and we are hardly to suppose, that they are all fools, or that none of them have the capacity to form such an account as this. But, on the other hand, I desire, and may very rationally request, " shew me by what means hops are profitable." If you ask my profit upon wheat, I shall state the expense, the average produce, and the balance between the two. If you ask Mr. Mayo the profit upon potatoes, he recites 1*l.* 14*s.* per acre; and he replies satisfactorily, because he does the same: he gives you the expense, and he gives you the produce; but when we come to hops, it is here as in Kent, in Essex, in Suffolk, and every where else: we have general assertions of profit, and when we come to examine, we find particular accounts of loss.

From this Battel account, let us deduct 1*l.* 5*s.* for oast, which I suppose is all an expense of the landlord; all the manure, or 2*l.* 3*d.*; also the interest of the first stock of poles, 1*l.* 2*s.*; and likewise that of stock in trade, 1*l.* 4*s.* amounting in the whole to 5*l.* 11*s.* 3*d.*; and then we shall have reduced the annual loss of an acre to 2*l.* 4*s.*

Upon twenty acres, a capital of 500*l.* in the landlord's pocket, and of 920*l.* in the tenant's, are sunk, without paying a penny interest; 600 loads of dung diverted annually from the profitable uses of the farm; and all this for the yearly loss of 44*l.* !

I take the fact, from all the information I have at different times, and at different places, received, to be this. Hops are the gambling of farmers: men put into a state-lottery, knowing that there is

a vast

a vast loss upon all the tickets, though immense benefits are made on some. And farmers are equally sensible, that if all hops are taken into the hop account, loss will be the balance of it; but they enter into the cultivation expecting the prizes of the hop-lottery. Others there may be, that will do the same thing, but upon more prudent principles: they will voluntarily submit to the annual loss of a few pounds, in order to have that certainty, which in some years confessedly arises, of a large sum at once from a great crop. This certainty, however, arises but seldom; for the great crop alone will do little: it must be a great crop when some considerable districts get a small one; that is to say, it must be a great crop, and a great price at the same time. To make hops answer steadily, several circumstances must unite; some other manure must be used than the dung of the farm, which cannot, consistently with profit, be thus diverted: they should be in espaliers, to save in the expense of poles, and to throw the binds nearer the ground, on the principle of vineyards, which never ripen well, nor yield plentifully, when the vines are suffered to rise high. Another reason is, the power of picking as early as you please, without cutting the binds. The latest picked hops will always give the best crop the following year.

The parish of Salehurst contains the largest plantation of hops in the county; between three and four hundred acres. Mr. Pooke, of this place, an intelligent and practical farmer, had twenty-five acres; the largest produce which he ever gained was twenty-one ton; 13 cwt. 24 lb. upon twenty-one acres and a half. During the time of picking, all the women and children are set to work; from 1000 to 1200 are employed,

ployed, one year with another, for three weeks or a month. In crop years, still greater numbers are employed; above 500 hands from other parishes find employ. The average produce, 9 cwt. per acre, at 4*l.* per cwt. One acre has yielded 100*l.* Mr. Pooke has sold a crop for 1800*l.*; but the same land, in other years, had only brought 40*l.* The crop was so deficient in 1793, that he set his on the ground that year at 1000*l.* less than a medium year. By the same method of calculation, the deficiency of the parish was estimated from 10 to 12,000*l.* The calculation of the annual expenses of planting and keeping up one acre of hops, which this sensible planter was so obliging to draw up according to a request I made him, I shall insert in this place, as it is done with exactness, and the result of more than half a century's experience.

" I begin the year with January; in which month, or the beginning of February, it may be proper to plough up land to plant with hops.

		£.		
Ploughing one acre nine inches deep, if the mould will allow it, set at	}	£.0	10	0
Harrowing the same very deep and fine, just before it is planted, ...	}	0	4	0
Making 1200 holes with a spade, to plant the hops in at six feet square, which is the quantity of hills an acre will contain at that distance (though they are planted at various distances, according to the humour of the planter); the holes to be made square, about nine inches over, and about nine inches deep, of which a man will make one hundred and a half in a day, which, per acre, is	}	0	12	0
Carry forward, ...		£.1	6	0

Brought forward, ...	£.1	6	0
Twenty-four cart-loads of dung and mould, well mixed together and rolled, each load to contain 12 cwt. at 1s. 3d. per load per acre, ...	1	10	0
If we allow one good shovel-full, or rather more, to each hole, the putting of which in the holes, and filling them up with the best mould that comes out of them, made fine by chopping it a little in the hole with the spade, ...	0	12	0
Six plants or sets for each hole, the cost of which will be, per acre, ...	1	16	0
Laying the plants out, and setting them in holes per acre, ...	0	4	0
Shimming the acre twice in each alley, four times over in the summer, ...	0	16	0
Hoeing round the young plants, where the shim cannot come, and weeding,	0	2	6
One shovel-full of the dung and mould laid on each hill soon after Michaelmas, which will take up 24 of our cart-loads, at 1s. 3d.	1	10	0
A year's rent for the land, ...	1	5	0
All sorts of tenant's taxes, ...	0	7	6
	£.9	9	0

" No produce from this first year's planting; and as in Sussex in general, it is better to re-plant the hops once in ten years on the same, or plant fresh land, a proportionable part of the above sum should be added every year, to the future expense of keeping the hop-ground up the ten years it may be supposed to remain.

" January,

" January, the second year, but in which the hops will be first poled.

Digging the young hops in this month, or February, at 1s. 4d. per 100, or per acre, ...	£.0	16	0
Dressing and mending the young hills with more plants, if needful, ...	0	2	6
Sharping the poles, and setting them up, at per acre, ...	0	15	0
Tying the vines to the poles, ...	0	10	0
Half-hilling, that is, laying about three good shovels-full of fine earth, taken out of the alleys, on each hill, at 2½d. per 100 hills, or per. acre, ...	0	2	6
Hoeing round the hills, there being more to hoe when poled, being farther round,	0	3	0
Shimming, as before, ...	0	16	0
Branching, ...	0	1	6
Whole-hilling, at 5d. per 100, or per acre,	0	5	0
Incidental expenses, such as tying, or setting up poles that may be blown down,	0	5	0
Stripping and stacking the poles, ...	0	4	0
One year's wear of the first poles, there being three to a hill, or 3600 on the acre, at 12 feet long, ...	4	10	0
Tithe, ...	0	10	0
133 loads of dung and mould laid on each acre, a load to every nine hills, once in two years, to be spread and dug in at 1s. 3d. per load, is 8l. 6s. 3d.; the half of that to be charged yearly,	4	3	1½
Rent, ...	1	5	0

Carry forward, ... £.14 8 7½
Brought

Brought forward, ... £.14	14	8	7½

All sorts of tenant's taxes, such as poor-tax, church-tax, way-tax, and hundred-tax, set at	0	7	6
Part of the expense of the first year's planting, with something for the interest of that money, set at, yearly, ...	1	5	0
Interest for the money laid out yearly on the acre of hops, ...	0	15	0
Dressing and mending the hops with plants, will in future be more than I charged for the young ones, by per acre, ,..	0	2	6
The hop-ground will in future require larger and longer poles, the yearly wear of which will be 7*l.* 4*s.* per acre, that is 2*l.* 14*s,* per ann. more than the wear of the first poles, which added to the above 4*l.* 10*s.* charged for the wear of the first poles, makes it in future 7*l.* 4*s.* per ann. Expense to pole the acre of ground, ...	2	14	0
The early expense of raising and keeping up an acre of hop-ground for 10 years in this place, is ....,.......................	19	13	1¼
To keep it in high cultivation, as it is generally done in this parish, it may safely be set at yearly per acre,...........	20	0	0

" Suppose this parish to average at 8 cwt. of hops per acre, one year with another, which I do not think it does, there would be little or no profit at 4*l.* per cwt. as I will endeavour to prove.

The

The expense of the hop-land yearly, as above,	£.20	0	0
Picking, drying, duty, and sending to London, being our market, and selling them, will cost 1l. 10s. per cwt.	12	0	0
	£.32	0	0

" I am persuaded in my own mind, if the hops should not keep up to a smart price, the plantation must be reduced. I have bought, in my time, 100 poles of the same sort, and at the same place, at 4s. 6d. as have been sold about two years ago at 21s. per 100. It was 55 years since I bought them at 4s. 6d*.

" The parish of Salehurst, which is usually called Robertsbridge, from the name of the village, is supposed to have the best plantation of hops in the county, the land being kindly for them, especially about the church, where it is rich. This county may average at from 5 cwt. to 6 cwt. per acre, one year with another, but not more. This parish has upwards of 300 acres of hop-ground in it at this time, but I cannot say exactly how much.

" In my calculation of the value of different articles in raising and keeping up hop-ground, I may have proba-

---

* The usual length of hop-poles is from 14 to 16 feet. Ash are the best, excepting chesnut, which are not commonly used : willow are good, if they are not set the first year; if so, they are apt to grow crooked : beech very bad.—*A. Y.*

Various and manifold have been the advantages resulting from the study of horticulture, but in no one instance would it prove more truly beneficial than in producing a dwarf hop. I propose it to be a subject for a premium : the art of the gardener would certainly accomplish it.— *Mr. Trayton.*

bly

bly set some rather too high, and others rather too low ; but, upon the whole, I believe it will be found to be nearly the truth, and not any great error, I presume, in any of it. Our county in general do not grow the crops we do, neither is it so well cultivated and manured ; consequently are not at the same expense : generally, they that look after it best make most of it ; but there are some pieces of land so unaccountably kind for hops, that they are better by nearly one-half the value of the expense of maintaining it, than others are adjoining ; and yet the best judges on earth could not have known which would have succeeded before they had been planted ; but there are infinite quantities of land where there is not any part of it ever can succeed. The great uncertainty of the crop, occasions hops to be a subject of gambling ; and so many people speculating on them, increases more the uncertainty of the price ; that some people will be getting money by planting and dealing, when others must be losing. Upon the whole, I really believe there are such great quantities of hop-ground planted where there must be money lost by it, that although many have undoubtedly been very considerable gainers, yet take the whole body of planters together, I have my doubts whether it has been of much advantage to them ; but as soon as one is tired, another will take it up ; for so long as there are a few go d prizes in the lottery, people will buy tickets, though the chances of gaining are against them."

V. CABBAGES,

Are little cultivated ; only by a few individuals is it that any attention is paid to them. In the strong
soil

soil which is ill adapted to turnips, cabbages would turn to great advantage. It is by interweaving such crops as these, by judicious management, with corn, that such an arrangement of crops will better support each other, and rear up a greater proportion of stock.

The culture of cabbages would perhaps be one of the greatest means of meliorating the husbandry of the Weald. But that they do not appear every where adapted to this adhesive clay, is seen from the cultivation of them in the Stag-park at Petworth. The preparation by manuring, tillage, hoeing, &c. were well attended to: there was a crop, and a tolerable one; but they burst, and rotted. Wet seasons and humid soils may sometimes cause this; but draining is a remedy; and from the capital manner in which the Stag-park has been drained, I have not a doubt of its now producing as fine crops of cabbages as need be expected.

Mr. Davis was for some years a cultivator; he sowed the seed in his garden in March; if the season was wet, later. May or June, he planted, either upon plain land or ridges; the latter, the best method. With regard to the comparative value ascertained with turnips, he estimates an acre to be more than equal to two acres of turnips. He first plants the *flat Dutch;* after this, the *drum-head;* then the *Scotch.* The Dutch are the best, as they stand hard weather better than any others, and weigh heavier; he has had them as high as 30 and 40 lb.

June, 1789, he planted one acre and a half of the flat Dutch (sent him by mistake for the drum-head); he horse-hoed twice, and twice hand-hoed. In January, having no other food for his cows than straw, he began cutting, and gave his cows three per day each,

with

with straw, for a week; he then increased the quantity
to six or seven: he found by a winter milch-cow he
had then, that they produced a great deal of milk;
his other cows gained flesh very fast. The beginning
of March, all the cabbages were consumed; he then
gave hay twice a day; but the cows fell off, and did
not thrive equally as upon cabbage and straw: three
cows out of six slunk their calves.

The next year he planted five acres with three sorts,
drum-head, long-sided, and Scotch field-cabbage, of
which the former were the best, although he was partial
to the flat Dutch, as hardier. In winter he began feeding his cows; and they did well, and found the supposition of cabbages causing the slinking their calves,
to be erroneous, having none that did it. He gave
cabbage to his fattening bullocks, which were then
upon corn, reducing it a third, as he found that
both together was rather too strong food: they throve
faster than before with corn. He had at that time
seventeen ewes (which a ram had stolen among), that
yeaned twenty lambs; as he had nothing but hay and
corn, except cabbage, he shut them up in a barn,
and fed with cabbage three times a day, and with
corn twice. The lambs soon began to eat; he then
weighed once a week, and they gained one pound and
a half for the first three weeks, and afterwards more;
at eight weeks he sold for 25s. per lamb, weighing
eight and nine pounds per quarter: the ewes went off
in July at 25s.

The following is the weight of a crop, communicated
by Mr. Kemp, of Coneyborough.

Seed sown middle of March; planted June 1, and
watered, and twice afterwards, the season being a dry
one.

Novem-

November 7, 1794, weighed 40 ; the average size (the whole weight 680 lb.) being 17 lb. each. Quantity of land, three roods sixteen perches. Number of cabbages 4116, at 17 lb. each, is 31 ton 4 cwt. 84 lb. (112 lb. to the cwt.) Sorts, an equal number of Scotch, and drum-heads.

VI. CARROTS.

Carrots are not cultivated to any extent, as food for cattle : they might unquestionably be of singular advantage, where the soil is light, and deep ; perhaps of all other applications whatever, they are the most beneficial for fattening bullocks : they are excellent for horses, and good for sheep. Mr. Kemp found them well-suited as food for horses ; his largest produce has been 280 bushels over two-thirds of an acre. It is greatly to be regretted, that the excellent qualities of so valuable a plant are not better understood. The Earl of Egremont, in the winter of 1796, fed his large dairy of cows with carrots, and with great success, and the butter excellent.

The following statement of a crop of his Lordship's, shews the vast productiveness of carrots, and of parsnips.

Roods.	Bushels.	Bushels.
43 of carrots	270, equal to 1004 per acre.	
53 of parsnips	203, ditto	613 ditto.
6 waste.		

102

And this too without manure. Certain it is, that after parsnips, they are the richest food that grows;
more

more valuable than potatoes, and much more so than turnips. Upon sandy loams, there cannot exist a doubt of the superiority of them (with good management, and judicious application) to every other food cultivated in England for stock; they are more productive than most other roots, are more nutritious; are drier, and more saccharine (excepting parsnips); and many thousand acres there are in this county, which might be cultivated to immense advantage in this manner.

VII. RHUBARB.

The Earl of Egremont cultivates the *rheum palmatum* for medicinal uses, and has it dried and cured in as good order and preservation, as any imported from abroad. It is taken out of the ground in autumn, after standing seven or eight years; it is then washed of the dirt, and dried, either in the sun, or laid over the flues in the hot-houses, after having been cut into pieces. Mr. Andre, the domestic surgeon and apothecary, uses no other, and finds no difference between this and the foreign. A considerable saving might be made in the importation of *rhubarb*, if others cultivated it for like purposes.

VIII. OPIUM,
(*Papaver somniferum*).

The largest quantity of this invaluable drug that was ever cured in this country, was raised in 1797 from the Earl of Egremont's garden at Petworth; and the fact now indeed thoroughly ascertained, that all the foreign opium is highly adulterated, renders it an
object

object of immense consequence to encourage the domestic growth. Mr. Andre is convinced, that in all his practice, he never made use of any of this drug that could be compared with this. The operation of collecting the produce, is effected by a gentle incision on the heads, as they grow, with a knife or other sharp instrument; which is frequently repeated; and the juices which exude from the wound, are scraped into an earthen vessel, dried by the sun, and preserved for use.

IX. SAINFOIN.

Very considerable tracts are peculiarly adapted to the culture of this invaluable plant.

The chalk-hills contain many thousand acres of land, upon which no other plant has ever yet been discovered to thrive to such advantage as this, and none which ever promised fairer hopes of success to the industrious farmer for the expense of cultivating it. Calcareous earths, of all other soils, are the best suited to the growth of sainfoin. But the exertions which have been made in this line, are weak and feeble, and certainly not commensurate to the merits of the plant.

Sowing.—It is usual to sow the seed with barley in the spring. Another, but a worse method, is either with wheat in autumn, or harrowed on it in the spring.

The soil should always be in good heart, clear of all rubbish, and always succeed a capitally managed turnip fallow. If the land is turniped for the two preceding years, a better crop will be ensured; and twice feeding enriches the land*. Six bushels are sown with

---

* Turniping land for two years previous to sowing it with sainfoin, is, in my opinion, unnecessary. If it is properly attended to one year, it will in general be found sufficient, and 3 or 4*l.* per acre saved in expenses, by a crop of corn gained.—*Mr. W. Dann.*

the

the corn, and perhaps not quite the usual quantity of barley, that the effect of too large a produce might not endanger the tender shoots.

Feeding.—Sainfoin, after having been mown, should not be fed till Michaelmas, when it will afford great plenty of grass till Christmas; it must then be laid by for the scythe, but never fed close: sheep, by close feeding, are apt to bite the crown of the root, which injures the plant. Many are decidedly of opinion, that it ought never to be fed. The duration, from eight or ten years to fourteen or fifteen years, if well manured with ashes. It is the best food which can be given to lambs, being sure to preserve them in a good habit of body; and they are particularly fond of it. It is equally acceptable to horses, and no hay comparable to it. Sheep will feed upon it till Christmas, without the expense of either turnips or hay; and there is no other mode of managing chalk-hills to such profit, as no substitute will maintain such a stock.

Mr. Pinnix, of Upmardin, has a thorough knowledge of the value of sainfoin; and the great breadth of it over his farm, is a feature in the economy of his business, which at once indicates his superior discernment in this branch of his profession. He has discovered, that the cultivation of sainfoin enables him to keep a far greater number of sheep than any other artificial grasses; and it is generally the poorest soils which are laid down with it. For this reason, upon a farm like his, valued at 7s. per acre, the land sown with sainfoin cannot be estimated at more than 5s. Now every possible expense in laying this land with sainfoin, is repaid by the two crops of turnips, and the barley.

barley. With such a preparation, so excellent, and so highly to be commended, the sainfoin will last good to mow full ten years, and be worth 50s. per acre, which is ten times the rent of the land ; and all this without any expense, either for seed or of tillage. For the next four years it may fairly be valued at 20s. per ann. The produce varies from 25 to 30 cwt.

X. LUCERN.

Lucern is commonly cultivated in the immediate neighbourhood of Eastbourne and Brighton, with a sprinkling in a few other places ; also about Chichester. About Eastbourne, sown broad-cast, and many pieces are very fine ; 20 lb. of seed to the acre upon the richest and deepest soils ; not answering to advantage upon any other. They mow it three times, to soil their teams in the stable. It is likewise made into hay.

The Rev. Mr. Durnford has for some years cultivated lucern in a small way at Berstead. He prepared the land with turnips and a fallow, marled and harrowed, and the seed drilled in eighteen inches from row to row, the beginning of June. The first year was once out ; the second year, three times ; the third, four times. It is now in its eighth year, but it begins to fail. Mr. Durnford stocks it at the rate of four horses, and as many cows, per acre in summer ; it gives excellent butter ; but it ought to be cut the day before it is given to the cows. After each cutting, it is hand-hoed.

The Earl of Egremont has cultivated it in the paddocks for soiling cattle, drilled. It was thrice mown, and wherever it failed, his Lordship scattered chi-

cory-

cory-seed; and the whole produced an abundant herbage*.

XI. CHICORY.

This plant, by the experimental improvement of the
age, has recently been introduced to the knowledge of
the farmer. For rapidity of growth, luxuriance of
burthen, nutritious qualities of the food, and duration, it stands unrivalled. All sorts of cattle and
sheep feed upon it with avidity. Where it has been
cultivated, it is usually sown with Lent corn, mixed
in a certain proportion of other artificial grass-seeds.
The Earl of Egremont, having ascertained the merit
of it, spread the cultivation over several acres, and
finds it a most useful and profitable plant. In 1798
and 1799, he had above 100 acres of it, and the use it
was of to him, in the support of an immense stock of
cattle and sheep, exceeded every thing that could have
been expected from the soil; and the benefit would
have been still greater, had that stock been still larger,
for much of it ran for seed; but this afterwards produced an evil, by ploughing it for wheat.

---

* By means of sainfoin and lucern, with the addition of clover, the
farmer is furnished with what he calls *artificial grasses*, suited, with
good management, to almost every kind of soil. The first, to the chalks,
gravelly and stony lands; the second, to light lands; and the third,
to clays. Few places indeed are so happy as to admit the cultivation of
all with equal success; and yet we observe these three growing side by
side, at the foot of the South Downs, near Eastbourne, seeming to vie
with each other which should flourish the most, and yield the greatest
crop. But this was in a soil wherein the calcareous and argillaceous
were so happily mixed, that almost any vegetable might succeed; and
yet here we sow them, ploughing up a stubble upon a level, where
there is not a stone to impede them, with eight stout oxen.—*Flor. Rustier.*

CHAP. VIII.

GRASS LANDS.

———

SECT. I.——NATURAL MEADOWS AND PASTURES.

THE management of the meadow and pasture lands
does not materially vary from those common practices
which usually govern other counties in this important
division of the work. Here indeed there is but too
much reason to complain of negligence, with respect
to the improvement of grazing land. Pasture over-
run with rubbish, or covered with standing water,
from inattention to draining, are the necessary consequences of such a slovenly conduct. Many opportunities of watering meadow land are at present lost
to the owners : a due attention to the principles of
irrigation, in converting the various streams which in-
tersect the country, to these useful purposes, would open
mines of inexhaustible treasure. Irrigation is but lo-
cally known. It is only in the western parts of Sussex, that any signs of it are conspicuous. The want
of a proper mode of managing pasture is the more reprehensible, because it is obvious, that the Weald in
general, from its natural quality for grass, as well as
from the uncertainty of ensuring the production of full
crops of grain, is far better adapted to the raising of
cattle, than corn. The tenantry here, from considering
the corn product as the main object in view, lose sight
of that arrangement which the nature of the soil should

ever

ever dictate to the farmer the system to follow. Wet and tenacious hungry clays seldom pay the owner the expenses of cultivation; and when, in addition, is considered the natural impediments to corn, which flow from a country so thickly interspersed with wood-land, we are surprised at that attention to tillage, which occupies the thoughts of those farmers, and characterizes all their measures. Hence it is we discover the difference between the circumstances of the farmers who live on the South Downs, and in the Weald. The former adapting the crop to the land, know the sensible effects of such a system; the other expect the same consequences, when working with different materials.

The following are the grasses which are found on the gohanna-ground in the neighbourhood of Petworth, in very good up-land meadow moderately moist; the most numerous first.

Trifolium pratense.
Festuca pratensis.
Cynosurus cristatus.
Holcus lanatus.
Ranunculus acris.
Heiracium spondyllum : sheep and hogs very fond of it.
Plantago lanceolata.
Lolium perenne.
Centaurea nigra.
Trifolium repens.
Anthoxanthum odoratum.
Poa annua.
Poa trivialis.
Rumex acetosella.
Poa pratensis.
Lotus corniculatus.

Lathyrus

Lathyrus pratensis.
Achillea millefolium.
Phleum pratense.
Avena flavescens.
Dactylus glomeratus.
Ranunculus bulbosus.
Ranunculus repens.

Marsh.

Besides the natural pasture and meadow-land, are several thousand acres of marsh-land, either situated along the coast, or in the neighbourhood of the rivers which empty themselves into the sea. These marshes perhaps are to be ranked amongst the finest of their kind that are any where to be met with; and the conduct of the grazier in the management of the fertile level, is the direct reverse of that unsystematic policy which is the guide of the up-land farmer in the arrangement of his grazing land.

Very considerable improvements have been effected of late years in the marshes. The brooks or levels have been, and are now, sometimes subject to be flooded with the violent rains which periodically flow from the hills, but more particularly in the winter. If, as is sometimes the case, these inundations take place in summer, the whole produce of the land for that year is lost by the stagnant muddy water; and no cattle will taste the herbage that year. The tide is another evil sometimes complained of, as the banks are not every where put into a proper state of defence against the incursions of the sea. An act was however obtained a few years ago, for widening the channel near Lewes, and making a shorter cut to the sea; and it has essentially benefited the Lewes and Laughton Levels.

SECT. II.—CLOVER, TREFOIL, RAY-GRASS.

THE artificial grasses in the highest request, and chiefly cultivated, are red and white clover, trefoil, and ray. These plants, which modern husbandry has brought into cultivation, must in every respect be considered as invaluable grasses, and adding in no inconsiderable degree to the wealth and prosperity of the farmer. In many places we find an almost universal growth of Dutch clover and trefoil. It is seen along the side of the turnpike-roads, in the lanes, and in every field on the south side of the hills; about Selsea, it springs up spontaneously in the greatest luxuriance; and by clearing the land of spear-grass, and other weeds, judicious management in a few years would convert these lands into the finest meadows in the world. There are no better plants than these; and the indigenous growth should excite farmers to cultivate these excellent plants, and obtain a fine fleece of clover and trefoil, where none is visible at present but the spontaneous growth.

The quantity of seed is various; but the following is considered as the proper proportion:

Dutch clover, ...	2 gallons.
Trefoil, ...	2 ditto.
Ray, ...	4 ditto.

This is for permanent pasture; but when the land is laid down for a layer of one or two years, it is then,

Clover, ...	1 gallon.
Trefoil, ...	1 ditto.
Ray, ...	3 ditto.

However,

However, there is no fixed rule in cases of this kind. The quantity of seed sown is of little consequence in any of the operations of farming, beyond the hedge which bounds the field. The course in which these artificial grasses are introduced, is generally with barley and oats; sometimes with wheat in spring. Clover is certainly the most valuable of any of our grasses; but land has been known to be surfeited with it, when repeatedly sown.

The cultivation of our best natural grasses has been long called for, and lately recommended by that elaborate botanist, Curtis, and by many others, as likely to turn up a very valuable acquisition. No branches of the art of agriculture are less understood, than a right knowledge of the properties of our grasses, and the soil congenial to each. Till very lately, they were entirely neglected, excepting ray, and one or two others, all of them inferior to many of those in a natural state.

As there is undoubtedly a particular period when the grasses are in a proper state for mowing, and as that state is most probably about the time of their flowering, should all the under-mentioned grasses be found, upon fair trial, to deserve cultivation, the following diagraph would seem to divide them into proper assortments to be sown together; supposing the fields or meadows where they are to be sown, to be principally intended for hay. If an assortment for *three* crops only be desired, the brackets on the right hand will shew the division. If *five* crops are required, the brackets on the left hand will direct to the assortment; in the division of *three* parts, the first crop will be fit to cut early in June; the *second* about Midsummer; and the *third* about the middle of
July.

July. In the division of five *parts*, the *first* will be
ripe about the latter end of May; the *second*, the
beginning of June; the *third*, about Midsummer; the
fourth, about the beginning of July; and the *fifth*,
the middle or latter end of July.

The *annual meadow, vernal, smooth-stalked meadow, small fescue, dogstail, yellow oat,* and *fine bent*,
seem to be best adapted for the feed of sheep; the
rest for the larger kinds of cattle;—the *soft brome,
smooth-stalked meadow, smaller fescue,* and *yellow
oat*, are partial to dry soils;—the *vernal, foxtail,
rough-stalked meadow, quake-grass, meadow-fescue,
soft grass, meadow-barley, catstail,* and *marsh-bent*,
flourish most in moist soils; and soils of an intermediate quality, as to moisture and dryness, will best
suit the remainder.

1 { Annual meadow (poa annua), flowers first
week in May.
Vernal (anthoxanthum odoratum), flowers second week in May.
Foxtail (alopecurus pratensis), flowers second
week in May.
Soft brome (bromos mollis), flowers third
week in May.

2 { Smooth-stalked meadow (poa pratensis), flowers fourth week in May.
Rough-stalked meadow (poa trivialis), flowers first week in June.
Smaller fescue (festuca ovina, rubra, durius-
cula), flowers first week in June.
Quake-grass (briza media), flowers second
week in June.

} 1

Rough

3 {
Rough cocksfoot (dactylis glomerata), flowers second week in June.

Tall oat (avena elatior), flowers second week in June.

Meadow fescue (festuca pratensis), flowers third week in June.

Darnel (lolium perenne), flowers fourth week in June.

4 {
Dogstail (cynosurus cristatus), flowers fourth week in June.

Yellow oat (avena flavescens), flowers first week in July.

Soft grass (holcus lanatus), flowers second week in July.

Fine bent (agrostis capillaris), flowers third week in July.

Meadow-barley (hordeum pratense), flowers third week in July.

Catstail, (phleum pratense), flowers third week in July.

Marsh-bent (agrostis alba), flowers third week in July.

2

3

In laying down land with artificial grasses, clover, trefoil, ray, burnet, &c. it has been supposed, that to feed the young layers the year they are sown, is prejudicial to the future growth of the plant. The Earl of Egremont has laid many acres with Dutch clover, ray-grass, and burnet, in one field, with red clover, ray-grass, and trefoil in another; and to discover whether close feeding was detrimental, his Lordship covered these layers with sheep in the autumn, and at Christmas, after having contributed to fatten many wethers

wethers for Smithfield, others were turned in, and nothing could be more favourable than the future progress of these layers. His Lordship is satisfied, that so far from its being injurious to the grasses, it is highly advantageous to feed them, as it enables the plants to throw out a thicker, more vigorous, and luxuriant herbage, the following spring; he therefore constantly pares them to the root in the autumn and winter, and again in the spring, and through the summer.

SECT. III.—HAY HARVEST.

In the operation of hay-making in Sussex, there are no particular features in the management which deserve commendation. If the season permits, it generally commences about the end of June: after it is cut, the swarth is shook out; it is then heaped into small cocks; the second day it is windrowed, and sometimes made into the larger cock, and the third day carried: this is when the weather is favourable. But hay-making so much depends upon the state of the weather, and the judgment of the farmer, that there is no fixed rule of proceeding, where the work depends upon contingencies. An improvement in making it would be, to have the hay always cocked at night. Meadow-land is mowed every year, and afterwards fed: upland pasture is cut every second year. The produce of the first, rises to two ton, and upwards, per acre; the other seldom, upon an average, exceeds a ton and a half. Clover yields from one and a half to two tons and a half.

The following singular and interesting method of applying linseed-oil on hay-ricks intended for fat-

tening

tening beasts, merits the attention of the curious. It
was communicated by the Earl of Egremont.

" SIR,

" I received your letter in regard to oiling hay.
I made practice of it about three years; but always
choose to do it when the weather is fine, and can get
it up without taking much rain. My method is, when
stacking the hay, to take a water-pot, and sprinkle
over every layer very lightly a quart of linseed-oil
to a ton of hay. I find that the hay comes out of the
rick very moist and very clammy: fatting beasts and
fatting sheep are very fond of it, and thrive upon it
very fast. I think it not proper to give it to horses,
or milch-cows, as I think it is too hot. I wish it not
to be reported in my name, as I did it for my own
security.

" Your most obedient humble servant."

Salt.—When the unsettled state of the weather has
damaged the hay in the field, salt has not unfrequently been used, by sprinkling it with the hay in
forming the stack. Mr. Edsaw, of Fittleworth, and
his father before him, have constantly adopted this
practice, in the proportion of a gallon of salt to a ton
of hay. Mr. Edsaw has applied it to the hay which
has been well made, as well as to that which has been
damaged.

SECT. IV.—FEEDING.

AFTER the hay is cut and carried, pasture-land is usually fed with cattle and sheep. Few traces of any well ordered systematic arrangement are here visible. In the Weald, where much of the land is under grass, the aftermath is pastured with bullocks, cows, young stock, &c. That admirable practice, of reserving the *rouen* for the pinching part of the spring, when all artificial provender fails, and before the young clover and other grasses have begun to throw out their shoots, is hardly known in the county. The Earl of Egremont has usually some portion of the Home-park wattled off for this reason, either for his Lordship's different flocks, or for the deer; and experience has declared the beneficial effects of it, for now he has it in his power to apply the hay for other purposes, and save a considerable consumption by the deer.

1799. His Lordship has continued this practice to the present moment, and with increasing success. He is now practically convinced, through a variety of severe and open winters, that the resource of *rouen* is one of the most important that can be secured on a farm. It is also a constant practice with Mr. Ellman, at Glynde, and Mr. Sherwin, at Petworth. Mr. Ellman usually saves 40 acres.

Upon flock farms it is usually important to ensure a provision of this nature, to supply the place which the deficiency of turnips, rape, rye, &c. unavoidably occasions at that season of the year.

In the marshes which border upon the sea, we find the grazier covering those fertile and exuberant levels with

with the greatest quantity of stock which the soil is
capable of bearing. The stock upon these marshes
consist of cattle as well as sheep. In the Level of Pevensey, cattle were universally preferred to sheep.
The marsh ground about Winchelsea and Rye, as it
wants fresh water, has been thought better calculated
for sheep : these grounds are universally stocked with
them ; and the general rule is, to have no more bullocks than what are sufficient to keep the pasture fine,
which is usually one to three or four acres. Pevensey
having plenty of water, was considered as better adapted for oxen. It should seem as if this circumstance
had governed the custom of the two marshes : the
soil and rent are nearly the same ; yet there are very
few fortunes made in Pevensey, but many about Winchelsea and Rye ; and this is attributed to sheep being
found to turn out so much more profitable than oxen.

But throughout the whole range of Pevensey
Level, it is to be observed, that the number of sheep
have been very much increased of late years. Graziers have now discovered, from the late rapid advances in the single article of wool, and the still increasing demand for it, that sheep pay far better than
beasts, whilst the loss is comparatively less.

It is not the usual custom to winter-graze cattle.
The land would be too much poached, and there is not
always a sufficient quantity of grass to feed them. In
warm weather the herbage grows so thick and luxuri-
antly, that the grazier's own stock, with what he is
able to procure from the hill-farms, is insufficient
to pare it down ; but then it is allowed, that when the
grass springs up in this very rapid manner, there is
little substance in it.

Men are generally employed in the Levels to mow
down

down the over-grown herbage, as it grows rank, and sheep or cattle are not inclined to feed upon it. An irresistible proof would this be, that they are greatly under-stocked (if superficially considered), and viewed only in May or June, when in the greatest luxuriance.

The increase in the quantity of sheep annually pastured, is to be accounted for from the good management of the grazier in laying his lands dry, by opening and keeping clean the ditches, and making drains at proper times to receive the superfluous waters. The introduction of sheep has also very much contributed to augment the fertility of the land, and with it, the quantity of stock; as sheep, by their close bite, pare down grass much nearer, and therefore leave no such waste as cattle: moreover, they expose the ground to be mellowed by the winter frosts, and which produces in the spring a much finer herbage, and a greater abundance of grass, which in itself is equal to manure; not to mention the amendment immediately flowing from the sheep.

The profit of these marshes is very considerable, as may be gathered from the following estimate.

The capital necessary for stocking 100 acres, will be at least 490*l*.: 250 sheep, and 150 lambs, and about a score of bullocks, of different ages.

Annual Expenses upon 100 *Acres.*

	£	s	d
Rent, at 25s.	125	0	0
Rates, at 4s.	20	0	0
Tithe, at 2s.	10	0	0
Scot, 1s. 6d.	7	10	0
Highways, at 6d. in the pound,	3	2	6
Carry forward, ... £.	165	12	6

Brought

Brought forward, ...	£.165	12	6
Church-tax, at 4d.	2	0	0
Labour, 1s. per acre; thistling, 1s.; } fences, 1s.	15	0	0
Draining, ...	2	10	0
Looker, ...	5	0	0
Lambing, ...	1	5	0
Clatting, ...	0	5	0
Washing, 3s. per 100, ...	0	6	0
Winding, marking, attendance of three } men, at 3s. per score, ...	1	18	0
Keeping 150 lambs thirty weeks in the } Weald, at 3s. per. score, ...	33	15	0
Driving into the Weald, and back } again, 12s. 8d. per score*, ...	4	8	0
Other incidentals, 6d. per acre, ...	2	10	0
Allow for six oxen, 8 cwt. of hay per } week, at 3s. 6d. for 24 weeks only, ...	33	12	0
	£.268	1	6

* High as this calculation may appear to some, it is accurate; for I send several hundred lambs into the Weald, and cannot estimate my expenses at less than what Mr. Young does, though they may be somewhat more. These must vary according to the country they are sent to, and the distance they are driven. Some may not drive them so far as I do; others I know drive them still farther, and at a greater expense. When the accommodations on the road, the many turnpike-gates they pass through, and the men employed in driving, &c. are taken into the account, the calculation of the author is founded in accuracy.—*A Romney-marsh Grazier.*

Produce.

Produce.

100 wethers, at 33s.	£.165	0	0
40 ewes, at 31s.	62	0	0
Six fatting beasts, 140 stone each, at 3s. 3d.	136	4	0
Eight two-yearlings, at 6d. per week for 20 weeks,	4	0	0
Six yearlings, at 4d. per week, for 20 weeks,	2	0	0
Wool of 60 wethers, 7 lb. at 10d.	17	10	0
Ditto from 40 ewes, 5 lb. at 9d.	7	10	0
	394	4	0
Expenses, £.268	1	6
	126	2	6
Per acre, £.1	5	2½

The profit of marsh-land will appear from the following account of 160 acres, Guildford's, near Rye: the original purchase was 7500l. It keeps six ewes per acre all the summer, and three per acre all the winter.

Expenses.

Interest of 7500l. at 4½ per cent.	£.337	0	0
960 sheep, at 20s.	960	0	0
40 bullocks, at one to four acres; besides the sheep at 8l.	320	0	0
Interest of 1280l. at five per cent.	64	0	0
Looker,	7	0	0
Carry forward, £.1688	0	0

Washing

Brought forward, ...	£.1688	0	0
Washing and shearing 48 score, at 2s.	4	16	0
Thistling, 8d. per acre, ...	5	6	8
Losses, two and a half per cent.	32	0	0
480 sheep put out to winter to May-day, at 3s.	72	0	0
Rates and taxes, ...	64	0	0
Expenses, ...	£.1866	2	8

Produce:

960 sheep, at 40s. 6d. viz.

Sheep, ...	£.1	3	0	....	£.1944 0 0	
Lamb, ...	0	15	0			
Wool, ...	0	2	6			

£.2 0 6

40 bullocks, ...	440	0	0
Produce, ...	£.2384	0	0
Expenses, ...	1866	2	8
Profit, ...	£.517	17	4

This is stated in a singular way, to shew the profit of these marshes. What common farm would bear to have the interest of the purchase of the fee-simple charged to it, as an expense in the same way as a West India plantation? Yet here is that and every charge paid, and above 500l. a year profit on it. It would lett at 40s. per acre. But to ascertain what it would very well answer to give for it, let us calculate as for a tenant.

Expenses,

Expenses as above, ...	£.1866	2	8
Deduct interest, ...	337	0	0
	£.1529	2	8
Suppose rent at 3*l.*	480	0	0
And additional taxes, ...	21	0	0
	£.2030	2	8
Produce, ...	2384	0	0
Expenses, ...	2030	2	8
Profit, ...	£.353	17	4

Which on the capital of 2000*l.* is 17*l.* 13*s.* per cent. ; from which it is plain, that when lett at 30*s.* and 40*s.* per acre; and farms rising from 1000 to 2000 acres, as beyond Winchelsea, there is no wonder that they should be esteemed good farms. Some years ago the rent was 22*s.* 6*d.* : now much higher.

CHAP. IX.

ORCHARDS.

IN the western parts of Sussex, are some considerable orchards; and where the soil is adapted to the fruit, the plantations are thickly interspersed, and the cider held in much estimation, as it makes a pleasant, palatable, and nutritious beverage; and as this county contains a soil well calculated for the production of it, there is no doubt but that new plantations might be made to considerable advantage.

The neighbourhood of Petworth yields the best liquor of any in the county. Lodsworth is noted for the excellent flavour of its cider. The Author has had the pleasure of tasting it at the Earl of Egremont's table, of a superior quality. The best in Sussex is produced on his Lordship's estates, and at Sutton, Bury, Bignor, &c. At Sutton, Lord Egremont constructed a press, which was obtained out of Herefordshire. It is only in a slip of land under the South Down hills, that the cider culture is in any request.

The soil which is considered as best adapted to the fruit, consists in a soft sandy rock basis, with a stratum of a light, but tolerably rich hazel mould. Strong clayey soils here are not suited to cider. The sorts of fruit are various, and the cider is compounded of different kinds; the chief of which are vulgarly called pear-apples, maiden-apples, cockle-pippin, &c. Sufficient regard to the choice of the fruit is not attended

to

to in the manner it ought to be. The harvest is at Holyrood; but the time much depends on the influence of the season. The apples are piled in the orchard for ten days or a fortnight, to mellow; and three or four months after being made, it is thought fit for the table; but this depends upon refining: fifteen bushels of fruit will make a hogshead of liquor. The market is chiefly at home. Some goes to Petworth, Midhurst, &c. but the greater part is consumed by the farmer's family and labourers. The price varies from 10*d.* to 15*d.* per gallon.

CHAP.

CHAP. X.

WOODS AND PLANTATIONS.

———◆———

SUSSEX is one of those counties which, from the remotest antiquity, has been celebrated for the growth of its timber, principally oak. Indeed no other part of England is able to vie with it in this respect, if we consider the woods, either in regard to the extent of them, or the qualities of the timber they produce.

The quantity of land cannot be estimated at less than 170 or 180,000 acres; and the quality of the oak timber may be collected from the circumstance of the Navy contractors preferring it in all their agreements, and stipulating for Sussex, before every other species of oak. The reigning feature of the Weald, is its timber, in which it is enveloped, and overspreads it in every direction, flourishing with great luxuriance, and so naturally adapted to the soil, that if a field were sown with furze only, and live-stock excluded, the ground in the course of a few years would be covered with young oaks, without any trouble or expense of planting.

Before the Norman Conquest, this part of the kingdom was one continued forest, extending from Hampshire into Kent; and the number of parishes ending with the Saxon word, *hurst*, or wood, are a strong presumption, that they were first cleared and cultivated by settlers from that nation. In the neighbourhood of

Salehurst

Salehurst we find no less than eight adjoining parishes ending with this word. At the Conquest, these woods were valued, not by the quantity of timber, but by the number of swine which the acorns maintained.

The great demand of late years for bark, has been one of the chief reasons for the extensive falls of oak, which, in consequence of the high price, has advanced so considerably, that the fee-simple of extensive and well wooded tracts of land has been paid by the felling of timber and under-wood in two or three years; and that upon several estates in the western part of the county, the value has increased full 100 per cent. in the space of twelve years. When we take into consideration the turn for improvement, and that spirit which has been so strongly exemplified in the addition which the highways have received, and the more easy communication to sea-ports than formerly was the case, by extending the inland navigation of the county; by improvements in the rivers, and by opening fresh channels: these circumstances thus connecting the interior with the coast, facilitate the transport of the timber to the dock-yards upon much easier conditions than what was ever before practicable. Consequently we find that the quantity of oak which has of late years been sent to Portsmouth and other places, has exceeded the amount which was transported twenty-five years back in the proportion of four to one; and from the survey which has been drawn of this county, as well as from the prevailing testimony of experienced Surveyors, it may be relied upon as a fact, that far greater quantities of oak timber have been lately felled and carried coastwise from Sussex, chiefly to the King's yards, than the country will in future be enabled permanently to supply.

The

The quantity now standing of a size for the Royal Navy, when brought into comparison with what has been within half a century, is indeed inconsiderable; and as there is but little regular succession in reserve, it follows that the annual supply will necessarily grow less. How far it is an object of importance to the go-vernment, and of profit to individuals, to promote the cultivation of oak, shall hereafter be considered. The subject is not an unimportant one.

Underwoods.

The mode of managing the underwoods is, to cut them from eleven or twelve to fifteen or sixteen years' growth; upon favourable well growing soils, from eleven to thirteen; upon poor grounds, from fifteen to eighteen. The age of cutting depends upon the qua-lities of soil, and the application of the crop, so that no fixed rule can be laid down, other than the gene-ral one above-mentioned, from eleven to sixteen years. The Earl of Egremont's underwoods are cut at twelve to sixteen years of age, where the growth consists of oak, beech, alder, and willow: the underwood is then the most valuable part of the conversion, except in the vicinity of hop-plantations, where the poles pay a much better price; woods which abound with birch, ash, hazel, and willow, of which hoops are usually made, at ten to twelve years of age; newly planted grounds are earlier cut; the shoots are more rapid.

It is worthy of remark, and deserves noting, that un-derwoods at twelve or thirteen years' growth, are as valuable upon some soils, as they would be if cut down at a later age, especially if they are advantage-ously planted in the neighbourhood of hop-grounds; as

poles

poles of that age and size are equally as good, and answer all the purposes of larger ; and when the underwood has exceeded the size of poles, its utility is no otherwise essentially serviceable than as it is valuable for fuel ; the younger therefore it is cut, if fit for market, the more productive it will turn out, and the sooner the succeeding crop will be ready for sale ; for when underwoods are left too long before they are cut, besides growing slower, the interest of the money is lost for which it might have been sold. The coppice upon the most growing soils (for considerable is the difference which exists in this respect) is worth from 8*l*. to 10*l*. or 11*l*. per acre ; but to gain such a product, the land must be exceedingly kind.

Application.

The purposes for which the coppice is converted are various—poles for the hop-plantations, bavins and spray-faggots for the lime-kilns, cord-wood for coaling, hoops for the use of the coopers, besides affording an abundant supply in fuel, and other purposes. Of all the various species of underwood (excepting perhaps alder), ash is the most profitable: the smallest piece is of use in some shape or other, and adapted to a greater variety of purposes than any other wood. Excepting Spanish chesnut, it forms the most durable poles for the hop-planters ; for whose use the various sorts of poles may be arranged in the following manner :

1. Spanish chesnut,
2. Ash,
3. Oak,
4. Willow,

5. Maple,

5. Maple,
6. Red birch,
7. Beech,
8. White birch.

This last is the very worst for poling to any size. But the light in which ash is considered as so valuable, is the application to which the shiverers convert it in quartering it into middling, long, and short hoops. In this respect its value is clearly ascertained. Birch on poor wet soils, pays well, and is rapid in its growth; but on all soils where the alder is in plenty, which, as it makes the best charcoal for the gunpowder-manufacturers, is the most valuable of underwood, is converted to patten-poles and powder-wood. Patten-poles cutting, are 2s. per hundred: they meet in general from three-fourths to one foot each, and sell for 5d. per foot. The powder-wood cutting and stripping is 3s. 6d. per load, which is sold for 1l 4s.

Value.

The value of underwoods, like most other productions, has advanced considerably in their price of late years. Those belonging to Battel-abbey, have, it is said, more than doubled their value in twenty years. The immense tracts belonging to the Earl of Ashburnham, have equally increased. Before his Lordship used them as fuel for his vast lime-works, they supplied several furnaces with charcoal in casting cannon for the use of Government; and although the demand was great at that time, still it was to be had at a moderate price. But when the art of extracting sulphur from pit-coal was first discovered (for that coal could

could not work in its natural state), very good cannon were made in Scotland, and in many other places; and the expenses in casting them having turned out so much lighter than would be the case at these furnaces with charcoal, of course the manufactory ceased in Sussex. And when these iron-works, which took off and consumed such prodigious quantities of wood, deserted the Weald, it was but very reasonable to conclude, that wood would then be procured in the greatest plenty, and consequently cheap; but the contrary has been the case: such a new demand has been created for the consumption of these extensive underwoods, in burning limestone for manure, and the great and still increasing call for hop-poles (3600 to the acre, and from 16s. to 20s. per hundred); all this, with an increased population, and a better system of husbandry, which every where pervades the whole country, are the reasons why wood-lands have been rising in value; and some people consider them as the most profitable of any land whatever.

Soil,

That the soil of this county is very congenial to oak, is apparent from the growth of it, which is in many places astonishingly clean and rapid. It is a weed which springs up in every protected spot. The best soil is a very strong and stiff clay: the red clay is well adapted to the growth. Lord Sheffield has young plantations which are remarkable for their quickness of growth; and the Earl of Egremont's Pheasant Coppice, which consists of several hundred acres, is another instance how well adapted the soil is to the production of this plant.

New

New Plantations.

In newly planted underwoods, it is to be observed, that in the first cutting, which is made at seven or eight years' growth, the profit is little or nothing ;—in the second it is still inconsiderable; so that for sixteen years the return from young plantations is trifling (not a very encouraging prospect to a planter) ;— the third is the most profitable cutting, as the plantation has now reached its ultimate perfection ;—the fourth equals the third ; but after this the coppice advances no more. The effect of the young timber is now visibly apparent to the prejudice of the underwood, which in sixty years, if the trees be left to stand thus long, is destroyed.

A fine nursery of young timber is rising in Stanstead-forest, which in another century will most amply contribute to fill up the place of that which has lately been felled. The soil of this forest (960 acres) principally consists of strong stony land ; on some parts of it there is found a deep and rich loam. The greater part of it has been replanted with oak ; some of which the father of the late Mr. Cathery (who was the agent and superintendant of the estates) planted, and the remainder by the son. The ground was turfy ; and the method which he adopted in sowing it was, to make a little hole in the ground with the broad end of a picke-axe, into which an acorn was planted, three feet apart from each other. In the course of twelve or fourteen years after the planting, the first thinning took place, and every four years after, this operation was continued, taking away those trees that impeded each other. The land,

prior

prior to its being planted, was valued at 4s. per acre to the landlord. A curious circumstance occurred the following spring after the acorns had been planted; for Mr. Cathery looking over the nursery, upon examination discovered, that the mice had eaten holes in the greatest part of the seed; still the trees grew up, and few if any of them failed.

Profit.

With respect to the profit of timber, Mr. Clutton is of opinion, that it pays 5s. per acre per annum, and the underwood the same, and as the rent of the county is about the same, the timber brings the woods to a par with the arable and grass. The value of oak at a hundred years growth, is about 7l. Lord Sheffield has sold out of 30 acres of wood, the fee-simple of which has not amounted to 200l. 100 trees for 1000l. and 100 more for 400l. But these trees had done so much damage, that the underwood which at fourteen years growth, before their being felled, had sold at 40s. per acre, at the same growth after felling, fetched 7l. 10s. per acre. This is a remarkable fact, and which deserves calculating: 1400l. from 30 acres, is 46l. 13s. per acre, and that divided by 100, the supposed growth of such trees, is 9s. which is the rent per acre per annum which the timber paid. The growth however, instead of 100 years, was more likely to be double, which diminishes the rent in the same proportion. But taking the account as it stands, the trees did mischief to the underwood, to the amount of 5l. 10s. in fourteen years: this is near 8s. an acre. Hence it appears, that the profit of the timber was only about 1s. an acre, even upon the supposition that trees of ten pounds each were no more than 100 years growth.

growth. Wherever one crop is made to grow to the prejudice of another, this will generally be found to be the case. Tellows are preserved as well from stubs as from seed, if not too old, but those which are preferred, spring from acorns. Woods well covered with timber, rarely have many thriving tellows that remain, since they are overshadowed, and find a difficulty in fighting their way through the branches of the other trees; the effect of which is, that a good succession of young oak seldom follows a fall of old timber. Stub-timber is by some people preferred to the growth from seed; for when a good stub is cut, the succeeding root springs up full three feet the first year, when an acorn hardly makes its appearance above ground; and fine oak timber, two loads to a tree, has been cut from stubs; though it is very seldom they make good timber trees.

But no where shall we find oak to flourish with greater luxuriance than in the neighbourhood of Petworth. The Earl of Egremont has felled several acres, which produced a profit of 560l. per acre. The circumstance of 3000l. being raised from the sale of 200 trees, cannot fail of proving how congenial the soil is to this plant. If, therefore, the advantage of timber is any where striking, when connected with profit, we expect it in those situations where it flourishes with the greatest success. But notwithstanding the above-mentioned products, great as they are, the crop is a losing one. Some of the best thriving woods on the Petworth estate, clear 12s. per acre per annum. *Raffolding wood* is of this description. It adjoins a farm of his Lordship's of 120 acres, which is lett at 50l. per annum, and it is very near the land in his own immediate occupation, and precisely the same soil

soil which yielded 373 sacks of wheat upon 54 acres in 1794.

Raffolding wood measures 23 acres of 14 years growth.

	Amount of Timber, Bark, &c. sold.	Expense of Barking, cutting up Wood, &c.	Tellows cleared.
	£. s. d.	£. s. d.	£. s. d.
Tellows, ...	80 15 0¼	23 12 8	57 2 4¼

Which is 2*l*. 9*s*. 8½*d*. per acre, or 3*s*. 6½*d*. per acre per annum.

	Amount of each Article at the Sale Price.	Expense of cutting and clearing the Coppice.	Underwood cleared.
	£. s. d.	£. s. d.	£. s. d.
Underwood,	248 10 11¼	112 0 0¼	136 10 11

Which is 5*l*. 18*s*. 8*d*. per acre, or 8*s*. 5½*d*. per acre per annum.

	s.	d.	
Tellows cleared, ...	3	6¼	per acre per annum.
Underwood, ...	8	5½	
Total, ...	12	0	per acre per annum.

So that the gross produce of the most flourishing wood, in soils which are singularly adapted for oak, is 20*s*. per acre per annum! whilst the produce of the same sort of land adjoining this wood, which is under cattle and corn, and calculated at no more than three times the rent, is above seven times as great; and if we raise it to four times the rent, which is then under the mark, the gross product of the same sort of soil is more than nine times greater in corn than in wood.

Age.

Age.

The opinion of surveyors, and other well informed persons, vary much as to the age at which oak arrives at perfection; and different periods have been assigned, from 100 to 180, or 200 years. Their opinions no less differ as to the mode of managing woods: some recommend the bringing up trees of different ages in the same field, and as often as any of them arrive at maturity, to fell those trees, and leave the rest standing, to be cut in succession; whilst other persons, of equal skill, advise, that when the great number of the trees in the same wood arrive at their full size, the whole should be cut down, and the ground completely cleared and replanted.

Felling.

An oak never comes to perfection under 100 years at least; and the fall of the best timber in the county will not average more than 4*l.* a tree, top, bark, &c. all included. When a wood is properly stocked, five trees may be taken down per acre at each felling of the underwood, and tellows saved in their place. If thirty trees are left upon each acre, it is generally supposed to be a quantity fully sufficient to arrive at any perfection. People however differ upon this point, and some think that not more than twenty or twenty-five should remain upon an acre; for it must be observed, that if the timber is too close, the underwood must be of less value, the price of which is raised in a greater ratio than that of timber.

When a tree is six inches in girth, it becomes timber, and when they are worth 40*s.* they never pay interest.

terest. For profit, they should be always cut whenever the tree fetches that price. A very considerable gain will then arise from the underwood; but when the timber is left standing 100 or 120 years, the underwood is effectually destroyed. When felled at an early age, the value per foot is certainly small, but if kept to such an age as to become an object of national defence, the value decreases in proportion to the age of it.

The following is a short calculation upon the utility of early cutting timber that grows in coppices, by the very ingenious Mr. Upton, of Petworth, the Earl of Egremont's timber surveyor, and the same person to whom the Society of Arts lately voted their gold medal, for the plan and model of a new barn, which he has lately invented.

Suppose one acre of coppice ground to contain 100 oak trees, at 55 years growth, worth	£. 100	0	0
At 100 years growth, worth (which will seldom be the case)	600	0	0
If cut down at 55 years growth, the 100*l*. by compound interest for the remaining 45 years, will accumulate to	940	0	0
Add to that 45 years growth on the ensuing crop, exclusive of the loss of underwoods, ...	80	0	0
	£. 1020	0	0

" But," says Mr. Upton, " if this mode become general, small timber would be of less value, and our Navy fall short of a supply of ship-plank."

The

The high price which of late years has been gained by timber, is as much to be attributed to the value of the bark, as it is to the worth of the timber: The season for felling is regulated by the time of barking: when the sap begins to rise, which usually takes place in April, the tree is then felled. Bark peeled from young tress, is much superior to that which is stripped from older ones ; it abounds more in sap, and there is no such waste in it, as the hard and dead part of an old tree is dressed, which is not the case with the other. In regard to the timber, the trees that are growing, and formed to make four-inch plank, are left standing for that use, and such as are growing and forming for three-inch plank, are frequently left for that purpose, without ever considering whether they will pay by standing for such scantling. The other classes of timber are large logs, to be hewed, small planks, and timber for carpenters' use ; but all timber that has finished its growth, should be immediately taken down, though no larger than a pole ; whoever keeps timber after it has stopped its growth, will lose the value of the timber in seventeen years, with the interest of the money it might have been sold for, and the injury done to the underwood by its standing, and preventing the succession of fellows.

Improvement suggested.

It must seem surprising, that in a country where the Navy is an object of such importance, and in a county like Sussex, where oak might be called its staple commodity, no complete trials have ever been made of increasing the duration of timber, so easily practicable, and so important in its consequences. Experiments have been incidentally tried by the Navy Board on
winter-

winter-felled timber, stripped of its bark in the usual season, and the tree left to dry till the following winter, before it is cut down. But stripping the tree of its bark, and allowing it to stand in that state three years to season, before felling it, has the same effect in converting the sap into useful timber, as allowing the tree to stand with the bark on it for twenty-five years longer, would have. In the spring the trees are teeming with vegetation, and their cavities overflowing with sap, which, if the tree is felled at that season, remains in the pores: thus it putrifies, leaving the tree full of cavities, and the timber weak. Besides all this, it breeds worms, and is liable to shrink: for these reasons it should seem, that during the winter, when the sap has retired, is the properest season for felling. The solidity, strength, and duration of the timber, is thereby increased, and being exposed to the effects of the sun and wind before felling, is so dried and hardened, that the sappy part in a manner becomes as firm and durable as the heart itself.

Upon the whole, it is a singular, curious, and interesting circumstance, and experiment has confirmed the beneficial tendency of the measure.

The Policy of encouraging the Growth of Oak.

I do not hesitate to question whether our policy in promoting the cultivation of timber is not erroneous. I am aware, that there does exist in the minds of some people a predilection in favour of woodlands, which arises from an idea of the superior comparative profit of them. Hence it might appear a degree of presumption in an individual to controvert the general sense of a county, when it carries along

SUSSEX.]   with

with it the fair and plausible appearance of being
founded upon experience, and the result of established
practice. Prejudice is not easily eradicated: to
combat it with success is no easy task. Systems taken
up from father to son, without any attentive investigation of debtor and creditor, in inquiries like the one
before us, are for ever occurring. If the appeal rests
upon facts, the question was decided before it was
proposed. Instances have already been produced,
sufficient to convince any unprejudiced person how
profitable is a timber estate, when put into competition with cattle and corn.

The statements above-mentioned proved an annual
loss to the public, upon the comparative gross receipt, to a considerable amount; and as it is a question too important to be passed over without some
further investigation, I shall throw together a few remarks upon a subject which has been much misunder-
stood, and involved in error, from occupiers, or their
surveyors, not having paid a sufficient attention to
matter of fact.

The arable and pasture in the Weald may be cal-
ulated at 425,000 acres; the woods, plantations,
coppice, shaws, &c. at 170,000. The exact amount of
either is not material in the consideration of the present
subject. The comparative produce by wood and by
corn, merits observation.

The common system of husbandry in this part of
the county, is a fallow, two crops of corn, and one of
clover.

Wheat,

Wheat, two quarters and a half, at 44s. £.5 10 0
Straw, stubble, chaff, ... 1 10 0
Oats, 4 qrs. at 18s. 3 12 0
Clover, twice mown, ... 4 10 0

£.4)15 2 0

Produce per acre per annum, ... £.3 15 6

Now, in order to draw up the comparison as favourable to woodland as impartiality admits, the following is a crop of 23 acres of wood growing in a highly favourable soil upon the Petworth estate, and exceeding very much the average value of the county.

Produce of Timber.

722 feet of timber, ... £.19 11 1
2 loads 37 cwt. 3 qrs. 19lb. of hatched bark, 51 3 3¼
7 stock 3½ qrs. from ditto, ... 5 16 3
Bavins, ... 4 4 4½

80 15 0¼
Coppice, ... 248 10 11¼

23)329 5 11¼

14)14 6 8

Per acre per annum, ... 1 0 5

Produce by corn, £.3 15 6
Ditto by wood, ... 1 0 5

Difference, ... £.2 15 1 per acre per annum.

Hence there appears a balance of 2l. 15s. 1d. per annum, estimating the gross produce, against woodland;

six

six times the rent in one instance, and not twice in the other. And be it observed, that these 23 acres of land grubbed up, would yield, like the land in the neighbourhood, *at least* a product of 3*l*. 15*s*. 6*d*. one year with another. If we call 12*s*. of this profit or rent, and put out to practicable compound interest in 100 years (and a less growth cannot be allowed for oak), and then see what it becomes. If we add another 12*s*. for farmer's profit, the difference is far greater. How much more then, if we ground the calculation on produce, rather than on profit?

The most material point in the present inquiry is, to ascertain the cause why arable and pasture are rated at so low a value, when it is known, that throughout the kingdom in general, the same sort of land would fetch a rent much nearer to 20*s*. than 12*s*. From what source is this difference to be derived? Is it to the soil, culture, or to the management? It is the effect of that influence so strikingly obvious to every observer, which the cultivation of timber inflicts upon the land adjoining it. I do not exaggerate when I state, that the culture of corn is in many places enveloped in a forest of timber. Viewed in a national light, its effects are sometimes distressing. Traverse this county ; remark the state of it in those parts which are teeming with timber : observe the corn surrounded by a forest in every hedge-row ; and then calculate the mischief : the damage it receives is hardly to be estimated. By aiming at too much, neither is gained. Without doubt, the considerations upon which tenants occupy their farms, are made compatible with such effects. It is the public who is the greatest loser ; but landlord and tenant come in for no inconsiderable share of the loss. I have viewed many fields of corn, which the finest

finest harvest weather would scarcely bring to maturity.
Perhaps the farmer is satisfied on this score, whilst
the proprietor indulges himself in a review, that shaws
are better things than close clipped hedges, and woods
than fields of corn. Such assertions are without number: but the present condition of the tenantry in
the Weald, is a refutation of such ideas.

The singular custom of *shaws* must be condemned;
broad belts of underwoods, and trees, two, three, and
four rods wide, around every petty enclosure. The
landlord is tenacious of preserving them, because they
afford protection to a quantity of timber; and the
tenant is allowed the underwood at the regular period
of cutting. The history of this custom is evident.
Long since the time of the Conquest, the whole
county was a forest: fields of grass and tillage were
opened gradually in the woods; and whilst land was
cheap and plentiful, no accurate attention paid to surrounding them with fences, the forest continuing to
form a sort of fence. Carelessness and ill husbandry
continued the practice, till at last the landlord, finding the sweets of great falls of timber from these shaws,
made it an article in the lease, to preserve them
against those encroachments, which an improved system
of husbandry would be for ever necessarily making
upon them. The country is very generally wet; the
means to air and dry it here used is, to exclude the
sun and wind by a screen of underwood, and a forest
round every field: these are small, so that a great
number are so wood-locked, that it is not surprising
when the corn is not ripened. At the same time
that this mischief is done, the wood itself is (timber
excepted) but of a miserable account, as any one
may suppose, when he is informed that these shaws
have

have a fence only on one side, and consequently
are exposed to be eaten by the cattle that graze in
the fields: hence we find an imperfect system of
wood, and an injured one of corn.

The arable and pasture in the Weald amount, as I
before remarked, to 425,000 acres. Of this, suppose
we strike out 125,000, as not materially affected by the
timber, &c. the remaining 300,000 acres are under the
full influence of it. Now, in order to bring this to a
par with the other land in the neighbourhood, the
tenant could amply afford to pay an additional 5s.
per acre, and be a considerable gainer by the bargain,
provided that the country was laid open, and the petty
enclosures enlarged. Here then is an annual loss to
the public of 75,000l. a year, resulting from the mischief which these shaws cause to the adjoining land,
which is a clear annual deduction to that amount from
the value of the woods; so that the damage which a
predilection for the cultivation of woods occasions to
the crop of corn, is nearly in proportion to the rent
which the plantation pays its owner.

The advocates in favour of this species of property
tell us, that it is the best land of which a proprietor can
be possessed; that although the estates may be of
large extent, he occupies them himself with success
and advantage equal to the most attentive and econo-
mical farmer; that to make himself master of the
business requisite in this line of rural economy, does
not require the labour which attends the cultivation
of an arable farm, or the management of live-stock;
that he has only to order and see that good fences
are made round the woods, to prevent their being damaged by the inroads of cattle; and when fit for the
market, if he does not chuse to sell them in the retail,

tail, by cutting the wood himself, he puts them up to sale by auction; the customary mode in many parts of the county; that he is sure, by this means, of receiving the real value. In short, that it is a safe, improving, and valuable treasury.

Consider the influence of woods and forests in a political light, as affecting the dearest and most important interests of the kingdom. Here the evil is flagrant enough. To encourage the growth of oak in a kingdom rapidly increasing in population, when we have so lately experienced to our cost, the sad effects of inability to feed our own people, is not giving encouragement to what most demands attention. Recent experience has taught a lesson of instruction, fresh in the memory of all. In order to ward off the apprehension of a famine, we have been under the necessity of importing corn to a vast amount; and in 1797, the balance against us was for no less a quantity than one million quarters of wheat, and very nearly another million quarters of oats! To have corn in abundance, we must, in the first place, lessen our forests and woods; for in proportion to the size and extent of them, and the waste land, must be our dependence on foreigners for a part of our food. Apprehensions of scarcity are periodical, and manifestly alarming to government. Butcher-meat, although it has lately declined (written in 1797), has been for some time at an unexampled price; almost all the productions of our soil have doubled their value. Under these and similar considerations, does it not argue a singular want of foresight, that people should be found who will stand up as advocates for a wilderness, on comparison with corn, cattle, and sheep; for the benefits which arise from raising wood, instead of feeding our own people. It is

is curious enough, that the woods in this county, if
grubbed up and planted with wheat, would add more
than 500,000 quarters of corn to the national produce;
and this is a quantity which, if we look at the sums
annually paid for the importation alone, is of *some
consequence*; compared with it, the produce by wood
is too small to merit observation.

But it will be said, that political views render the
production of timber necessary. Such views should
be explained; they will in all probability discover
themselves to be rotten. " Convert your timber into
corn, and the nation is undone. What becomes of
the Navy—the wooden walls of old England? Granting that it is for the interest of individuals to grub up
their woods, still oak must be raised, and that too in
quantities, or the means of national defence disap-
pears."—As if the additional wealth created by the
conversion of timber into tillage, would not be able
to command the most unlimited supply for the Navy.
Deal we find to be a commodity as essentially necessary in the construction of our houses, as oak in the
building of our ships; yet where is the inconvenience
of importing it? The North of Europe, and Ame-
rica, hold out such inexhaustible stores, that any apprehension of scarcity is unfounded, and what proves
the scarcity to be ill founded is, that the contract
price for oak in the King's yards, has not advanced for more than forty years; a decisive proof,
either that the quantity has not declined, or that the
foreign growth is every way adequate. Scarcity is
complained of; the scarcity of timber is unexceptionably the most convincing proof of national prosperity. To complain, is preferring a produce that
yields 20s. to another that at least pays 4l. In proportion

portion as our woods and wastes are made to vanish before population and corn, must be the scarcity of wood.
As the kingdom advances in cultivation, woods and
forests are made to disappear ; and if our enemies make
use of foreign oak in building their navy, we may
surely do the same on at least equal terms.

In whatever light this subject is considered, whether in respect to the landlord, or his tenant ; to individuals, or the public, the woods are inferior to
corn ; and the first step to an amelioration of the
Weald, would be the diminution of them*. By properly

---

* I cannot altogether agree with the Reverend Author, in his ideas of
the Weald being so much enveloped with woodlands and large hedge-
rows. In some particular places it certainly is, and would be an improvement to grub and clear the land from hedge-rows and timber.
Thus far I agree with the ingenious Author; but surely this practice
ought not to be followed, except where the soil is kind for corn and
grass. Let the bad soils remain, by all means, in the same woody state,
with the addition of a much greater quantity of it being planted with
underwoods, as a nursery for timber. The greatest improvement that
can be made (and done with the least expense and trouble in that district), would be for landlords to take away from their tenants from ten
to sixty acres, in proportion to the size of their farms, of the very
poor land, which in its present state nobody receives any benefit from ;
and I am satisfied the tenant would be glad to get rid of the good-
for-nothing land, as they generally call it, for a small compensation ;
and this land, which now is looked upon as not worth cultivation,
would most probably in a few years be valued from 6_s._ to 8_s._ per acre.—
_G. F._

All ideas of the present value of such land, derived from the application of it in its _unimproved state_, is liable to error. Where is the _good-for-
nothing land?_ I am acquainted with little land in the Weald, properly
so called, and the region of timber, which nobody receives any benefit from ; for the great tracts of waste-lands form no part of the present
question. But however, taking them into the account, and connected
with the farmer's other land, they pay some rent, not less than 2_s._ to
4_s._ per acre. Taken at 2_s._, the produce is 6_s._ Now, should this planting

perly lessening them, the improvement of such heavy
soils would already be more than half carried through,
and the consequent success great, rapid, and effective.
Corn and cattle, mutton and wool, would mark the
progressive improvement of the county, and the
Weald, in lieu of being covered with woods, would
smile with plenty and prosperity.

To those gentlemen who are such sticklers for encouraging the production of timber, it will be very
satisfactory to observe, that the Sussex woods, under
proper management, would more than supply the
whole Royal Navy.

---

ing speculation of raising coppice, as a nursery for timber, succeed, let
him calculate the progressive increase of 6*s.* per annum, at compound
interest, during the term his trees are to stand. Such a calculation
will not turn out any inducement to convert whatever live-stock it is that
feeds upon them to timber, by way of a *great improvement.* If it be said,
that planting is preferable to the present waste state, the comparison
is admitted: enclosed and divided, they will be fit for any application.—
A. Y.

CHAP.

CHAP. XI.

WASTES.

———

THE tracts of land which come under the description of mere wastes, in Sussex, are very considerable; they chiefly occupy the northern side of the county: out of a portion containing, by computation, 500,000 acres, these almost desert tracts take up no less a space than 110,000 acres of it; and what renders it the more singular is, the apparently beneficial circumstance, that this great range is within the distance of 35 to 45 miles of the capital; and all might not only be converted to the great benefit of the county, of which they compose so large a part, but be likewise highly productive to the empire at large. It is not a little curious, that such immense tracts of land should be still left in a desert state, when they are every where intersected by turnpike-roads, and in the neighbourhood of such a market as London! These are surely advantages great enough to recommend the culture of them. At first sight, the soil is a discouraging prospect; generally a blackish sand, ferruginous, poor, and frequently very wet, over a bottom composed of an earth resembling marl in colour, though not in quality. Under this comes a sandstone; and over the whole tract, iron-works once flourished. To this ferruginous quality of the soil, its poverty has been ascribed.

Paring

Paring and burning, would be the making of this
soil: it is the abuse of this excellent practice, which
calls for condemnation. The soil, though poor, is
susceptible of considerable improvement. And be it
remembered, that some of the greatest exertions that
have been undertaken in this island, have been on
soils poor and sandy, some of which have not exceeded these in fertility, and without possessing any
of those advantages which arise from the vicinity of
London. It is idle to say, that such a soil is too
poor for profitable cultivation. That such great
tracts in Surrey, as well as in Sussex and Hampshire,
should be suffered to remain in their present state, is
a most unaccountable negligence, and to a superficial
observer, a motive for concluding, that vast cities,
instead of shedding a benign influence over the neighbourhood, have a tendency to the reverse. The
wastes, only within 40 or 50 miles of London, would
supply that city with bread.

The greatest improvement that I know undertaken in this county, has been effected on the Stag-
park at Petworth, some years ago, by the Earl of
Egremont. Previously to its being improved, it was
an entire forest scene, overspread with bushes, furze,
some timber, and rubbish; of no kind of use, if
we except a few miserable and ragged young stock
which it annually reared; and would not have lett for
more than 4s. or at most 5s. per acre. The undertaking of converting between 7 and 800 acres of land,
was an exertion to be expected only from an animated
and enlightened improver. It was begun about sixteen or seventeen years ago; the timber sold, the underwood grubbed, and burned into charcoal upon the
                                                spot;

spot: and every part of the park has been since
drained in the most effectual manner: the whole of
it enclosed and divided into proper fields, and planted
regularly with white-thorn, all of which has been
trained in the neatest manner. All the crops upon
the ground succeed each other in a system of correct
cultivation, and so luxuriant, that few tracts of 20s.
or 30s. per acre, can be said to be more productive.
Extraordinary fine crops of wheat and oats are raised,
as high as five quarters of the one, and ten quarters of
the other; fine crops of barley and tares, and vast ones
of turnips; and artificial grasses; clover, ray, chicory,
rib, &c. in great profusion.

It is thoroughly well stocked with Sussex, Devonshire, and Herefordshire cattle; flocks and fatting
sheep of the South Down and Spanish breed, Leicester and Romney: the whole of it is a garden.

Since the first edition of this work has been published, some considerable tracts of the poor sandy soil
lying along the northern side of the county, have been
brought under a course of improvement. Mr. William
Seaton is converting a part of Tilgate forest, by unremitted exertion, into a well ordered and systematic
arrangement of crops, by denshiring the forest (here-
tofore no other than a rabbit-warren); and it well merits attention, that no plant or root that has yet been
tried upon this land, seems so well adapted to the soil as
potatoes. An account of the expenses and produce of
six acres of the forest, very recently enclosed from
the warren, was sent by him to the Society of Arts,
for which the gold premium was adjudged last year.
These six acres, the rent and all the taxes of which
amounted but to 6s. an acre, gained him a produce by
potatoes of 80l., which is upwards of 13l. per acre;
a proof

a proof how well suited is the culture of potatoes to this
land, more particularly when we take into consideration
that the soil does not appear to have been chosen from
any circumstances of superiority over the remainder
of the warren: the preceding crop was oats, and they
yielded so badly, that the piece was then considered
by Mr. Seaton as not worth the expense of cultivation,
and for two years it was thrown open to the rabbits. In
October, 1796, it was enclosed and ploughed; March,
1797, harrowed, and soon after cross-ploughed; the
beginning of April harrowed again, and soon after
ploughed a third time; in ten days harrowed a third
time, and ridged up for dung (12 cart-loads per acre);
and before April 25th, 20 bushels of potatoes planted.
In June they were hoed and earthed up, and in
October 250 bushels per acre were taken up.

CHAP. XII.

IMPROVEMENTS.

SECT. I.—DRAINING.

THIS operation is not yet thoroughly understood; the practice is confined to a few spirited individuals. Hollow-draining is the first improvement
wanted; though it is rendered difficult to execute by
the nature of the soil. The tenacious properties of the
clay very greatly retards, and in some places absolutely prevents, the subsiding of the water. In this
case, surface-drains only can be of any use; but
wherever the upper soil is formed of a greater proportion of loam than of clay, the water will pass through
it with ease, and the operation may be attended with
great success. The trenches are made three spit (two
feet) in depth, and from four to eight, or ten inches
wide, at bottom, and eighteen inches at the top:
besides the spade, the trunking-tool, and the scoop,
are used. The small spray of bush-faggots is trod in,
to prevent the materials (as sea-beach, stone, or
sand-stone) from settling at the bottom. In the neighbourhood of the sea, beach is used, and it serves for
excellent drains, and lasts for ever. It is commonly
laid in the drains about 10 or 11 inches thick, over
it a small quantity of stubble or straw.

But the art of draining has lately received a rein-
forcement of knowledge, from an important discovery
of Mr. Elkington. His system is not so much the
constric-

construction of drains, to draw off any wetness occasioned by rain, or. overflowing, &c. as the more
complicated operation of draining lands, rendered wet
by subterraneous waters originating in hills and rising
grounds. To discover the heads of these springs,
is the main point of the work. His knowledge and
experience in draining boggy land, brought him to be
employed in various parts of the kingdom. Among
other places, he came to Petworth, where the Earl of
Egremont soon cut out work for his ingenuity.

Lord Egremont wanting a supply of water for his
lake, Mr. Elkington was of opinion that it might be
gained from a large hill of sand-stone, which had already been drained into several small reservoirs, from
which the water was conveyed to the lake but in a small
quantity. Undertaking to procure a much larger
stream, not only by discovering more water in the hill
than was at present known, but also by diverting some
springs which break out in a common on the other
side of the hill, he agreed to convey the water into the
park, by cutting his trenches on the east side of the hill,
in order to draw the water which issues from the north
and west parts of the hill upon the common. Mr.
Elkington pronounced, that by boring, the water
should be made to boil in his trenches like a fountain.

When his drains were finished, it appeared that
no water was gained by them. Mr. Elkington said,
that this was no fault of his, as the springs were not
perpetual, but dried up in summer. In very dry
weather, such as 1797, this might be the case; but
the drains have been running from that time to this
very strongly; all of them in their original channels, and none in Mr. Elkington's.

The

The plan upon which he proceeded, was to sink a ditch from the level at A

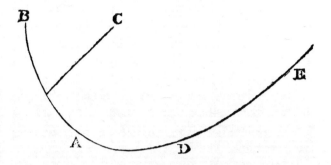

up to B, where it is eleven feet below the surface of the ground; he likewise bored down eight feet lower, and found nothing but clay. At C, he did not carry up a ditch, but sunk a well six feet deep, and bored down seven feet lower, where also it is all clay. He then sunk a ditch from the level A, round the end of the hill to E; the water oozed out a little all the way; and at D, and a few yards each way, there is a smart spring, as much, or rather more, than the well above; but the soil is composed of such loose stones and sand, that it immediately sinks, and runs into the ground. At E, there is another little run: the bottom of the ditch at D, is six feet under the surface, and the hill rises very fast. At B and C, where Mr. Elkington expected to find stone, and a hogshead of water every minute, he neither met with stone, or one drop of water. There is a well, and a spring which constantly overflows about five feet perpendicular, and about ten yards as the hill rises above B. Mr. Elkington said that his trench would lay the well dry; but the spring flows over the well as much as ever. His charges for this were 30*l.*

SUSSEX.]   Another

Another attempt was made by Mr. Elkington, to drain a meadow called Budham, lying below a gently-rising ground, and along a river. Mr. Elkington conceived that this meadow was wet from springs in the hill, and that cutting a trench above 500 yards, would cut off these springs; and as the water in the river was higher than the meadow, he laid the mouth of his trench into the river, two feet below the surface of it; he contending, that the water of the springs would run into the river, without the river running into his drain.

He was told, that the wetness of the meadow was owing to a mill-head penning up the water above the level of the meadow. There was an old ditch for carrying the water from the meadow into the river, when it was low; this he said might be stopped up, as his drain would answer the purpose. When about 200 yards of his drain had been finished, he found that it did not answer his purpose, and he deepened it two or three feet; but when finished (at the expense of 100*l*.), the effect was, that when the ditch was stopped up, the meadow was flooded by the springs, as his drain did not answer the purpose of carrying the water off, although altered backwards and forwards several times; and when the ditch was open, the meadow was flooded by the river. When Mr. Elkington was last at Petworth, he said that some of the springs were beyond his work, and he could only recommend to bank the river out at the ditch, and to build a wind-pump to pump the ditch dry.

After this unsuccessful undertaking, Lord Egremont took a quite different method. The level of the meadows on each side of the river having been taken, it appeared that the meadow-ground on the opposite side

side was beneath the level of the meadow which required to be drained, and consequently that the draining could be effected by a trunk laid across the bed of the river; a wooden pipe was therefore laid at the bottom, to receive the water of the ditch, and it was carried on by an open drain passing through this other meadow, on the opposite side (which it also drained) up to the bridge, under which it passes, close to the turnpike-gate, by means of a pipe, and it empties itself into the river at the mill-head. This has answered most effectually, so that the water in the old ditch now stands always a foot below the surface of the meadow; and more than one hundred acres of contiguous meadow have been highly improved by these new drains: much of it a mere bog before being drained, is now converted into a fine water-meadow, and worth full 3*l.* per acre. These grounds are at any time capable of being flowed by the means of sluices made through the towing-path, which acts as an embankment; and in summer, if the river is too low, by fresh streams which flow into it from the upper grounds; and the water can at pleasure be drawn off by drains into the lower level below the locks, and sometimes, where particular circumstances render it necessary, by the means of culverts carried across the bottom of the river. The failure of Mr. Elkington seemed to proceed from a want of that theory and principle which might have been looked for in an experienced drainer.

A third failure of Mr. Elkington's occurred also at Petworth. Lord Egremont has a forcing-engine, worked by a water-wheel at the river, for raising water 178 feet high, in order to supply the town of Petworth. This is an expensive machine, so that his

Lordship

Lordship would have been at the expense of 1000*l.* to procure an equal supply (a hogshead a minute), by bringing springs from distant grounds. This Mr. Elkington undertook to do : but as his Lordship had seen the preceding failures, he declared he would be at no uncertain expense. Mr. Elkington offered to procure the requisite supply for 1000*l.* if he succeeded, and to pay 1000*l.* if he failed ; afterwards reduced to 500*l.* and the water to half a hogshead each minute. But upon Lord Egremont's having the agreement drawn up by a lawyer, Mr. Elkington declined it. He went however to work on his own account, in order to retrieve his reputation as a drainer. The hill from which he expected to be able to draw the water, is of a large dimension, spreading a circumference of several miles, and is formed of whyn and other stone : springs of no great account, break out all around it at different levels. Mr. Elkington took the level of the Petworth reservoir, and fixing on a spot on the side of the hill above that level on the gohanna ground, where a parcel of these small springs break out on the sides of the hill, cut a very deep trench, and bored ; but all in vain. He then tried at another place, where two small springs broke out. Here he worse than failed ; for he not only found no water, but actually lost one of the old springs, which supplied two cottages with water, and did not even catch that of the other spring, though close to his trench. In one part of this trench he told Lord Egremont's director of similar works, that he would, at such a spot, find stone, and a spring that would run a hogshead in a minute ; but they found no stone, only clay, and no water. There is also a well and spring at another spot, which Mr. Elkington said

his

his drain would lay dry : it had however no such effect, and flowed afterwards as much as before.

Having thus described his failures, it is necessary to observe, that he drained an acre of boggy meadow very well and successfully, though at the great expense of 40*l.* Lord Egremont considers him as a very good common drainer, though a very expensive one; but without any particular skill or knowledge not possessed by any other good drainer.

I have thought it proper to insert the preceding details, not by way of prejudicing any man against a person who has certainly performed, in other cases, great and singular improvements; but merely to caution the world against an appearance of mystery and intuitive knowledge, which a certain degree of success may have given to Mr. Elkington's manner. It goes only to prove that he is very far from being infallible, which I have heard some persons very nearly declare him to be. He has executed works sufficient to prove his merit, and wants nothing of that sort to add to his reputation.

SECT. II.—PARING AND BURNING.

This is one of the greatest improvements which land is susceptible of receiving. It converts an old worn-out turf into corn and grass; it adds new life and vigour to the soil, and changes the nature of it: but in the hands of a needy tenant, it is almost certain destruction to the soil; instead of improvement, in his management, it ends in impoverishment. He breaks up, pares and burns, and drives the land with three

or

or four crops of corn, and then lays it down again.
The fertility of his new land tempts his rapacity to repeat his crops, till the soil is exhausted of every particle of fertility; and when it is so reduced that no
corn can be made to grow, can it be wondered at that
landlords should object to a system which is entitled
only to execration? It is the gross abuse of a practice,
which, when properly conducted, is an admirable
improvement to any land. If paring and burning
exhaust the staple, the rent of it, so treated, certainly
would not have advanced in a few years from 50 to
100 per cent. If it extracted the nutrition or food of
plants inherent in the soil, it would have had the
effect of destroying the productive properties of earth
long since, in those countries where the practice has
been a favourite one for many centuries.

In 1763, the late General Murray laid a field down
to grass till he returned from Minorca, and it was not
one penny the better in all that time; he then pared
and burned, and limed it with the ashes, ploughed,
and laid it down again directly, without sowing any
corn; and in all the uplands of Sussex, there is not a
finer piece of grass than it has been ever since. This
is a remarkable experiment; and we may draw a conclusion from it in favour of the practice, that it is
only the abuse which merits condemnation.

When a farmer pares and burns, he knows that
he is in possession of a dunghill, and his first business is to get the heart and blood out of it as soon as he
can, by corn-cropping: with such management the
practice is execrable; but if applied with proper caution, there is no safer or better husbandry. This trial
might have succeeded better if a crop of turnips had
been taken after the paring; these fed off with sheep,
and

and then laid down with barley, which is husbandry
(and I name it for that purpose) that is applicable to
common management: whereas, farmers will not hear
of 30s. or 40s. to pare and burn, and 60s. or 80s.
more in lime, in order for grass only, however excellent the husbandry, which this undoubtedly is.

This husbandry has been practised in Sussex by
Messrs. Seaton, Dixon, and Bradford, &c. Great
success for a time attended it; but from the want of
sufficient capital in some cases, and too much corn-
cropping in others, the final result was not such as it
would have been with different management.

SECT. III.—MANURING.

THE manures used in Sussex, besides common dung,
are;

1. Chalk.
2. Lime.
3. Marl; and in a small
   degree,
4. Sleech.
5. Soap-ashes.
6. Wood-ashes.
7. Peat-ashes.
8. Coal-ashes.
9. Rags.
10. Sheep-clippings.
11. Pilchards.
12. Paring-dust.
13. Gypsum.

The three first are used in great abundance, the rest
partially.

1. *Chalk.*

This is in great request, and used in quantities from
800 to 1600 bushels per acre.

Mr. Peachey, of Chichester, spreads 8 bushels to a
perch. Mr. Gell, at Applesham, lays it down as a
rule,

rule, that it should be exposed to the air for a year or
two before it is ploughed in, for the frosts to pulverize
it, in order to unite it the better with the soil.   He
manures with 140 cart-loads to the acre, each load 30
bushels; and he estimates the expense at 5*l.* per acre.
This energetic and spirited farmer has already covered
his well cultivated farm with 20,000 loads of this manure: and what appears to be the singularity of the
circumstance, it is all done upon a chalk farm.   His
exertions, in this respect, have been unusually great.
In the operation of such interesting experiments, he
applies chalk in union with lime; first, 120 loads of
chalk upon a layer; two years afterwards, lime for
rape; a kiln of 12 loads to two acres and a half. The
wheat-stubble, with clover amongst it, sown on this
preparation, marked a crop of extraordinary goodness,
and the clover a very superior crop.   The expense,
however, of the improvement, enormous.

120 loads of chalk, dig, fill, and spread,
 at 4*s.* per score, ... } £. 1 4 0

Four carts and 16 oxen, and four dri-
 vers, 40 loads per day, three days, } 3 16 0

Lime.—1500 furze fag-
 gots, at 6*s.* per 100, } £. 4 10 0

Six loads chalk, labour,
 &c. at 5*s.* .... } 1 10 0

Burning, ... 1 1 0

Beer.—Emptying kiln,
 and spreading, ... } 1 10 0

Repairs of kiln, ... 0 5 0

Divided by two and a half, £. 8 16 0, gives 3 10 4

£. 8 10 4

Which

Which expense is invested by a tenant upon land, the fee-simple of which would not sell for more than 4*l*. What say the farmers of Europe to this, English ones alone excepted! It is impossible not to admire the spirit which animates such improvement.

" Mr. Lickfold, of North Chapel:

A broad-wheeled waggon, eight horses and two men, eight miles out and eight home, two days, at 15*s.* per day,	£.1	10	0
Turnpikes, ...	0	6	0
Chalk digging, ...	0	10	0
800 furze bavins, the produce of three quarters of an acre of three years' growth, two acres and a half, ...	2	0	0
Burning, ...	0	10	0
Emptying, ...	0	1	6
Carriage of nine cart-loads to the field, ...	0	9	0
For two acres and a half, £.5	6	6	

The two acres and a half of furze just as good land as the two and a half it manures.

Many of the farmers carry the chalk twelve miles, and through very bad roads."—*Annals*, 38, p. 660.

Chalk we see highly contributing to the melioration of different soils; but variety is as essentially necessary in manure, as it is indispensable in seed : hence it is, that in land repeatedly limed, the effect is no longer visible.

Chalk should undoubtedly be substituted in lieu of it, in all those districts where the land has been repeatedly limed. Soft, soapy, and free chalk, might be tried to very great advantage, and marl likewise.

The navigation of the Rother, effected by the Earl

of

of Egremont, has had the good effect, among many others, of dispersing great quantities of chalk in the line of country through which it passes, at a much less expense than is effected in the transport of this commodity by land. At least 40,000 tons are dispersed in the neighbourhood of the Rother and Arun.

2. Lime.

This is an article of the greatest consequence where chalk is procured in such abundance, as all the farmers use it very plentifully to manure their crop, chiefly for wheat. But the present use of it renders the expense so heavy, and the repetition so rapid, as to put the effect of liming in a very questionable point of view*. The farmers generally lay it on their fallows from 80 to 120 bushels, every fourth or fifth year, and some use it every third year. The effect of lime is unquestionably great, more especially upon lands lately broken up, and by a prudent and judicious disposition in the management, it will turn out an excellent manure; but repeated so often, it answers no longer. Indeed, sensible farmers have discovered this to be the case by long experience, and they mix it with other manures, or mould, or no longer use it.

As it is chiefly with a view of ensuring full crops of wheat, that we see such exertions effected in liming, I shall in this place enter rather more at large into this

---

* " A very just observation. Lime, as a manure, certainly benefits land in some degree; and so it ought, otherwise the expense is certainly thrown away. Ask 99 farmers out of 100, whether it pays or not: they cannot tell you, for they never calculated the great expense of manuring their land with lime. The general answer is, that it is an old established rule, the custom of the country, to lay lime on to their fallows; but my opinion entirely coincides with Mr. Young's, that it seldom answers the expenses."—W. F.

practice,

practice, and describe the structure of the kilns in this county, with the method of burning as practised, as well in the *tunnel* as in the *flame* kilns.

As the chalk-hills extend no further than East-bourne, in order therefore to supply the rest of the county, the chalk is shipped in sloops from the Holy-well pits at Beachy-head, from whence it is carried to the Bexhill, Hastings, and Rye kilns : here it is burnt into lime, where the farmers come with their teams and take it away at 6*d.* per bushel. In this trade 16 sloops are considerably employed from April to the month of November. Nine of these belong to Hast-ings, and seven to the port of Rye. The total quan-tity consumed at these kilns, for one year, amounts nearly to 633 sloop-loads of chalk, each containing 550 bushels, or about 350,000 bushels.

That the public may have all requisite informa-tion respecting the burning, I have inserted the account of a kiln, and process of burning, which I had from a lime-burner of Hastings, who has been employed in the trade for many years. The kiln is seventeen feet in the clear, at the bottom, nineteen in depth, and fourteen over ; 70,000 bricks were used in constructing it, which, at the time of building (25 or 26 years ago), were 25*s.* per 1000. It has four eyes at bottom, each 21 inches wide in the run of the shovel, and the same in length. These are situated at the op-posite sides of the kiln, and are used for drawing out the lime.

The arched way round the kiln, is eight feet wide in the ring, clear of the buttresses, which are three feet thick. The whole circumference of the inside circle is 90 feet. The conveniences are all excellent, as a waggon with one horse can stand in the porch, clear of of the door-way. The kiln contains about 1200 bushels of chalk, proper coal-measure; and the draught, in full work, is 300 bushels of lime every day. To burn one kiln, requires six chaldron of coals, Welsh, or Hartley.

The process in burning one of these is, to lay at the bottom about 30 faggots, and upon that a small quantity (about half a cord of wood, covered with straw); upon this is laid coal, and upon the coal, chalk, continued in this manner till the kiln is three quarters full, when the faggots are lighted at the bottom; and as quick as the chalk is converted into lime, it is drawn from the bottom, and replenished with chalk at the top, the kiln being always full. A chaldron and a half of coals is the usual quantity to 300 bushels of lime. The chalk wastes one-fourth in the operation. They think that the lime is much stronger when burnt with coal, as the chalk is always cut into small pieces before it is put into the *perpetual* or *tunnel* kiln; whereas, in the *flame* kilns, the chalk is put into the kilns in large pieces, the size of a man's head, and larger, without any breaking; and in the act of burning, it must happen that some of it will either be too much or too little burnt; that, for instance, which is placed at the bottom directly over the fire, will be unequally burnt, whilst that which is at a greater distance from the fire, will not receive its due portion of the heat; for, in these flame-kilns, the heat being forced upwards through the chalk, it generally happens that the lower part is burnt more than the other.

In the coal-kilns, the fire is continually advancing upwards, and the fire spreads more equally: they possess an advantage in the quick dispatch of drawing the lime; but in the flame-kilns, after the chalk is burnt,

burnt, much time is lost by waiting till the lime is
cold, and by emptying it at the mouth.   Last year, at
Hastings, the price of lime was 50s. per hundred bushels, and a drawback allowed of 5s. per cent. to
those who bought 500 bushels.   This year (1797) the
price is advanced to 2l. 14s. 2d.   The demand for lime.
from these kilns rather decreases.

The account of a lime-burner at Hastings :

	Bushels burnt.
1788,	70,000
1789,	80,000
1790,	98,000
1791,	103,000
1792,	80,000
1793,	60,000

This decrease is caused by the erection of two new
kilns, in opposition to those from the proprietor of
which this account is extracted.

The lime-burners at Hastings, Rye, and other places
along the coast, prefer chalk to stone lime, as being
softer and more yielding : but those farmers who have
been in the habit of manuring their land for a number
of years with one sort, derive benefits from a change.

The price of 100 bushels (the medium quantity for
one acre) at any one of these kilns was, in 1793,
2l. 14s. 2d.; the year before, 2l. 10s. and a drawback
of five per cent.   The price is now advanced.   The
season for burning is all the summer.

Besides the lime burnt from chalk, another great
supply from limestone is drawn from the bowels of the
earth, in the Weald.

Of this the Earl of Ashburnham is almost the sole
proprietor, and the greatest lime-burner in all the
                                    kingdom ;

kingdom ; the spray-faggot of all his extensive woods
being cut down as fuel for his kilns. These lime-
works are situated in a valley surrounded by woods ;
and as they are of a different construction to the foregoing, I shall in this place insert the following account of one of them, with the process of burning with
faggot-wood, accompanied with the plan, elevation,
and section of one of his Lordship's lime-kilns, for
which I am indebted to the spirited and enterprising
superintendant of the lime-works.

The plan of the lime-kiln, drawn by a scale, and
shewing the appearance at different heights, will enable a bricklayer to build one. It must be set into a
bank of earth, and care taken that no wet can lodge at
the bottom, which must be paved with brick ; the
breast-wall above the throats, may be done with stone,
laid without mortar ; and the bricks in the inside of
the kiln, may be laid either in loam or mortar. It
will be necessary to have a rim of iron, about two
inches wide, round the top and inside of the throats,
to prevent the lime-burners from loosening the bricks
as they put in the fuel. The bench is used to form a
steady base for the arch to spring from ; and when
done with stone, it is never liable to be burnt, as the
embers lie as high in the kiln, whilst burning, as the
bench ; and if the stone is of that nature which retains
its shape during burning, without cracking or opening, it does not get sufficiently done. It has a hatch,
merely for the convenience of taking the lime out ; and
the size of it is not material, as, of whatever size it may
be, it must be closed up with earth and stones during
the burning of the kiln. The first operation is the
filling, done by forming the arches of the kiln, which
are a continuation of the two throats, to the far end ;

and

and they are turned higher and lower, according as it is intended to have more or less stone in the kiln ; but they generally stand hollow about four feet. The arches spring from the benches, and care must be taken to fill up the sides as the work advances, and also the space upon the middle bench, or the arch would not stand. There is no occasion to be very particular as to the size of the stone in the arch, but it may be put in as large as a man can readily lift. The arch being turned and safe, the largest stones, about the size of a man's head, are placed nearest the breast of the kiln ; when it is filled within two feet of the top, smaller stones are put in ; and within six inches of the top, the smallest of all, and as small as possible. The kiln being now filled level with the surface, it is then covered over with bricks ; care having been taken, during the operation of filling, to place the limestone adjoining the sides and back part of the kiln, *hollow*, which assist the flame in penetrating through the stone, and meeting with some resistance from the closeness of the smaller pieces at the top, is, by that means, thrown more into the body of the kiln. This finished, a gentle fire is kindled, which is kept up with a moderate degree of heat for fifteen hours ; by which time the kiln becomes thoroughly heated, the limestone has done cracking, and the inside of the arch assumes a pale red colour. At this time the work goes on as quick as possible, there being now little fear of the arch failing. It is to be observed only, that towards the conclusion of the burning, when the kiln necessarily becomes very hot, for ten minutes in every half hour the lime-burner may stop, and put no fuel into the kiln, and the operation will proceed on with the same expedition. When the limestone is thoroughly

roughly burnt, there is a clear red fire at the top, and an appearance of sulphur upon some of the bricks may be generally seen in the hollowest parts of the lime-kiln. It is then necessary to throw a little clay upon the tops of those bricks, in order to choke the fire, and force the heat elsewhere, and, by covering the surface with dirt, the heat is gradually conducted over the whole. When cool, the bricks and dirt come from the lime without the least injury; but it must remain thirty hours before it can be emptied. The tools necessary are:—a prong, to push forward the faggots, and sometimes to lighten them up in the throats; a long pole, reaching to the further end of the kiln, for stirring up the embers, to make them throw out a fresh degree of heat; a large hoe for raking the embers; and a large iron shovel-pan to carry them away. In putting the fuel in, the stronger end of the faggot is first thrust forward. The ashes are worth as much per bushel as the lime, either for the use of the farmer or soap-boiler. The two sorts of limestone in use are very different in the effect which the fire has upon them. The one, a grey stone, is a mass of marine shells, and the exuviæ of sea animals: this will at first bear the necessary degree of heat without danger; is very tough, and will open a little without flying; but, upon fire being continued too long, will vitrify. The other is a blue stone, very much inclined to crack and fly to pieces, and requires great attention, to prevent the stone forming the arch, from breaking and letting in the kiln. By continuing fire too long, and too fiercely, it runs into a powder, although it does not vitrify like the other: it is a much stronger cement than the grey, or chalk. At first, difficulties may arise in the burning, and the stone may tumble in; but be the difficulty
culty

culty what it may, care and perseverance will over-
come it. It may not be worth while to bind the furze,
when used as fuel, in faggots; but whether it shall
be burnt as faggots, or loose, it should be stacked
when cut, to retain its strength, and it may be used
in its dry state; this mode, therefore, should be
adopted. There should be water near the kiln, for
the convenience of wetting the iron over which the faggots are put, and also for wetting the tools, and the
ground round the kiln, to prevent the scattered faggots
or furze from taking fire. The top of the kiln should
be level with the surface of the adjacent ground; and
a drain should be made from the hatch round the kiln,
to carry away any wet that may fall, and which would
otherwise keep the kiln cold, and therefore waste the fuel.

The hole for the reception of the embers will be most
convenient on the left hand side of the mouths of the
throat, at the distance of five or six yards, so as neither to give much trouble in conveying them from the
kiln, nor reflect too great a degree of heat on the
burner. For burning coal, the *tunnel*-kiln is superior
to the *flame*-kiln, for no heat is lost. In a flame-kiln
this is not the case, since a great degree of heat, and
much time also is consumed, before it can be emptied.
Chalk loses one-fourth in the kiln. Those farmers
who for many years have limed with chalk till it is
useless, by changing it for the stone-lime, have reaped
great benefit; and so, on the other hand, with stone-
lime. Variation is necessary*.

* Might it not answer, to use the manure of chalk and stone, without
decomposing and destroying what may be its most essential virtue by fire?
For instance, if it was, in its natural state, pulverized by large mills
prepared for that purpose, would it not, thus prepared, prove equally,
if not more efficacious? The experiment however is worth trying,
whatever may be the effect of it.—*W. Fox*.

SUSSEX.]   The

The great demand for lime in the eastern parts of the
Weald, induced the Earl of Ashburnham, a few years
since, to set about a method of drawing up the limestone from under ground, for the supply of the neighbourhood. This great undertaking he has most successfully accomplished, and the neighbouring farmers
for many miles round, are now supplied from his
works.

The lime-works are situated in a valley in the centre
of Orchard-wood, Dallington-forest, &c. The shaft
by which we descended is four feet by five, boarded,
with ladders for the men to go and return from their
work, which is 80 feet deep, more or less : through
this the stone is drawn up in barrels, of 5 cwt. to each,
one descending while the other ascends. The whole
machinery is moved by a horse, and is the same with
that generally used in collieries. Drains are constructed at the bottom to take off the water, by means
of a level, continued as the work moves on, and
serves not only for conveying away the water, but also
for bringing air to the different works. The process
in separating the limestone from the solid bed is, to
blast it with gunpowder : a hole is bored in the rock
with an auger; a pricker is put into this whilst the
powder is ramming down, and when this part of the
operation is finished, the pricker is taken out, and a
wheat straw filled with powder is put into the place of
it, and a small piece of touch-paper to the top of the
straw, so as to communicate with the powder within,
and give time to the workmen to seek a place of safety.
When the rock is blown up, the stone rolls down in
large blocks, which are broken to a portable size, and
then conveyed in barrows or little waggons, on roads
framed for the wheels to roll along, to the foot of the
shaft.

shaft. A boy fills the bucket, which is drawn up, and stacked into square yards, being previously cleansed of all dirt and shale, which would otherwise vitrify and injure the lime. Each stack is five yards in breadth, and ten long: from thence it is taken to the kilns as wanted. In general, it is much better that the limestone should remain for a time in this state, that any remaining dirt which adheres to it, may peel off with the weather. The situation of the kilns is close to the pits, and lower in the valley, so that the limestone is carried *down* to the kiln, and the labour facilitated. When burnt, the farmers come with their waggons to carry it away. His Lordship has opened a communication with London, and now sends from Hastings by water. The kilns begin working in April. In 1792, the account stood thus:

April,	6000 bushels.
May,	8000
June,	26,000
July,	35,000
August,	21,000
September,	10,000
October,	9000
November,	6000
	121,000 bushels.

Respecting these lime-works, it is impossible not to admire the spirit with which his Lordship entered upon this arduous undertaking, by sending for miners and artificers skilled in the operation of mining: his success has corresponded to the spirit which first animated his endeavours, and he now reaps the fruit of his labour, in creating a supply for the neighbour-

ing

ing farmers, which before was to be had but in small
quantities, and that at a dearer rate, or it was obliged
to be brought from a distance.

3. Marl.

In the maritime district, this excellent manure is in
great abundance a few feet under the surface. It is to
be preferred when it contains much of that greasy kind
of soapiness, which has worked such wonders in various parts of this district. Great exertions have been
used in marling these fertile soils. It is, I believe,
more or less, found every where on the south side of
the Downs. Great quantities are dug out of pits on
the sea-shore, which are generally covered at high-
water mark. Near the sea, at Ford, &c. white marl
is dug out of the ditches, and spread with great success. Mr. Milward has greatly improved his estate at
Hastings, by marling. In three years he raised 60,000
loads; dug and spread at the rate of 5s. 6d. for 800
bushels.

The farmers spread it upon their lands according to
circumstances—from 10 to 20 waggon-loads (800 to
1600 bushels). Wherever the soil tends to a reddish
loam, or inclines to be sandy, here it is that marling
is practised with the greatest success. With regard
to the season of laying this manure upon the land, the
most proper is in winter or autumn, upon a clover
ley, for the frost to pulverize it: the field is fed in
the following spring, and then bastard-fallowed for
wheat. This is considered as the most judicious way,
but the more general rule is that of spreading it in
summer, and then ploughing it in*.

                                                    The

---

* Marl should always lie on the land six or eight months before it is
                                                    turned

The following analysis of the calcareous soils, &c. of the neighbourhood of Petworth, was made at Lord Egremont's,

---

turned under; but the longer it lies, the better it will answer, for when it is immediately worked into tillage, for the want of sun and frosts, &c. it goes to a clay, and is longer before it shews its good effect: properly applied, it will be beneficial the first season, and when it is worn out, if repeated again, will still answer the better, if properly applied.—*Mr. Harper.*

Do they use *marl* or chalk for meadow and pasture? Is the chalk of the hard or soft nature?—*H. Strachey.*

Chalk rubble is used upon meadow land, marl upon arable: chalk is used, of both kinds, but the soft greasy, soapy, by far the best.

With respect to marl, I shall point out an egregious blunder, which some time ago came accidentally within my observation. Being on a visit to a relation who had lately taken a farm in the eastern part of Sussex, on walking over his land, I observed several pits, out of which I supposed marl had been dug, and as the land was inclining to a light sandy loam, I thought marl might prove a valuable acquisition, and digging in the pits, I found a soft substance underneath, which looked like marl; but on trying it with acids, I discovered it to be a soft clay. On this disappointment, I inquired whether any body in the neighbourhood had ever found and tried it. I was answered, that it had been tried, but was not found to do any good, and that it was entirely left off as a useless practice, and that there was of course a general prejudice against what they deemed marling. Being desirous of inquiring farther, I sent to a neighbouring marl-pit, which was said to have been a good one, and used not many years before. I got some from thence, and on trying it, found it to be mere clay, nor had it any marl at all mixed with it; and this farmer's land being stiff clay, no wonder that adding clay to it, did not answer; nor was it any wonder that, under the influence of such an error, it should be looked upon that marling such land would not pay for the labour and expense. But it by no means follows, that marling, or even claying, with such clay as that was, such light land as that I walked over, would not have been a most beneficial practice; and I am inclined to think, that in some former times it had been used for that purpose, or I cannot account for so many pits dug in various parts of a farm. An error of this nature, attended

with

Egremont's, at Petworth, by Mr. Marshall, the well-
known agricultural writer : it will throw considerable
light upon the subject of chalk, marl, limestone, &c.

Petworth, April, 1791.

Hard Marl of Duncton.

100 grains yielded in one experiment 76 gr. calc. mat.

$$\begin{array}{r} 24 \text{ residue—a fine} \\ \underline{} \quad \text{silt.} \\ 100 \\ \underline{} \end{array}$$

In another experiment, 78½ calc. 21½ residue.

100 grains soft marl of Duncton, 80 gr. dissolv. mat.

$$\begin{array}{r} 20 \text{ resid. as above.} \\ \underline{} \\ 100 \\ \underline{} \end{array}$$

100 Grains Chalk of Duncton.

First trial, ... 73 dissolved.

27 resid.—fine tenacious silt.

Second trial, ... 75 dissolved.

25 resid. as before.

Tillington Whinstone.

First trial, ... 75 dissolved.

25 resid. principally fine sand.

with pernicious practical consequences to so capital a branch of agriculture as manuring land, deserves animadversion, and the ignorance
on which it is founded, deserves to be exposed. The detection of error
is always a capital step towards finding out truth.—*Rev. Mr. Davies.*

Second

Second trial, ... 74 dissolved.

26, as before.

—

Marble of the Weald.

First trial, ... 92 calcareous.

8 resid. blue silt.

—

Second trial, ... $92\frac{1}{2}$ } as before.

$7\frac{1}{2}$

———

The hard sandstone of Petworth-park, non-calcareous, and the very hard ragstone non-calcareous.

———

Petworth, 30*th April,* 1791.

Blue Stone of Sutton.

64 Calcareous.

36 An ash-coloured friable earth.

Chalk Marl of the sea-beach, near Middleton-church, dug up in a state of paste, as the tide was leaving it.

96 Calcareous.

4 Brown slime, with a non-calcareous gypsum, like fragment.

———

Petworth, 19*th May.*

Chalk Marl of the sea-beach in Middleton, picked up in knobs on the beach in the tide's way.

$98\frac{1}{2}$ Calcareous.

$1\frac{1}{2}$ Brown slime, with some minute fragments.

Chalk Marl of Deanswood.

98 Calcareous.

2 As above.

May

Limestone of East Sussex (brown part.)

91 Calcareous.
 9 Rusty-coloured friable earth.

Limestone of East Sussex (blue part).

95 Calcareous.
 5 Black, gunpowder like silt.

Efflorescent Matter of the New Road.

64 Solution precipitate white.
36 Residue, friable earth.

Limestone of Tillington Street.

81 Solution white.
19 Residue, fine sand and friable silt.

Hardham Blue Marl.

$8\frac{1}{2}$ Solution, a purple tinge.
$91\frac{1}{2}$ Residue, grey, smooth, tenacious, subsoluble in
    water.

Houghton Chalk, middle strata.

99 Solution.
 1 Residue, a brown matter lodged in the pores of the
    paper.

Houghton, lower strata.

97 Solution.
 3 Residue, grey subtenacious silt.

Houghton, upper strata.

99 Solution.
 1 Residue, as the middle strata.

Duncton,

Duncton, East Pit, upper strata.

96 Solution.

4 Residue, grey silt, as the lower strata of Hough-ton pit.

Duncton, East Pit, lower strata.

97 Solution.

3 As above, except the colour somewhat darker.

Duncton, West Pit, upper strata.

$93\frac{1}{2}$ Solution.

$6\frac{1}{2}$ Still browner and more tenacious, but perhaps discoloured in drying.

Duncton, West Pit, lower strata.

96 Solution.

4 Light coloured silt, with some white fragments, apparently of plaister stone.

" Marl Flour" of Duncton, West Pit.

41 Solution.

59 Residue, tenacious, impalpable, resists water, like fullers'-earth, but somewhat darker.

" Maamstone" of Duncton, Bury, &c.

20 Solution.

80 Residue, resembling the residuum of the grey chalk, but more friable, and somewhat sandy.

Limestone of Rotherbridge.

68 Solution.

32 Residue, a fine loose sand.

4. Sleech.

4. *Sleech.*

Sleech, or sea-mud, is not uncommonly used as manure in the neighbourhood of the sea: they spread
from 12 to 1300 bushels of it for wheat ; but the land
has been too frequently dosed with it, to render it any
longer answerable. It is inferior both to marl and
chalk.

5. *Soap-Ashes.*

These are used upon pasture-land : they mend worn-
out grass, by killing the moss and other rubbish, and
produce a fresh layer of white clover. Mr. Clutton
bought these ashes at $2\frac{1}{2}d$. per bushel, at Cuckfield, in
1793, and spread 200 bushels to an acre : the improvement great.

6. *Wood-Ashes.*

These, like the above, have very much improved
cold and wet pasture-land in the Weald, where they
are an excellent dressing. The virtue residing in ashes
is very great, and not having their manure exhausted
like the above, are much more beneficial. Small spots
of poor hungry pasture have been very profitably benefited by wood-ashes, to near twice their value ; and
great advantages might be made of this manure in a
forest country, if attention was paid to the saving of
them. The Weald is a forest, and the consumption
of wood abundant, and selling any part of his manure
is no advantage to a farmer. Mr. Mayo, at Battel,
mends his pastures, in low and spungy situations, with
these ashes, and nothing can equal them.

7. *Peat-Ashes.*

Mr. Gell, of Applesham, is certainly one of the
greatest manurers in Sussex. With lime and chalk he
has

has made very powerful improvements. He has tried peat-ashes for various crops, and undertakes to say they are good for pease, turnips, clover, and sainfoin.

8. *Coal-Ashes.*

These too are a great improvement of grass. Lord Egremont has doubled the value of his park with draining and coal-ashes, which before was covered with moss, rushes, and rubbish. The difference of the grass, where the land has been covered with ashes, and where it has not, is most striking. They have created a sweet bite of white clover and trefoil, liked much by sheep.

9. *Rags—Sheep-Clippings.*

These are of service chiefly in the hop-grounds, for which they are thought an excellent manure. Great benefit is said to be derived from the application of these rags and clippings, in contributing to preserve the plantations in a state of constant moisture and vegetation in the driest seasons, when grounds which have been manured with dung, have been dried up, and the hops failed.

10. *Pilchards.*

Fish have always been known to contribute greatly to the melioration of land, by the quantity of obaginous matter with which they abound. Mr. Milward has manured with them, but found no benefit.

11. *Paring Dust.*

A fellmonger of Petworth tried this as a manure in his garden for potatoes and cabbages. For potatoes, the experiment was very efficacious. After the trench

is

is made, and the potatoe-cuttings placed in it, he covers with dust, and a great produce gained.

12. *Gypsum.*

Among other manures, gypsum, or plaister of Pa-
ris, has been tried in Sussex, but none of those marvellous properties have been discovered to reside in it,
which some experimenters have supposed that it contains. Mr. Pennington, of Ashburnham, gives the
following account of it: — " Having procured equal
quantities of French and English gypsum from Mr.
Green in the Borough, the following trials were made
of it; in every instance as much ground was covered
with the one as the other, on distinct, but adjoining
spots. On the 14th of June, 1791, in six different
fields, portions of from 40 perches to four perches,
which were accurately measured, upon natural grass,
beans, potatoes, pease, and barley, were covered, at
the rate of eight bushels to the acre; the soil a sandy
loam, but in which the sand is of so fine a grit, that
every shower makes it poach in winter. On the day
it was strewed, it was showery, and on the 15th it
rained from ten in the morning till evening. Though
I attentively observed those spots through the summer
of that year, I could never perceive on them the least
appearance of greater luxuriance than on the surrounding ground. I hoped to see some effect in the year
following, but was disappointed. On the 13th of
April, 1792, I strewed in the morning, whilst it rained
heavily, two square perches of red clover, sown in
1791 with barley, and which fully planted the ground;
and in the afternoon of the same day, a quarter of an
acre in the same field, at the rate of six bushels to the
acre. The next day was showery, and in the following

ing week a great deal of rain fell. I could not at any time in the summer perceive the least improvement from the gypsum. The 29th and 30th March, 1792, I sowed patches of wheat, spring tares (at the same rate of six bushels to the acre) : the 30th, and five following days, were showery, but none of these crops were benefited by my making use of it."

Besides the above manures, which are partially used, there are others, though the quantities are too small to specify particularly. Sea-weed is collected, and mixed into compost, oil-cake, &c. &c.

SECT. IV.—WEEDING.

THE best farmers on the South Downs eradicate the kilk and poppy by constant attention to ploughing and weeding; whilst the fields of others may be seen in a perfect blaze, as it were.

Charloc is a very pernicious weed ; it is moreover a great enemy to lambs : when they are turned into it at two or three months old with the ewes, they will frequently die suddenly by eating it. Mr. Ellman observes, that the real cause of such a luxuriant herbage of weeds as is too often beheld in the Downs, must be attributed to the practice in vogue amongst farmers, which is, that as soon as the corn is off the ground, they plough in the charloc, which vegetates in the ground, and the land becomes stocked with it. Now the turnwrest-plough, from the nature of the share, does not cut the earth, and throw it up so well or so evenly, as the round plough : consequently, the ground not being properly turned over, the kilk has a fair opportunity of of vegetating; so that if we examine only a small piece of land under these circumstances, it will in all probability contain several seeds of this destructive plant.

Wheat is generally hand-hoed in the spring ; if foul, the operation is repeated. Women and children are employed, at 6*d*. and 8*d*. per day. Mr. Woods hoes his pease by fixing together two five-inch hoes at three inches asunder (between which a drill passes), in such a manner, that a man draws it after him. Of this work one man will hoe an acre per day.

Upon dry soils subject to poppy, Mr. Ellman, of Shoreham, ploughs his tare and rape land for wheat the beginning or middle of September, to sow the wheat the middle of October : the harrowing kills this noxious weed ; and in putting in the seed, he likes to tread much with oxen or with sheep. A neighbour treads his with oxen in March, which he thinks better against the poppy than doing it at the time of sowing.

SECT. V.—-WATERED MEADOWS.

On the western side of Sussex, that admirable practice of watering their meadows in a regular manner, is very well understood, and successfully practised. The course of the Lavant river, from its spring-head to Chichester, waters the finest and most productive meadows in the county. The water is let on the grass in December, when it waters for three weeks : this three weeks is equal to all the rest of the year ; for at this time the moss is entirely killed by it, and the young grass will then begin to shoot out in a very luxuriant manner.

manner. In spring-watering, it is usual to let the water over the land twenty-four hours each time; and in May the watering ceases altogether. In July, from two to three tons of hay are mown per acre, and the rouen fed with cattle till Christmas, but seldom with sheep, as they are found to rot. If wethers or ewes, before lambing, were turned in, they would certainly die. Eighty ewes bought at Weyhill fair, were turned into some land adjoining a watered meadow: it happened that a score of them accidentally broke into the meadow for one night; taken out the following morning, and kept till lambing: the score that had broke loose produced twenty-two lambs, all of which lived, but every one of the ewes died rotten before May-day; the remaining sixty made themselves fat, nor could a rotten sheep be discovered among them: several of these were put into the meadow with their lambs, but received no injury. The soil of this meadow ground is either peaty or gravelly; it is cut into lands of thirty or forty feet width, with a drain and water-carriage to each land.

Stock and Product.—As soon as these meadows are mown, oxen are turned into them, at the rate of 100 to 140 stone to two acres, till Christmas; which at 3s. 6d. per head per week, the accustomed valuation, is 1l. 8s. for September, October, November, December. They are taken up to the stalls for winter-fattening, and during the three succeeding months of January, February, and March, the same ground is stocked with two couple of ewes and lambs per acre, which at 6d. per week, each couple for twelve weeks, is 12s.: this, in April, is increased to five couple for six weeks,

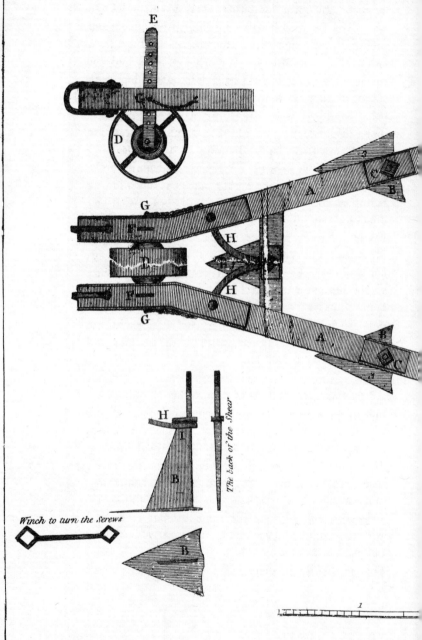

AN *ENGINE* to clear Land of Weed

E

D

G

F

G

H

H

A

B

C

A

B

C

H

I

B

The back of the Shear

Winch to turn the Screws

B

1

Ja.ˢ Lambert del.ᵗ Lewes

nly called in Sußsex A NIDE OT.

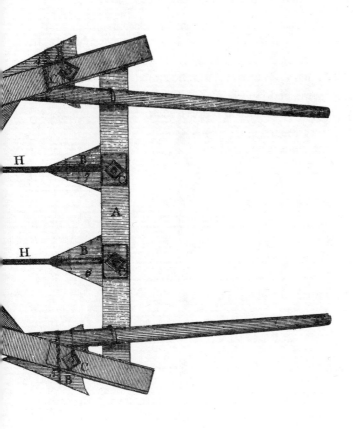

e *of one Inch to a Foot*

| 3 | 4 | 5 | 6 |

weeks, which amounts to 15s. more. The hay is mown in July, and the ordinary crop two tons.

Together, the account will stand thus :

September, October, November, December, half an ox per acre, at 1s. 6d. per week, ...	£.1	8	0
January, February, March, two couple ewes and lambs, at 6d. each couple,	0	12	0
April, and part May, five couple, six weeks, ...	0	15	0
Two tons of hay, at 40s.	4	0	0
Produce, ...	£.6	15	0

The expenses are :

Rent, ...	£.1	10	0
Labour, ...	0	4	0
Watering, ...	0	2	0
Rates, ...	0	6	0
Tithe, ...	0	4	6
Expenses, ...	£.2	6	6
Produce, ...	6	15	0
Remains a profit per acre, of	£.4	8	6

The waters which have been collected from some hills about Petworth-park, in order to form a sheet of water in front of the mansion, after passing through the lake, are let off upon a slope of the park, and may irrigate five or six acres, to the great benefit of the grass; and though it is only in the winter season of the year that any flow takes place, yet in the dry time of the summer, the herbage upon this irrigated

gated part of the park is more luxuriant than elsewhere. Indeed the advantages of irrigation are too
clearly seen, to admit any doubt of its effects. But
we have a notable instance in Burton-park, where it
has worked little, if any sensible improvement, though
conducted in a very skilful manner.

The Rev. Nicholas Turner, who lives near the spot,
and is thoroughly acquainted with the ground, is
clearly of opinion, that watering this land has not
done one atom of good to it. The soil is an extremely
poor blackish sand. Mr. Turner says, that the slimy
particles of the water which issue from the chalk hills
are beneficial in the winter months, but that in summer
the heat so acts upon the water, that it deposits its
earthy base, which adhering to the blade, prevents
the growth of the plant. Whatever be the cause, it
is clear that no improvement by irrigation is visible in
Mr. Biddulph's park.

CHAP. XIII.

LIVE STOCK.

SECT. I.—CATTLE.

WHOEVER has given much attention to husbandry, and practised it for any length of time, well knows, that of all others, the profitable management of cattle and sheep is the most difficult branch of farming. It is here that improvement is slow and tardy in its growth ; and success is least to be expected, and late before it comes. The improvements by manuring and draining, with a right application of the course of crops, have generated great alterations for the better in those branches more immediately connected with the plough. A new turn has every where been given to the face of the country. The return is speedy and certain in tillage; in live-stock it is distant and uncertain.

The breed of Sussex cattle and sheep, and the system upon which they are founded, forms the most distinguishing feature in the husbandry of this county. The cattle must unquestionably be ranked amongst the best in the kingdom ; and had Bakewell, or any of his associates, adopted the middle horned breed, either of Sussex, Devonshire, or Herefordshire, in preference to the inferior stock which the reputation of his name, and the mysterious manner in which his breeding system was conducted ;—had he, I say, gone to work with any of the above-mentioned breeds, it would have

contributed

FAIR MAID. *SUSSEX COW.*

of the Glynd breed, in the possession of M.ʳ John Ellman.

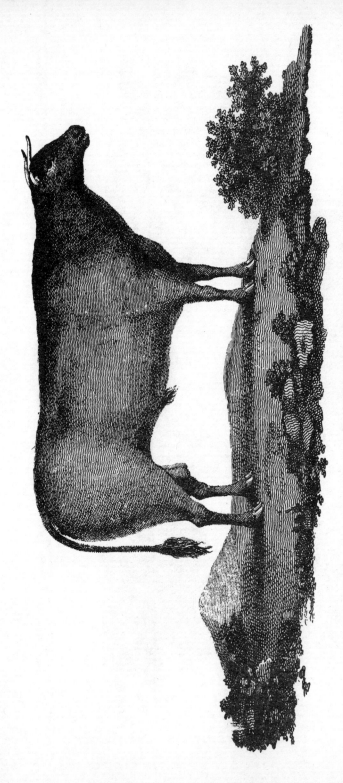

MERCURY. SUSSEX BULL.

bred at Glynd, by M.ʳ John Ellman.

contributed to exalt the superiority of his stock beyond
the power of local prejudices to remove. Ideal merit
would have fled before real improvement. A good
thing will force its way, in spite of the steadiest opposition : it stands in no need of assistance from monopoly, to make its way. Improvement carried on in
the dark, must be of temporary duration. Secrecy
excites curiosity. To price, it seems, merit is attached : it wears an imposing aspect ; it commands
the attention of the farmer : he never questions the merit of a breed, when others are as the dust in the scale.
Justice however demands that we acknowledge, that a
high money-price has had an effect as certain in its
consequences, as if the same had absolutely and *bonâ
fide* been given. We are witnesses of the rivalship it
has excited. Other districts are fast copying the example which Bakewell has set.

The celebrity which Bakewell has gained, and the
reputation of his name, stand deservedly high. To
his knowledge and experience it is, that the breeding
world are indebted for whatever exertions have animated the kingdom in advancing the merit of other
stock. If we examine into the state of our cattle and
sheep before he turned his attention to the subject, and
again view it at the time of his decease, the inquiry
would be so favourable, that we cannot but confess,
that such a weight of improvement as has been created
in England, in consequence of his extraordinary exertions, is one of the most singular circumstances in the
annals of our husbandry. It is not a short-lived improvement, which might be thought to exist in a few
districts, and without extending its influence beyond
the interested views of any private society. No such
thing. This I consider as the least part of its benefit :
                                                    it

it is a great national concern, which must in its nature
be lasting. The circulation of inquiry, the exciting
emulation, and instigating others to a rivalship; the
encouragement of merit by a great money price: these
are circumstances that have contributed to raise the
merit of our cattle and sheep: the progression is rapid,
though still in its infancy. The whole island is elec-
trified. The age of frippery is fast sinking into con-
tempt and oblivion, and another, truly great, because
nationally advantageous and conducive to the happi-
ness of thousands, has arrested the attention of mankind. Men of the highest rank and fortune no longer
keep aloof from rural concerns; they professedly take
a pleasure in zealously contributing to promote the
study of these important designs, and rival each other
in restoring the plough to the rank and estimation
which it so deservedly claims. As *nihil agricultura
melius est,* so *nihil homine libero dignius.*

In treating of the management of Sussex cattle,
with a view to greater clearness, the first object to be
considered is, the division of the subject; and this
raises an inquiry into the purposes for which cattle
are bred in this county. This is the leading question,
and it will materially tend to elucidate the arrangement, by considering cattle under the three purposes of,

I. Beef.
II. Dairy.
III. Work.

I. *Beef.*

Under this head may be classed the description of
Sussex oxen, in relation to,

1. Colour.

1. Colour.
2. Shape.
3. Fatting, age.
4. Food.
5. Thriving disposition, handling.
6. Flesh, offal.
7. Sale, price.
8. Weight, profit.
9. Compared with Devons.
10. Compared with Herefords.

1. *Colour.*

Sussex cattle are universally red; for wherever any other is found, it may be depended upon that the breed is stained with foreign blood. Many farm-yards in this county have a mixed colour of black and red and white, and all black, but they are a cross from Wales and other parts. This beautiful breed has been very much tarnished in this manner; but the nice breeders adhere as steadily as they can to the dark red. Sussex ideas run strongly in favour of this co-lour: it is a point, they say, of considerable import-ance, as the beasts are more kindly, and have a better disposition to fatten. To retain old customs is very natural to man; but prejudice in favour of colour, when opposed to shape and make, is carrying a man's peculiarities a little too far. It is repugnant to com-mon sense : where the points of an ox are good, colour is a perfect non-essential. It is readily admitted, that the thorough-bred beasts are a dark red ; but crossing has much altered the stock : it should never be at-tempted, without a certainty of mending the breed.

It has been asserted, that white is an infallible cri-terion of degeneracy in all the animals of the creation, and

and that oxen and cows that partake of it are weak and tender; that in the health and vigour of the forest, no animal is found of this colour. In confirmation of this, the Rev. Mr. Ferryman* remarks, that the white colour is the test of degeneracy : that the colours of the animal world which contain the most vivid hue, and are of the most shining lustre, are endued with the greatest portion of health and strength : that birds, when confined in a cage, shift their plumage to the sign of degeneracy : that the common poultry with high coloured plumage, are every way superior to white and its kindred ; and among mankind, a healthy habit is visible in the floridness of the constitution; as sickness is seen in the paleness of the looks, and grey-headed age in the whiteness of the locks. So far for Mr. Ferryman. But let me ask, what traces of weakness are visible amongst the frozen inhabitants of arctic circles ? What signs of any want of spirit, vigour, or health, is observable in the forest bordering upon the Pole, the residence of bears, black and white ? Many species of the inhabitants of Siberian forests regularly change their colour to a white, on the prospect of winter, and return to their original colour in the spring. But on what principles of degeneracy? The European swan is a pure white, the New Holland black; but neither are a degenerate race.

The Hereford breed of cattle are among the best breeds of England ; yet it has been the fashion of late to breed for white faces, and the stock is considerably improved. Formerly the true and high bred Herefords

---

* This gentleman has formed at Petworth a museum of British birds, which promises to be very complete.

were brown*. The pure white breed was probably the parent stock out of which all others have derived their origin. At this day we see them wild in Lord Tankerville's park; and as an instance that white appears the predominating colour, from a great number of calves which Lord Egremont has reared from Hereford bulls and Sussex cows, and the contrary, and Devonshire cows and Hereford bulls, and *vice versa,* all these calves have uniformly had white faces and bellies. A few years since, Mr. Davis, of Glynde, was in possession of a black ox with a white face, out of a red cow by a red bull; which shews, that in crossing various breeds, the properties of cattle will be dormant for many years, and then shew themselves in their posterity†.

2. *Shape.*

Mr. Ellman, of Glyd, has given us his experience of cattle, summed up in the following description of a thorough-bred Sussex ox. It should be observed, that these points were approved by several other intelligent breeders. A thin head, and clean jaw; the horns

---

* That white was thought no proof of degeneracy in stock, may be gathered from what Mr. Campbell says upon that subject in the Annals of Agriculture, where speaking of this vulgar error, he says, in a letter to the Earl of Egremont, " As to matter of fact, I can assure, that among other breeds of cattle which I have tried (and wherever I did try, I did it thoroughly, and not slightly), I have had bulls, oxen, and cows, of a white breed, as healthy and hardy as any others."— This is the opinion of an eminent breeder, founded upon extensive practice, and who was allowed by all who knew him, to have possessed a great share of knowledge with respect to cattle.

† It must however be admitted, that colour is at present of importance, as a mark to indicate what now must be considered as the true Sussex breed; with this view it may safely be attended to.

point

point forward a little, and then turn upward, thin tapering, and long ; the eye large and full ; the throat clean, no dew-lap ; long and thin in the neck ; wide and deep in the shoulders ; no projection in the point of the shoulder, when looked at from behind ; the fore-legs wide ; round and straight in the barrel, and free from a rising back-bone ; no hanging heaviness in the belly ; wide across the loin ; the space between the hip-bone and the first rib very small ; the hip-bone not to rise high, but to be large and wide ; the loin, and space between the hips, to be flat and wide, but the fore-part of the carcass round ; long and straight in the rump, and wide in the tip ; the tail to lay low, for the flesh to swell above it ; the legs not too long, neither thick nor thin on the thigh ; the leg thin ; *shut well in the twist ;* no fulness in the outside of the thigh, but all of it within ; a squareness behind, common in all long-horned beasts, greatly objected to ; the finer and thinner in the tail the better. Of these points, the Sussex beasts are apt to be more deficient in the shoulder than in any other part.

A well made Sussex ox stands straight and nearly perpendicularly on small clean legs ; a large bony gummy leg, a very bad point, but the legs moving freely, rather under the body, than as if attached to the side of it ; the horns pushing forward a little, spreading moderately, and turning up once. The horn of the Devonshire, which very much resembles the Sussex, but smaller and lighter, is longer, and rises generally higher. The straightness of the back line broken in many very fine beasts, by a lump between the hips.

Such are the observations of Sussex men upon their own breed of cattle : in addition to which, some further

ther remarks will be occasionally necessary, either in illustration of the foregoing, or to point out where different opinions are entertained upon the stock of this celebrated district. With regard to those parts which constitute the offal, head, bone, leg, &c. these are generally clean, fine, and light, proportioned indeed to the carcass; leg not gummy; head and neck small; chap clean; ribs springing; full flank; great breadth before and behind loin and shoulder. He who breeds cattle with points like these, cannot fail of obtaining a valuable stock*.

3. Age.

The age when the generality of farmers turn off

---

* In our tour through the county (1791), my fellow-traveller, whose opinion I cannot but mention as entitled to a considerable degree of attention and confidence, as being founded upon an early and long experience, had no hesitation in pronouncing this breed to be comparatively much inferior to a good selection from the Lancashire, Hereford, or Shropshire breeds. We had an opportunity of examining some heavy and well fed oxen at Lord Sheffield's; the deep red colour, and cockhorn, seem the distinguishing characteristics. I have often noticed them since in Smithfield. They have apparently too large a proportion of bone and offal; and I cannot help thinking them, on comparison with some other breeds, though a weighty, yet an uncouth and coarse animal.—*W. Pitt, Pendeford, Staffordshire.*

Mr. Pitt taking exceptions to Sussex cattle, on account of the too large proportion of bone and offal, is curious; still more so the observation, that " thorough bred Sussex oxen are uncouth and coarse animals." There is no saying exactly what is intended by such expressions. If it implies a *coarse*, and *thick*, and *rough* hide, or a *hard* and *coarse-grained* flesh, nothing was ever farther removed from fact than such an assertion. Sussex oxen are as remarkable for the fineness of their hides, as they are for the closeness and delicacy of their flesh. In his own Staffordshire long-horns there does not exist any shadow of comparison, for feeding, grazing, working. In quality of flesh, thriving disposition, &c. both the Sussex and Devons exceed you, and Hereford leaves you far behind.

their

their oxen from work, is at six years old. They will
sometimes work them another year; but if we take the
county upon a medium, we shall find that the far
greater proportion send them to the grazing account at
six. The proper time when cattle fatten to the best
advantage, is a matter undecided in the opinion of
some of the best judges in the county; yet it is a point
that deserves ascertaining, for it is very necessary to
know at what age they decline in the working state.
It is affirmed, that young beasts are much more profitable for fatting than older ones, as the union both of
growth and fat pays better than fat alone. We know
that cattle will continue their full work long after
the time they are usually sent to graze, and perhaps to
greater profit than when they were young. At six,
they are turned off; at seven, slaughtered; although
at ten or eleven, he is greatly to be preferred to the
work of a four or five-years old steer, and the additional labour would probably out-balance any injury
he might receive by more than three years' work; not
that the cattle would be damaged by any alteration of
this kind, for the greatest injury is effected in their
growing state; and this ceasing at six, an ox cannot
receive any damage which is not over-paid by his
work. Lord Egremont has a pair of Sussex oxen in
the eleventh year of their age, which, for seven years,
have done as much ploughing and carting as any two
horses in the county. His Lordship is now fattening
those beasts, and they thrive very kindly, and more
so than younger beasts in general. With half a summer's grass after taken from the collar, and an autumn's
rouen, they were, without other food, sent to Smithfield, and sold for eighty guineas: a remarkable fact,
bearing directly on the question of age and long work.

4. *Food.*

4. *Food.*

After spring-sowing is over, it is the common practice, about May, to turn the oxen into the brooks, pastures, or marsh-lands, during the summer, in order to prepare them for stall-feeding in the winter. Where the meadow is rich and the herbage luxuriant, one acre will readily support one ox in a thriving state for six months, turning them in the middle of May. Many farms on the South Downs have considerable tracts of marsh, which very much contribute to carry on a system of fatting, not perhaps applicable to districts less favourably situated. But let it be observed, that in the Weald, where there are no marshes, the cattle are fatted to perhaps equal advantage with those who are better situated in respect to meadow; so that it is obvious that the goodness of the cattle does not depend upon any extraordinary richness of the county.

Stall-Feeding.—It was once a question, whether cattle could ever be profitably fed, if tied up to a stall. Experiment, full, ample, and satisfactory in the highest degree, has now convinced the world, that it is the most profitable manner of feeding; and indeed the practice has been found so beneficial, that summer-feeding after the same manner, will by degrees gain ground. There is every reason in the world for believing that all sorts of cattle, as well as horses, should seldom, if ever, be allowed to graze, either in summer or winter.

It is an erroneous, and certainly an expensive method, to keep a farm under-stocked. Soiling is unexceptionable; but in winter, to allow cattle of any kind

to

to range over fields, treading and poaching the ground, and losing sight of all the advantages which confinement creates in respect to manure, is surely not a very beneficial system to pursue. The more cattle are confined, the sooner will they fatten. It is the same with horses, and holds good with every species of live-stock upon which experiment has reached.

The Earl of Egremont's dairy, of between twenty and thirty cows, and all the oxen upon the farms, are constantly tied up for the greatest part of the year; and such is the advantage attending it, that one-third of the food is saved, the oxen are on the spot for their work, the cows are milked with a fourth part of the usual trouble, more dung is made; no poaching and treading the ground, &c. &c.

Stalls.—Ox-stalls are generally ill-contrived in Sussex, exposed to the winds and rain, which checks the thriving of cattle more than we at first imagine. It may be safely said, that in proportion as the cold is excluded, will the ox get fat: warmth is almost as essential as food itself. We see little attention paid to consult the constitution and habit of the ox, nor to allow him the full indulgence of all conveniences and comforts, as well as bare necessaries. Mr. Ellman, of Shoreham, in the stall-feeding his cattle, has keel-ers in every stall, for the purpose of watering, with troughs of communication, in order to convey the water from a pump in the farm-yard to a general trough at the outside of the ox-house, which is again separately carried to each keeler; so that all the trouble of untying and driving to water, is wisely avoided. These keelers or troughs are placed even with the manger, and are of the same size and dimensions. In

each

each stall he fattens two oxen, and allows five feet room for each, and has enough for forty head of cattle. When the oxen are at first brought to the stalls, it is only during the nights; but about the middle of November they are regularly tied. Mr. Ellman has found that nine oxen, fed loose in a yard, by eating as well as destroying, consume as much hay as twelve tied up. Surely this is an important point. The waste of food, when the cattle are fed loose, is prodigious; when stalled, the food left by the fatting cattle, serves to support the lean stock which are tied in the same yard.

Mr. John Ellman has an ox-house upon a similar construction, and a very complete one.

In constructing these stalls, it ought to be remembered, that the bottom should be gently sloped, to carry off the urine: and a step is necessary at the heels of the cattle, for cleanliness, and in order to carry the urine to a reservoir.

Each manger should possess the conveniency of a separate partition for meal, bran, chaff, oil-cake, &c. without mixing with turnips, potatoes, cabbages, or hay; a pump conducting to each trough, to supply the cattle with water; and a weighing-engine at hand—a most useful, necessary contrivance in the fatting of cattle, that the owner may instantly know what is the state and progress of the beasts, and that he may compare the improvement and the expense together; the flesh, and the food necessary to acquire that flesh. Weighing-machines are amongst the greatest improvements which have lately been thought of; for they lay open an inquiry of the last importance, and which has been strangely neglected—to find out that breed of cattle

cattle which gives *the greatest quantity of flesh with the smallest quantity of food;* and this is only to be discovered by weighing-engines, which are essential for the purpose; and no complete ox-stall should ever be without such an apparatus.

Potatoes.—Under the article *potatoes*, the application of this root to the feeding of bullocks, was remarked upon. Fatting with this food is largely practised in Sussex, and a very interesting feature in the husbandry of the county, first introduced by Mr. Mayo, of Battel, and it is very generally approved. The mode in which they are given is of course various: washed, and unwashed; cut, and uncut; steamed, boiled, and raw. The common practice is that of giving them raw, at the rate of a bushel to a bushel and a half per ox per diem, besides eight or ten pounds of hay, more or less. The oxen take very readily to the potatoes; but they should be cut, and washed, if dirty. The full quantity is not given at first, but dealt out in proportion as they agree with the cattle, always observing to regulate the hay, or other dry food, according to the effect which the potatoes have upon the oxen. The smaller the quantity that is given at a time the better; at least three times a day, giving hay between each feed of potatoes. It takes from two to three months to finish an ox in this manner, who consumes 100 to 120 bushels, and often more, besides seven or eight hundred weight of hay. This quantity will fatten any tolerable thriver.

The most interesting inquiry in the feeding of bullocks, is that of ascertaining the profit of potatoes, and the value, compared with other food. So few are the expe-

experiments which have been registered upon this part of the subject, that little additional light can be thrown upon that division of it, which is undoubtedly the most interesting : it is only to be done by weighing.

The merit of potatoes for cattle has lately been questioned, and there are grounds for believing, that, compared with some other sorts of food, their value is not so great. Accurate and multiplied experiments are necessary, before any decisive conclusion can possibly be drawn. But it is never to be forgotten, that that very accurate experimenter, Mr. Dann, was for some years largely in the practice of fatting oxen on potatoes, but gave it up, from the conviction that, with every advantage of breed and attention, warmth and cleanliness, they would not pay more than 4d. per bushel.

Turnips are chiefly cultivated upon the flock-farms for sheep ; but it is a practice in various parts of the county to draw the largest for the bullocks. When they are given upon an empty stomach, the cattle will blow; but never when hay is mixed. Mr. Milward observes, that when they are given to the cattle fresh drawn from the field, his oxen are liable to the flux ; but taken up before-hand, no food is better, as the watery nature of it is removed ; but, in order to carry this into execution, he adopts the method of stacking, taking special care of guarding against the frost; and this very spirited cultivator never found his oxen thrive so well as on dried turnips.

The custom of drawing them two or three days previous to giving them to cattle. is very prevalent about
Lewes,

Lewes, Brighton, Shoreham, &c. Mr. Ellman constantly practises it.

Oil-Cake.—This food is much used in Sussex, to finish the cattle, and they thrive extremely well with it: though it is expensive, it finishes a beast for Smithfield in a shorter space of time than any other food.

Corn.—In seasons when beef fetches a very high price, and corn is very low, Mr. Ellman has fatted with it; and as cattle half, or three parts fat, sell badly; in such a case it has answered with him; but the mere weight of beef gained, rarely pays the expense. When corn is malted, it is allowed to be a most nourishing food; and Mr. Davis, of Glynde, remarked, that one bushel of barley malted, went farther in fatting, than double that quantity of oats unmalted. Mr. Dunn has also fed horses upon boiled barley and wheat-straw, with great success.

Lintseed and Barley.—Mr. Bridger, of Tillington, after soiling his cattle in summer upon clover and tares, till the rouen is ready, which is the end of July, and holds till the beginning of November; takes them into the yards, and generally fattens upon lintseed and barley, mixed and ground together: two pecks of lintseed to two bushels and a half of barley, and wheat-chaff given with the meal, besides hay. Of this mixture each ox consumed two bushels per week.

This food is used in large quantities by Lord Egremont. His Lordship has lately fattened several large Hereford and Sussex oxen, and Devonshire cows, upon it.

It was a very interesting experiment, and expressly under-

undertaken with the view of discovering the thriving progress and ultimate profit which large cattle made, when compared with smaller, fed upon the same sort and quantity of food, distributed to each in equal proportions. We have yet very few experiments registered, of the *proportion of food* to the *weight of meat*. All other points, however excellent in their way, are no otherwise satisfactory than as contributing to this essential end. And it is proper to remark in this place, that the standard of merit, as laid down in the greatest weight of flesh upon a stated quantity of food, will be in proportion to the good qualities in other respects. That breed which experiment says gives the greatest net profit in money, from a given quantity of food, must at last be allowed to contain the sum-total of merit. An ox, for instance, is comparatively good for nothing, which requires an expense in food to the amount of 7*l.* to gain fifty stone of flesh, whilst another only consumes to the value of 5*l.* to lay on the same quantity of meat. Again, whether two beasts, each, for example, of 160 stone weight, consume more food than two others of only half, or two-thirds that weight? Whether 10*l.* expended in raising a small beast up to 100 stone, will not equally raise a large one up to 200 stone? The following experiment, as far as it was carried, besides proving the merit of the Hereford cattle, goes to prove this reasoning to be founded in fact.

Nov. 26, 1797.—Three Hereford oxen, two spayed heifers, a Hereford cow, and two Sussex cows, were tied up in the Stag-park farm-yard, to a mixture of barley-meal and flax-seed. For the first seven weeks each had three gallons every day, of which one quarter was flax-seed, making three-fourths of a gallon. For

SUSSEX.] the

the rest of the time they were fattening, they had three
gallons, of which one-third was flax-seed, allowing to
each beast two gallons of barley and one gallon of
flax-seed, ground together, and mixed up with some
wheat-chaff, to facilitate the digestion, and prevent it
sticking in their throats. The flax-seed was six shillings per bushel. Besides this, they had hay in equal
proportion, weighed to them three times each day,
and a little more when they were shut up for the night,
which amounted in twenty-four hours to 13 lb. of hay
for each ox. During the time that they were fattening,
it was very evident that the large oxen throve much
faster than the rest ; and so fat were the three Hereford
cattle, that it was thought a dangerous experiment to
send them to Smithfield.

It ought to be observed in favour of the Hereford
cattle, that they were sent from that county into Sussex for working, and not for fattening. Of the eight
beasts, kept upon equal quantities, one of the Hereford
oxen, killed at Petworth, weighed 170 stone; the other
two, sent to Smithfield, weighed 200 each ; the two
spayed heifers, 90 to 100 stone each ; the Hereford cow
105, and the other two 150 stone each. The two largest
Hereford oxen, each weighing 200 stone, throve best,
excepting one of 150, which was by far the best thriver
of any. The cow of 105 stone was from Hereford, and
bought by the late Mr. Campbell, for the excellency of
the breed, and selected with much care, as were the
oxen, which were chosen by the same hand. The ox
killed at Petworth certainly throve better than the cow
upon the same food. Each came from the same stock ;
and there was a difference of 65 stone, butchers' weight,
besides the proportion of the fifth quarter, in favour of
size; the ox weighing 170 stone, and the cow 105.
                                                        The

The ox was indeed particularly fat, more so than was necessary; but this was the merit of the breed, and not any extraordinary time or method of fattening. It has been remarked by the late Mr. Campbell upon this subject, and may serve as a corollary to the above, that the thorough-bred Hereford cattle that have attained their full size, require a less proportion of food to make them fat, than others of the breed which are not so highly bred, nor so handsomely formed; and that the quantity of food to produce this fat, so far from being greater, they will consume a much less proportion than other smaller oxen of the same sort not so highly bred, and much less than cows not of the Hereford breed, which when fat, do not weigh above one-half as much.

The following is the weight of the eight bullocks when put up to fatten, and the weight of them fat.

	Nov. 27, 1797.			March 19, 1798.			Gain in 16 Weeks.			
	Cwt.	qrs.	lbs.	Cwt.	qrs.	lbs.	Cwt.	qrs.	lbs.	
Duke, -	17	0	7	20	2	0	3	1	21	} Hereford.
Drummer,	15	3	25	18	3	14	2	3	17	
Merry,	15	0	11	17	2	0	2	1	17	
Lock,	14	1	21	17	0	0	2	2	7	} Sussex.
Spot,	14	0	25	17	0	0	2	3	3	
Lion,	14	0	25	17	0	0	2	3	3	
Short,	13	2	7	16	2	0	2	3	21	} Devon.
Foot,	13	0	14	15	0	0	1	3	14	
	117	2	23	139	1	14	21	2	19	
				117	2	23				
				8)21	2	19				
				16)2	2	23				

19 per week, live weight.
4d.

6s. 3d. per week--10¾d. a day.

The

The barley and flax-seed which was ground for the oxen, from November 27, to March 19, was,

		£.		
Barley, 34 qrs. at 27s.	£.	45	18	0
Flax-seed, 11 qrs. at 48s.		26	8	0
Hay, 4 tons, at 4l.		16	0	0
	£.	88	6	0

Mr. Mayo, of Battel, who fed bullocks upon potatoes for sixteen years when he made the observation, asserted, that a Sussex cow of 80 or 100 stone, eats as much as an ox of 140 stone, and that the quantity eaten is by no means to be estimated in proportion to the weight of the animal; and his bailiff, a very intelligent man, was decidedly of the same way of thinking, and gave it as his opinion, that if he were to take in beasts to feed by the week, he would be paid as much for small as for large oxen, since experience had satisfied him that the one ate as much as the other.

How one experiment begets another! It is thus that improvement is accelerated. Mr. Dale, the miller at Petworth, got 120 stone of pork with two-thirds of the food, from two hogs, each 60 stone, that he did in six hogs, each twenty stones.

January 2, 1799, three bullocks were killed for the poor-house of Petworth.

	Carcass.		Hide.	
	St.	lb.	St.	lb.
Leicester brook horn, four years old, weighed	82	2	10	1
Sussex steer, three years old, ...	116	2	11	3
Bull stag, half Hereford and Devon, ...	107	3	13	2
	305	7	34	6
15 st. 7 lb. rough loose fat 3 5½ cauls	19	4½		

The

The fat therefore of these three beasts being 19 st. 4½ lb. and the carcass 305 st, is one-sixteenth of the carcass.

5. *Thriving Disposition.*

One of the good qualities in the Sussex cattle is, the propensity they have to fatten kindly. The hide of the best sorts yields a mellowness in handling; and it possesses a fineness and sleekness, which is the characteristic of good cattle, and an infallible criterion of a healthy habit; though we find some of them with coarse, rough, thick hides, which are a mark of hardness of flesh, as fineness and closeness of grain convey to the feel a fine texture in the hide. The condition of Sussex cattle is very much to be known by attending to this appearance in the hide; and it depends in a great measure upon the pile and growth of the coat: the shorter and sleeker the coat, the more thriving the beast; as, on the contrary, in proportion to length and hardness, is unthriftiness.

The Sussex breed would undergo a considerable alteration for the worse, if breeders followed the directions of those who teach, that one quarter of an animal is to be neglected, because it has a tendency to gather fatness on other parts, or tallow in the inside, and that this cannot be effected without a deduction from the more valuable quarters.

6. *Flesh Offal.*

The Leicestershire school, which teaches that cattle should be bred with a view to meat, not offal, is *prima facie* evident; and that all those parts, as the head, neck, legs, bone, tallow, considered as offal, or of inferior value, should of course be fine and light. In the prosecution of these ideas, they lay down rules, that the

the surloin, rump, and buttock, and, in fine, the hind quarters, are main points of attention.

Thus they talk of loading what they deem the valuable parts, at the expense of every other, and would breed a beast as heavy as possible behind, and as light before, because the one sells for rather more per pound than the other.

Sussex men breed their beasts with weight before as well as behind, in equal proportions, or proportion is no longer preserved, and the beast is mis-shapen. Nor is it often that an ox can be made to fatten upon the valuable parts, that does not in some measure fatten upon the coarser quarters. Instead of being light and narrow, Sussex cattle are broad and weighty; instead of hollowness in these parts, when they are low in condition, good Sussex thrivers are well filled up.

7. Sale and Price.

Smithfield is the greatest market for the sale of Sussex cattle, which is well supplied with fat oxen from this county, where they are deservedly held in the highest estimation. They go at all times of the year; but the grazier endeavours to bring his beasts to the pitch, so as to be enabled to meet the demand when it is probable there will be the greatest call for prime beef, which is from Christmas till May. After this the market declines.

8. Weight and Profit.

The mean weight of Sussex cattle, when brought to Smithfield, will be 140 stone. Many farmers breed for size, and others again pass by weight, and bring in runts, which devour as much as large oxen. The late Mr. Edsaw, of Fittleworth, was particularly in the breed

breed of large oxen, and has had them six feet high behind, and five and a half before, and six feet and a half girth over the heart. They have sometimes been slaughtered up to 216 stone, at which weight Mr. Ed-saw killed two. Mr. Ellman fed one at Shoreham a few years ago, which measured as follows:

	Ft.	In.
Length from crown to rump,	9	6
Girt before shoulders,	9	5
—— behind,	9	0
—— round the middle,	10	0
—— at flank,	9	0
From nostrils to tip of tail-bone,	14	8
Length of solid sides,	6	7

Weight, 214 stone. Sussex breed seven years old. In years when prices are moderate, lean oxen are bought in at 2s. to 2s. 3d. per stone, of that weight which they will be when fat, or from 3d. to 3½d. per pound. Supposing the beast fat to be sold at 4½d. (and it is more), this makes the profit by fatting on lean weight to be 1¼d. per pound, or 17½d. per stone of 14 lb.; equal to 9d. per stone live weight. When prices were moderate, they were sold fat at Smithfield from 3s. to 3s. 6d. per stone: now it is 4s.

9. Compared with Devonshires.

The Devonshire stock are nothing more than a variety of the same species as the Sussex. The parent stock was certainly the same. Other circumstances have in the course of time introduced an alteration. By Devonshire, is to be understood the red, so much resembling the Sussex; for Devonshire contains other, and very inferior sorts of cattle.

The

The thorough Devons are a bright red, neck and head small, eye prominent, and round it a ring of bright yellow; the nose round the nostril has the same colour: the horn clear and transparent, upright, tapering, and gently curved, not tipped with black; their bones are well proportioned and light. On comparison with Sussex, they do not rise to so great a weight, and consequently have not so great a strength for work. They are thinner, narrower, and sharper, on the top of the shoulder or blade-bone; they drop behind the shoulder: the point of the shoulder generally projects more, and they usually stand narrower in the chest; their chine is thin and flatter on the barrel, and they hang too much in the flank: they are as good feeders as Sussex, and as profitable to the grazier, their hides thinner and softer, and they handle as mellow: they are wider on the hips, and cleaner in the neck, head, and horns, and smaller and lighter in the bone.

Upon the whole, they are a very valuable breed of cattle; and the distinction between them and Sussex is not so striking, as to make a more minute description necessary.

10. *Compared with Herefordshire.*

The true-bred Hereford cattle, in respect to the kindly disposition for feeding and fattening, are equal, if not superior, to either Sussex or Devonshire-bred beasts, and, in point of working, by no means inferior to either. They are a large and weightier breed, yet as complete in their make; generally wider and fuller over their shoulders or chine, and the breast or brisket; also in the after part of the rump, which is much oftener narrower in the Sussex than in the Hereford. By naming the Sussex, the Devons may be included:
they

they are each of them a most excellent breed; but if the Herefords, upon comparison, excel the true-bred Sussex, it must include the other.

In the true Hereford we find no projecting bone in the point of the shoulder, which in some breeds forms almost a shelf, against which the collar rests; but, on the contrary, tapers off: they have a great breadth before, and equally weighty in their hind quarters; the tail not set on high; a great distance from the point of the rump to the hip-bone; the twist full, broad, and soft; the thigh of the fore-legs to the pastern joint tapering and full, not thin, but thin below the joint; the horn pushes sideways a little, and then turns up thin and tapering; remarkably well feeling; mellow on the rump, ribs, and hip-bone. The quality of the meat not hard, but fine as well as fat; little coarse flesh about them; and a high disposition to marble with fat. With respect to the kindness of their disposition upon comparison with Sussex and Devonshire, we have perhaps the most decisive proof of it that was ever brought forward, in Lord Egremont's experiment, and which tends to shew that these large cattle will gain their fat sooner than either of the other breeds. It has been said, that the Herefords are better made for fatting than for working; that they are large-boned, thick legs, quite a contrast to the clean thin leg of Sussex, and consequently that their motions are slower; but the trials which have been made of these cattle in the same teams, ploughing and carting, with a great number as well of Sussex as Devons, at Lord Egremont's, has ascertained the merit of the breed in this respect. Mr. Campbell thus writes to Lord Egremont upon the subject: " To the article of draught, which your Lordship mentions, I can answer from
much

much experience, that the Herefords are completely fit. I remember Mr. Young, in one of the Numbers of the Annals, speaking of Devons and Somersets, on comparison with Sussex, said that they did not come to so much weight ; therefore, he said, not so fit for draught : in that, the Herefords come nearer to them. Most certainly any of those breeds are easy to be had of a weight sufficient for the purpose. With respect to the general weight, as they are to be found on an average, the Herefords are known to outweigh the Sussex, and (being, I believe, of nearly the same specific gravity, which in some cattle makes much difference) to do so, forms the circumstance which is the chief reason of my giving them the preference. They are generally more completely formed, their several parts brought up more to a level, and the intermediate spaces more completely filled. In my humble opinion, a middle-sized ox of such shape, free on every point from any hardness of flesh, and the legs free from gumminess, from knee and hock inclusively downward, clean bone and sinew, will be the fittest for work, as well as most generally saleable when fat."

11. *Other Sorts.*

The other sorts of cattle to be found in Sussex, are a small breed brought by Welsh drovers, which, by crossing with the native breed, have very much injured the Sussex stock. The Alderney, Norman, and Jersey breed of cows, are to be found all over this county. Lord Egremont has some of the cows, and a very fine bull, of that breed.

II. *Dairy.*

II. *Dairy*.

Under this head, our information may thus be classed :

1. Shape of the cows.
2. Milk.
3. Butter.
4. Food.
5. Breeding, how practised.
6. Crossing.
7. Rearing.

1. *Shape of the Cows*.

As the shape of the cattle has been noticed, little remains to be said upon the form of cows.

However, it may be necessary to observe, that breeders aim at a clean and thin head, neck, leg, and bone, a fine shoulder and chap, and, generally speaking, for the same points in the heifer as in the ox. The true cow has a deep red colour, the hair fine, and the skin mellow, thin, and soft; a small head, a fine horn, thin, clean, and transparent, which should run out horizontally, and afterwards turn up at the tips; the neck very thin, and clean made; a small leg; a straight top and bottom, with round and springing ribs; thick chine; loin, hips, and rump, wide; shoulder flat; but the projection of the point of the shoulder not liked, as the cattle subject to this defect are usually coarse: the legs should be rather short, carcass large; the tail should lay level with the rump : a ridged back-bone, thin and hollow chines, are great defects in this breed.

The Earl of Egremont has in his possession, among many others, a heifer, which for beauty, proportion, and symmetry, may challenge the whole country, without finding her equal.

Length

	Ft.	In.
Length of back,	3	11
Cross hips,	1	5½
Fore-quarters,	1	7
Loin at six inches from hip,	1	4½
Hip to first rib,	0	7½
Girth, centre,	6	11
Chine,	5	10½
Fore-leg,	0	6½
Horn,	0	6½
Collar,	4	0
Neck,	2	8
Withers, to horn,	2	3½
Height, centre,	3	11
Width, natche,	0	10½

A heifer of Mr. Ellman's, measured as follows:

	Ft.	In.
Wide, from the centre of one hip to the centre of the other,	1	9
Length of the rump from centre of hip,	1	6
Centre of hip to perpendicular of fore-legs,	3	2½
Girth at chine,	6	4
Centre,	7	8
Girth, neck,	2	10
Leg,	0	9
Horn,	0	7½
Middle of tail,	0	4½
Height to hip,	4	4
Thick chine,	1	9
Thick centre,	2	6
Width six inches below rump,	1	1

2. *Milk.*

2. Milk.

The material object in the cattle system of Sussex, is the breeding and rearing of stock for working and fattening. The concern of the dairy is but a secondary object in this system.

Upon many farms, nearly as many fat oxen are annually sold as there are cows kept. 3*l.* or 4*l.* in the product of the dairy, had much better be lost, than an indifferent ox bred.

In quantity of milk, they are not to be compared with some other breeds, as the Holderness, Suffolk, &c. which two breeds are the greatest milkers in the kingdom; and therefore with these the comparison is unfair; but if a money value is the object, the difference will not be merely so great, if any at all. This indeed is a circumstance in the breed of cows, deserving the most attentive experiments. If the *profit* of a cow was in proportion to the *quantity* of her milk, there would be a much greater disproportion between them than really is the case. The Suffolk cow is usually a poor and miserable-looking one: the Sussex cows keep themselves almost beef whilst they give milk. If this fattening disposition was not an indication of rich milk, those cows would be so unprofitable, that nobody could afford to keep them. The best cows of the Duke of Richmond's dairy (all Suffolk polled), in May and June will give two gallons at a meal; but if they are averaged, about one gallon.

But what the Sussex cows lose in quantity, they make up in quality. Upon this subject Mr. Campbell observes to Lord Egremont: " The Holderness cows, and their relations the Fifes, give the greatest quan-

tity

tity of milk of any in the kingdom; they are also the coarsest and most open-fleshed beasts in it. The fine fleshed cattle give milk of a better quality, and a higher and richer flavour. The Guernsey cows, and your Lordship's East India cow, mentioned in the Annals, confirm what I say as to milk."

In point of butter and cheese, none beat Suffolk cows. In May and June, in the great dairy districts, all the large dairies have cows which yield eight gallons a day; and in May, June, and July, whole dairies of forty, fifty, and sixty cows, that give four gallons a day; yet they are an ill made uncouth animal.

3. Butter.

A good cow will give 5 lb. of butter in a week in the height of the season; and six will make from 30 to 40 lb. of cheese in a month, of skim-milk.

The following is the product of the Duke of Richmond's dairy at Goodwood. The breed is of the Suffolk sort—twenty in number.

Six Summer Months, 1792.

	lb.	s.	d.	£.	s.	d.
April 6.	57½	at 1	0	is 2	17	6
13.	47½	— 1	0	— 2	7	0
20.	53½	— 1	0	— 2	13	6
—	18	— 1	1	— 0	19	6
27.	48¾	— 1	0	— 2	8	9
—	14½	— 1	1	— 0	15	8½
May 4.	56	— 1	0	— 2	16	0
—	20¾	— 1	1	— 1	2	5¾
11.	31	— 1	0	— 1	11	0
—	15½	— 1	1	— 0	16	9½

May

	lb.		s. d.		£. s. d.
May 18.	88½	at	1 0	is	4 8 6
25.	36	—	0 11	—	1 13 0
—	47	—	1 0	—	2 7 0
June 1.	87¼	—	0 11	—	3 19 11¼
8.	40	—	0 10	—	1 13 4
—	40½	—	0 11	—	1 17 1½
12.	8	—	0 10	—	0 6 8
15.	74	—	0 10	—	3 1 8
—	13	—	0 11	—	0 11 11
22.	125¼	—	0 10	—	5 4 4½
29.	82¼	—	0 10	—	3 8 6½
July 6.	83	—	0 10	—	3 9 2
13.	45	—	0 10	—	1 17 6
20.	36	—	0 11	—	1 13 0
—	26	—	1 2	—	1 10 4
27.	50	—	0 11	—	2 5 10
—	33	—	1 0	—	1 13 0
Aug. 3.	90	—	1 0	—	4 10 0
10.	66	—	1 0	—	3 6 0
18.	59¾	—	1 0	—	2 19 9
Aug. 24.	60	—	1 0	—	3 0 0
31.	67½	—	0 11	—	3 1 10½
Sept. 7.	58	—	0 11	—	2 13 2
14.	54¼	—	0 11	—	2 9 8¾
21.	46	—	0 11	—	2 2 2
28.	46¾	—	0 11	—	2 2 10¾
	1816,	—	0 11½		£.85 15 2½

Which gives a product of butter per cow, for the summer half of the year, of 90¾ lb. at 11½d. or	£. 4 6 11½
Per month, 15 lb. 2 oz.	0 14 4
Per week, 3 lb. 12 oz.	0 3 7

Product

Product of butter from the same dairy by fifteen cows, during the six winter months.

		lb.		s.	d.		£.	s.	d.
January	5.	32½	at	0	11	is	1	9	9½
	12.	34	—	1	0	—	1	14	0
	19.	34½	—	1	0	—	1	14	6
	26.	29	—	1	0	—	1	9	0
February	2.	25¾	—	1	0	—	1	5	9
	10.	22¾	—	1	0	—	1	2	9
	16.	22	—	1	0	—	1	2	0
	23.	19	—	1	0	—	0	19	0
March	2.	20½	—	1	0	—	1	0	6
	9.	19¾	—	1	0	—	0	19	9
	16.	26½	—	1	0	—	1	6	6
	23.	31¼	—	1	0	—	1	11	3
	30.	42½	—	1	0	—	2	2	6
October	6.	36½	—	0	11	—	1	13	5½
	13.	30¾	—	0	11	—	1	8	2½
	20.	24	—	0	11	—	1	2	0
	27.	21¾	—	0	11	—	0	19	11¼
Nov.	3.	19¼	—	0	11	—	0	17	7¾
	10.	17	—	0	11	—	0	15	7
	17.	16¾	—	0	11	—	0	15	4½
	24.	19½	—	0	11	—	0	17	10½
Dec.	1.	24¼	—	0	11	—	1	2	2¾
	8.	29½	—	0	11	—	1	7	0½
	15.	29¼	—	0	11	—	1	6	9¾
	22.	29	—	0	11	—	1	6	7
	29.	31½	—	0	11	—	1	8	10½
		689	—	0	11½		£.32	18	10¼

Per

Per month per cow, 46 lb.	£.2	4	1
Per month, 7 lb. 10 oz.	0	7	4
Per week, 1 lb. 14 oz.	0	1	10

Annual produce of each, 136 lb. 6*l.* 11*s.* 0½*d.* in butter. From August 18 to April 20, no butter sold.

An extraordinary instance occurring, to shew that the Sussex cattle, though they have a great disposition to fatten, are yet valuable for the quantity of butter they give, it is proper to note it.

The gardener of Lord Hampden, at Glynd, had a red cow of the Sussex breed, which in one year, two or three weeks after weaning the calf, gave 10 lb. of butter per week for some weeks; the next year the same cow gave 9½ lb. per week for several weeks; nine for several more; and then for the rest of the summer 8 lb. to 8½ lb. per week ; and till the hard frost set in, 7 lb. ; and during the frost, 4 lb. per week. That summer cheeses were also made of her milk, about 6 lb. each. When she gave most milk, two per week were made; so that at the height of milking, she gave 10 lb. of butter, and 12 lb. of cheese, each week.

She never at any time gave more than five gallons of milk in a day. Towards winter she had a bushel of bran twenty weeks, which the profit by pigs more than paid for.

Four or five years before, the same person had a fine black cow from Lord Gage's, which gave also in the height of the season, five gallons per day ; but no more than five pounds of butter in a week was ever made from it; and they remarked as a fact, that they had often noticed, that the milk of a black cow never gives so much butter as that of a red one.

The owner above-mentioned paid the farmer 5*l.* per

SUSSEX.] ann.

ann. for the food of the cow in question, and said the
value of the produce was 8*l*., the calf selling at 8*s*.
I went to the yard, in order to handle this cow, and
found her to feel very well, though out of flesh : she
has the disposition of the breed to be fat ; her carcass
very large. Observations of this sort should always
be noted, because a great number of results will furnish something better than conjecture. We shall however be very much in the dark, till experiments are made
on the quantity of *food eaten by all sorts of cattle**.

4. *Food*.

---

* I am inclined to think this useful aliment of butter suffers greatly
in its quality and durability, in the ordinary process of making up. The
error I would point out is, the admission of water (warm or cold), both
into the churn, and in the heating and making up. Water is well known
to be a great dissolvent ; at least if it be not essentially so, it serves *in
vesiculi* as a conductor to air, which is universally such. Fresh butter
then, in consequence of imbibing water, and water being saturated with
air, is always in a progressive state of decay. Not so when water is excluded : its obginous parts are admirably calculated to secure it from pu-
trefaction ; and I am almost positive, that butter might be made with as
little trouble as the present method, to keep the whole year fresh and
sweet, without the least particle of salt, solely by the exclusion of water.
I was witness some years ago to a piece of butter being taken out of the
churn in very warm weather : there might have been water put in previous to the churning, and I believe there was, but it had none afterwards : a part of this butter was used for making ointment, the remainder was set by and forgot ; a fortnight afterwards, it was discovered to
be as fresh and sweet as ever, though it had never been salted. I have
heard it spoken of a notable old housewife famous for good butter, that
she always kept the floor of her dairy dry. The custom is exactly the
reverse at present in those parts, many pailfuls being thrown down in
the hot weather, which will assuredly rise again in steam, and affect the
milk with its humidity. I propose the following observations in the
treatment of the dairy concerns : A spacious room with a north aspect ;
wide airy lattices, with trees planted before at a convenient distance, of
a kind that yield no effluvia. Trees thus situated, will draw a current.
and ventilate the air ; the floor, stone or brick, washed clean, on the re-
moval

4. *Food.*

A variety of experiments, highly interesting to a dairy farmer, have been lately made by the Earl of Egremont, with a view to discover what food keeps cows in the best condition, and gives the greatest product in butter and milk: certainly an important inquiry, and the scale upon which it was conducted, was of that magnitude, as to make it in some degree complete and satisfactory : the result was, that potatoes, boiled or raw, was a very improper food for cows. In these experiments, the green food was potatoes and carrots ; the dry, hay, chaff, and oil-cake.

Ten cows were tied up to potatoes and hay, and ten to hay only, for a month : they kept themselves in good order upon the hay ; but those on potatoes wasted much.

The difference in the quantity of milk, inconsiderable; boiled potatoes and wheat chaff, equally unsatisfactory ; steamed, not much better.

---

moval of each successive mess of milk, and kept perfectly clean and sanded, which will absorb all humidity. The vessels used for holding the milk, to be washed clean, and afterwards rinced a first and second time with sweet milk; the churn served in the like manner, and all the dairy implements. A cruet washed ever so clean with water, will cause vinegar, if put into it, to become dreggy; but when rinced with a little of the same, will always appear limpid and clear. No water put in with the cream when it is churned. As the butter is taken out, put it into a tray full of holes, placed over any other vessel; avoid squeezing it into lumps; it will drain the better for being loose in its texture; remove it to a large tray without holes; recover all the crumbs that have run through the strainer; knead it well with your hands, previously rinced with the whey, and form it into a flat cake, the thinner the better; sprinkle salt over it, and leave it in that state half an hour, by which time the salt will extract all the whey; make it up the usual way, but use not one drop of water in the whole process.—*Mr. Trayton.*

Ten

Ten cows upon raw potatoes and hay, and ten upon carrots and hay, for a month; the carrots much the best; but the potatoes reduced them so low, though plenty of hay was given, that the consequences might have been worse, had they continued on this food any longer. December 10, 1795, the cows were tied to raw potatoes and some hay; in three weeks a change of food was absolutely necessary. January, 1797, three pecks of boiled potatoes to each, mixed with wheat-chaff.

November 28, 1796, ten cows to carrots, ten to potatoes: the carrots kept them in hearty condition; the flavour of the butter and cream nowise affected by the food.

5. *Breeding.*

The breeding system of this district is entitled to considerable attention, and is a most profitable branch in the management of live-stock. The cows are in proportion to the farmer's occupation, and all, or nearly all the calves are reared, which are kept in succession for work; so that a farm of eight cows will have six calves, six-year olds, as many two-year olds, four three-year olds beginning to work, four four-year olds, as many five-year olds, and as many six-year olds. Upon some farms, the calves reared are, loss excepted, equal to the number of cows: females are sufficient to keep up the stock of cows; and if other females remain, they perhaps change them with a neighbour for males. Others again spay the females, and work them as oxen. It will of course be observed, that the variations in the rearing must depend upon circumstances. If the cows continue good to an advanced age, fewer are weaned to supply their places. A difference of opinion exists relative to the best time for calves to be born.

born. Some prefer early ones that come in January; others think those of March and April the best. They universally suck the cow from ten to thirteen weeks, are cut at seven weeks, and are weaned by being shut up; and having a little grass given them, till they have forgotten the dam, are then turned out to pasture. The first winter they are well fed with the best hay; after that with straw, except after Christmas, while working, when they have hay, but straw alone till they begin to work.

Variations of no great account are found : many do not let the calves remain at night with the cows, till they are five to eight weeks old, according to the quantity of milk the cow gives. Sometimes one cow will suckle two calves. They sometimes lose calves by a distemper they call the *husk*, which is occasioned by little worms in the small pipes on the lights. A good cow will suckle two calves for the butcher to a considerable profit, after her own has had her milk for ten or twelve weeks.

Mr. John Ellman's succession system is :

14 calves, of which nine male; eight for oxen, and one for accidents: not taking to work.

14 year olds.

14 two-year olds ; of which eight worked a little at two years and a half.

14 three-year olds ; part of which taken for cows, and others, if not good, fattened.

14 four-year olds, eight worked.

14 five-year olds, eight worked.

14 six-year olds, fattened.

24 oxen worked in common, eight three year, eight four year, eight five years old ; intending to have eight
every

every year for the team; but rearing nine, prepares him for the chance of a steer not taking well to work, and consequently fats one every year at three years old*. His cows, upon an average, pay 4*l.* each by suckling for the butcher, besides rearing the calves as above, which suck twelve or thirteen weeks. The cow-calves that are reared, bring a calf at two years old, which runs with the dam all the summer, for seven or eight months. Such calves they call a *hurter*.

Lord Egremont always fattens several calves, and has tried for that purpose skim-milk mixed with lint-seed, boiled to the consistency of a jelly; in the proportion of a pint of jelly to a gallon of milk : it did not seem to answer†. Of several that were killed in 1797, the following appeared to be the live and dead weight.

Born May 27 ; killed August 5.
Alive, 211 lb.
Dead, 133 .... 140 lb. would be two-thirds.

Born May 27 ; killed August 8.
Alive, 209 lb.
Dead, 136 .... 140 lb. would be two-thirds.

Born May 27 ; killed August 15.
Alive, 235 lb.
Dead, 154 .... 156 lb. would be two-thirds, if
it had weighed 234 lb.

Born June 18 ; killed September 15.
Alive, 240 lb.
Dead, 160, ... two thirds.

---

* Instead of fourteen, all these numbers are now become eighteen, having increased his stock.

† Mr. Milward fattens with balls of flour mixed with rum.

The

The general advantages of this breeding system, are obvious to those who have been long in the habit of experiencing the inconveniences of other methods. It is a grazier's own fault, if ever he attempts to fatten an unkind beast; let him only take care of his stock, and he will need no apprehensions of that sort. Those who trust to fairs and markets, know that they will sometimes unavoidably have either ill-conditioned beasts, or be forced to give prices too high to answer, not to mention the uncertainty of fairs, and the enormous price lean stock frequently sell at.

***

The Earl of Egremont's Cattle System for Work.

The calves are dropped from December to the end of February; they are weaned immediately, never letting them suck at all, but the milk given for three days as it comes from the cow. But for weaning on skim-milk, they ought to fall in December, or a month before and after, and then they should be kept warm by housing; and thus they will be equally forward with calves dropt late in the spring, that ran with the cow. With the skim-milk some oatmeal is given, but not till two months old, and then only because the number of calves are too great for the quantity of milk; water and oatmeal are therefore mixed with it, to make it go farther. (Heifers with their first calves are exceptions; such do not become good milkers, if their calves do not suck for the whole season; but with the second calf are treated like the rest). In May, they are turned to grass; the first winter, beginning in November, they feed upon rouen, or aftermath, as called in some places; the following summer, that is, from May, they are at grass; the second winter on straw, but eat very little,

little, as they run out on short rough grass. They have been tried on hay alone, but straw and grass do better. The following, and every other summer, on grass, and are broke at Christmas, being three years old : they are lightly worked ; the only object is to break them in, in order that their work may begin in the spring. From this time their winter-food is straw, with the addition of a ton and a half of clover-hay, by estimation, and reckoning at the most ; but not being trussed, the utmost is taken. It is given between the finishing of straw and going to grass, that is, during the season of spring-sowing and a month before it, in order to prepare them for that work about the 10th of February.

His Lordship works them three, four, or five years, that is, from three years old to four, from four to five, from five to six, from six to seven, and from seven to seven and a half, being in this last case put to fatten after the wheat season. But his more common system is, to work them four years and a half, and then fatten.

The breed is Hereford, Sussex, Devon, and a mixed breed between Hereford and Sussex. The Hereford breed appears to be the best of the three, when pure, for the two objects combined, of working and fatting ; but the mixture of half Hereford and half Sussex, are equal. But with all crosses, the Hereford white face is sure to come.

When at straw in the winter, they work three days in a week ; for instance, his Lordship has now thirty-four, being twelve three-year olds, ten four-year olds, and twelve of all ages above that, as they happen to be good for work. And here it is to be observed relative to turning off from work, that when an ox will not bear hard work and hard food, he may on an even

chance

chance, if put to feed, fatten as well as one that would stand work and hardship much better, as the qualities of fattening well, and bearing hard work, are distinct. But the perfection of breeding is to have such as will do both; and the free temper and willingness in work of an ox, may make him be thought tender, and unfit for labour, if due attention is not paid to this circumstance.

Those thirty-four oxen are at oat-straw, with no other food, and sixteen of them are worked every day; and I could not but remark the very good order they were in; none of them complaining, by their appearance, of any want of better food. This straw system holds till about the 10th of February; then hay is given, to prepare them for the fatigue of spring-sowing; the hay system lasting till May, when they are turned to grass.

In forming an estimate of the value of their work, Lord Egremont, in order to be within the truth, takes to account only a part of their time; from three to four years old, two days a week, at 6d. a day; from four to five years, three days a week, at 7d.; from five to six years, three days a week, at 8d. These rates during eighteen weeks at straw in winter. But for thirty-four weeks in summer at grass, the beasts from three to four years, four days a week, at 6d. from four to five years, five days, at 9d.; and from five to six, five days, at 10d.

Years old.	Days.	d.	Weeks.	£.	s.	d.	
From 3 to 4	2	at 6	18	 0	18	0	Straw.
4 to 5	3	— 7		 1	11	6	
5 to 6	3	— 8		 1	16	0	
3 to 4	4	— 6	34	 3	8	0	Grass.
4 to 5	5	— 9		 6	7	6	
5 to 6	5	—10		 7	1	8	

£.21 2 8

Let us now calculate the expense and return of these oxen, as they are managed, taking it for a single one.

	£	s	d
Expense of weaning, ...	2	0	0
Six months' good grass, at 10*d.* per week,	1	0	0
Six months' grass, being the first winter, at 1*s.*	1	6	0
Six months' grass in the park, being much fed with sheep, deer, &c. at 1*s,*	1	6	0
Six months winter, at 9*d.*	0	19	6
Six months' grass, at 1*s.* 6*d.*	1	19	0

£.8 10 6

At this time Lord Egremont could sell them, or indeed so early as in August, for 11*l.* or 12*l.* lean;

	£	s	d
Say	.11	10	6
He has cost, ...	8	10	6
There would here be profit, ...	£.3	0	0

Supposing his food charged to November. On the contrary, if put into good grass in August, for fattening,

ing, he will by the end of the grass season, about the time when he would be broke, sell fat to the butcher at 15*l.* or 16*l.*

Bring down	£.8	10	6
Twelve weeks' grass, of additional value (that is, 3*s.* 6*d.* instead of 2*s.* 6*d.*), 2*l.*	1	4	0
	£.9	14	6
Sells for	15	10	0
Expense, ...	9	14	6
Profit, ...	£.5	15	6

This is on an average; but I was present at Petworth, and partook of one : a Sussex, turned to fatten, because a coarse ill-looking-beast; came fatting, between August and January, to 116 stone, which at 4*s.* would be 23*l.* 4*s.* and was excellent beef.

This is a considerable profit; but as it would, if generally practised, exclude the system of working, and force the use of horses for the entire work of the farm, it does not seem to be so large as to form inducement sufficient to change the system; accordingly **Lord Egremont** does not stop here.

Bring down	£.8	10	6
Six months' straw, at 1*s.* the third winter, ...	1	6	0
Six months' grass, at 2*s.*	2	12	0
Three months' straw, fourth winter, at 1*s.*	0	13	0
One ton and a half of clover-hay, ...	3	0	0
Carry forward, ...	£.16	1	6

Brought

Brought forward, ...	£. 16	1	6
Six months' grass, at 2s.	2	12	0
Three months' straw, fifth winter, ...	0	13	0
One ton and a half of clover-hay, ...	3	0	0
Six months' grass, at 2s.	2	12	0
Three months' straw, sixth winter, ..	0	13	0
One ton and a half of clover-hay, ...	3	0	0
	£.28	11	6

If the account is here stopt, by the sale of the ox lean (not Lord Egremont's practice), we may, for the better comparison of keeping and fatting, thus state it:

The ox would now sell for	£. 19	10	0
He has earned, ...	21	2	8
Expense, ...	28	11	6
	£. 12	1	2

This is selling in the spring; but if in the following autumn, then the lean account will stand thus:

Bring down	£.28	11	6
Six months' grass, at 2s.	2	12	0
	£. 31	3	6
He earned before, ...	£. 21	2	8
Six months' work, 5 days a week, at 10d.	5	8	4
	26	11	0
Sells for	19	10	0
	46	1	0
Expense, ...	31	3	6
Profit, ...	£. 14	17	6

But

But at this period his Lordship puts some of them to fattening; the account therefore goes on thus:

	£	s	d
Bring down, ...	31	3	6
Six months' rouen the seventh winter, at 2s. a week; if not rouen enough, then bad hay, two tons and a half, at 30s.	2	12	0
Six months and 3 weeks' grass, at 3s. 6d.	5	1	6
	£.38	17	0

	£	s	d
He may now be sold for	30	0	0
He has earned	26	11	0
	£.56	11	0
Expense, ...	38	17	0
Profit, ...£.	17	14	0

And here it should be observed, that I saw four oxen, two of which are ten years old, which his Lordship bought of my father; and two others coming eight years old, feeding upon rouen in January, through the severe frost of the end of December, and without having a mouthful of any other food, and thriving well: a very satisfactory proof how much rouen is to be depended on, even in such a season, and of the great profit attending it. The advantages of kept grass can hardly be exemplified in a clearer manner than in this practice; for no slight portion of the profit throughout the scale, arises from the cheapness of this food. The calves entirely depend on it for the first winter: they have some the second also, though at straw; and the winter previous to fattening, the oxen are put to it, to improve them. Its value is best

ascertained

ascertained by supposing its absence; for then hay
must be the substitute; and the expense of that food,
if reckoned at what it would sell for, every one knows
to be extremely great. I had the pleasure of seeing
Lord Egremont's whole crop of lambs thriving admirably on this food also, without the addition of
any other; a very severe frost leaving his turnips rotten, and yet the farmer free from all anxiety. Rouen
defies the season, and places the flock-master on velvet.
But to return:

	£.		
Bring down	£.38	17	0
Three months' oil-cake, eight cakes a day, at 8l. a thousand at London; carriage to Godalming, 15s. 8d. a thousand; a team of six horses 2500 cakes thence to Petworth, two men, and 3s. turnpikes — say 21s. for 2500; carriage in all, 24s. per thousand; 90 days, at eight cakes, 720, 17s.	6	12	0
Three months' hay, with cake, at 8 lb. per day, 720 lb. say half a ton, at 3l.	1	10	0
	£.46	19	0
He then sells for	32	10	0
His work, ...	26	11	0
	£.59	1	0
Expense, ...	46	19	0
Profit, ...	£.12	2	0

Difference in profit, lean or fat, 2l. 15s. 6d.; against
which there is the value of his dung.

But

But Lord Egremont's more common practice is, to keep them longer before fattening; in which case, the account goes on thus:

	£	s	d
Bring down	31	3	6
Three months' straw, seventh winter,	0	13	0
One ton and a half clover-hay,	3	0	0
Six months' grass, 2s.	2	12	0
	£37	8	6

If then sold lean, the account will be:

	£	s	d
His former work,	26	11	0
18 weeks, at three days per week, at 8d.	1	16	0
And 34 weeks, at five days per week, at 10d.	7	1	8
	£35	8	8
Sells for	19	10	0
	£54	18	8
Deduct expenses,	37	8	6
Profit,	£17	10	2

And here therefore it deserves remarking, that the profit on keeping him at work while he is in full strength, is an object; for his labour amounting to 8l. 17s. 8d. and his food only 6l. 5s. leaves a profit of 2l. 12s. 8d. per annum. Nor has Lord Egremont observed, that by thus keeping him through the summer which follows his seventh winter, that he is worse in any respect, either for fattening, or selling when fat.

If he is now fattened, the account goes on thus:

Bring

Bring down	£.37	8	6
Six months' rouen, as before, eighth winter, at 2s.	2	12	0
Six months and three weeks' grass, 3s. 6d.	5	1	6
Oil-cake, as before, ...	6	12	0
Hay, ditto, ...	1	10	0
	£.53	4	0
His work has been	35	8	8
He now sells for	32	10	0
	£.67	18	8
Deduct the expenses, ...	54	18	6
Profit, ...	£.13	0	2

Upon this account, which, it should be observed, *is* an account, and not a calculation, for it is a transcript of practice, Lord Egremont observes, that it seems remarkable, that the farmers are generally fond of fattening their oxen, by which they lose, and that nine-tenths of the kingdom are unwilling to work them, by which they as clearly gain; partly to be accounted for, in the first case, by the vanity of having fat oxen, which is an object amongst them of a sort of pride, as if a certain degree of respectability attached to the practice: not entirely ill-founded, as far as manure is concerned.

From the whole of this account, the advantage of working oxen on Lord Egremont's farm, is clearly manifest: but it should be observed, that this is to be extended no further as practice, than what he actually practises; that is, keeping both horse and ox-teams. Much of the soil of his farm, especially the arable, is a

strong

a strong clay; upon which it is of great consequence not to trample in ploughing, and of equal importance in spring and autumn sowing times, to be very quick for catching the right moments: horses trample less, and are more expeditious. He therefore finds it particularly serviceable to do all sorts of carting with oxen, except the longer journies, and at such seasons to keep the horse-teams uninterruptedly at plough. It must not however be hastily concluded, that oxen do not plough well, for they certainly do, and on his Lordship's farm; but in spring and autumn seed-times, horses are more expeditious. Six oxen in stiff land, for the first earth, plough but three-fourths of an acre a day, on the average of the several ages at which they are wrought. Those who, upon lighter soils, can use smaller teams, will of course find them more beneficial than here stated.

6. *Crossing.*

Crossing is universally practised. It is very strongly believed, that without this custom, the breed would infallibly degenerate; and in conformity to this notion, the Sussex breeders every year or two change their bulls; consequently this practice is in vogue, for the mere sake of crossing; and it has contributed to the deterioration of the stock. Bulls are seldom to be met with above three years old; so that, with this system, a man scarcely knows what his young stock will turn out. Mr. Ellman, in support of this opinion, gives it as the result of experience, that it is necessary in all kinds of animals: that it is, of course, better to cross from a finer stock than their own; but, if they have been long in one blood, it will be better to

SUSSEX.]   take

take a cross from a worse breed, rather than not
change, as the mere crossing will be advantageous
enough to induce this conduct. And it is thought
that this observation goes more pointedly to the means
of improving the health of the animal, and the disposition to fatten, than either to shape or colour.

Of the same opinion was Mr. Allfrey, of Friston, one
of the most experienced and sensible breeders that the
county had to boast. From his own knowledge, he was
decidedly for crossing. His father had many years ago
a most beautiful breed of beasts, and he bred in the
same strain for upwards of twenty years, when, as he
states, they were very much degenerated, and their
constitutions so bad, that it was no unusual thing to
lose four or five in a year. His flock too suffered
from the same inattention; but on changing his rams,
the alteration in a few years was really astonishing.
In the breed of his greyhounds, in which he was always
curious, observation confirmed him in the opinion.
Mr. Allfrey bred from dogs in the same strain, that
were very capital, till they could not run a mile; but
by crossing with others, they again improved.

In opposition to these Sussex facts, it was the practice of the greatest breeder the world has produced,
deduced from long and attentive experience, that to
cross with a breed which was not decidedly better than
the breed to be crossed, ought never to be attempted.
But when this is the case, Mr. Bakewell thought it a
necessary measure, but in all others a most mischievous
one.

The late Mr. Campbell, of Charlton, coincided in
these sentiments: he too was known to have possessed
great experience, and was a consummate judge upon
the subject.

                                              " As

" As to the art and mystery of generation or conception, all that I pretend to know (and that I do by many experiments to a certainty) is, that ill shapes, and properties of particular breeds, when introduced in others, even by a single cross, will continue to have effect, sometimes more, sometimes less, and sometimes lurking for generations, scarce perceivable, or even totally out of sight or feel, and then break out on some individual as strongly, and with as bad effect, as if there had never been any further mixture or addition of the blood on the other side. I therefore consider crosses to be a matter requiring the greatest caution, and what I should never choose to do, if there was one bad property in the proposed cross; and I am of opinion, that the surest and best means of improving a breed, is by constantly and completely weeding the original stock and nursery, and securing the opportunity of advantage from particular extra individuals which may happen to be produced in it; and in every respect availing oneself of all the use it may afford, and carefully preserving the continuance of it as long as possible, or until a yet better comes."—*Extracts of Letters from Mr. Campbell to Lord Egremont.*

III. *Work.*

The third division of the subject embraces the description of the draught cattle of this county, in respect to,

1. Training.
2. Yokes.
3. Collars.
4. Yokes and collars compared.

5. Pro-

1. *Training.*

The common practice is to yoke the steers in the double yoke, which is generally performed with the use of a rope, to confine them whilst the yoke is fixing on. Mr. Ellman observes, that this should be done between two pair of old steady oxen; one pair before, to prevent the steers from flying back, and one or two pair behind, to prevent them from pushing forward. In this way they are put to the plough, and the next day they are generally yoked with less trouble: sometimes, and which Mr. Ellman thinks a better way, we yoke, he says, one steer with one of our old steady oxen, one that is not very free to work. This last mode of training the steers to work, is less liable to hurt them in a warm day, as the ox will prevent the steer from fatiguing itself, which is often encouraged by the other, when two are yoked together.

At two years and a half old the oxen are broken to the yoke; at the outset the work is gentle, so that the young cattle are trained to the labour with other steady ones in the team: whatever is the work of an ox, it is made consistent with the progression in his value; for the breeder knows that the system would otherwise be deprived of the principal part of its merit; consequently the work is at all times so proportioned as not to affect the growth of the animal, which continues till the sixth or seventh year. This then is the reason why such numbers are coupled in a team, that, at first sight, there appears an absurdity in working their cattle in such numerous teams. Eight great oxen

in

in a single plough, is the common allowance upon almost any soil; and if the nature of it is heavier than ordinary, the number of them rise up to ten or twelve. It is not an unusual circumstance to see thirteen or fourteen pair, and ten or a dozen horses, in a field of less than twenty acres. When more than eight are used in one team, it should be observed that the rest are in training.

The customary load for a team is from eight to ten quarters of wheat. Now if this weight is divided among the eight oxen, each draws ten bushels for his respective share. So jack-asses would be more useful, as the expense of maintaining them would be far less, and the heavy weight of oxen upon wet land in rainy seasons, would be avoided. Ten bushels of wheat per ox; yet common Sussex cattle draw just so many sacks several miles to market, harnessed in single ox-carts, at Bradfield-hall, and are afterwards fattened to great profit. The necessity of the case then is not admitted, and most certainly *one-half at least* of the number, might readily be spared, without injury to the growth of the cattle. This system of ploughing with eight or ten oxen in yokes, is the more reprehensible, because the strength of an ox for labour is well known. For many years Sussex oxen have been used at Bradfield, and, in point of draught, found equal to the best Suffolk horses for all sorts of work, and especially for that labour which is not much understood in Sussex, the transport of heavy loads of corn to market. Upon a wet and adhesive clay loam, the daily work of two in a team, is equal to that of two Suffolk punches, an acre in a day; but then the oxen are in harness. Here they are used three years; in Suffolk we use them twice that time.

Another

Another circumstance, is the nature of the soil in that part of the county where ox-teams are chiefly in use. It may be characterized as a strong clay loam, wet and tenacious to a degree, and therefore a soil very difficult to plough to advantage; and as the usual custom is double yokes (though single ones are certainly to be preferred), when eight, ten, or twelve are used, four, five, or six, must necessarily follow each other upon the unploughed land, as many walking at the same time in the furrow; the latter is bad enough, but the former, it would seem, in any season upon such a soil, especially in a wet spring, would be destruction itself, when trampling should be avoided with the utmost solicitude.

Certainly, by a different arrangement in the system, a considerable expense of food and labour might be saved to the farmer.

3. *Yokes—Collars.*

The mode of working their oxen in this county has, from the earliest ages, been the established one of bows and yokes, both single and double. Oxen in collars are a late improvement. A wide difference of opinion amongst practical men exists, with regard to the best method of using oxen, in yokes, or in collars. Some very sensible men, who have worked them in yokes, and afterwards with collars, have gone so far as to say, that three in harness are competent to as much work as four the other way. This indeed is a point of the greatest consequence. To talk of men not liking innovations, and revolting at a change, is a mere apology for idleness. If this is the case, they should unquestionably be adopted.

The Rev. Mr. Davies, of Glynde, some few years ago,

ago, worked oxen singly in collars, and found it to
answer exceedingly well. He worked them gently
at first, and five in collars did the work of eight in
yokes, and with equal ease. This gentleman found
that he could make two teams out of one; but the advantage of the greatest consequence upon his farm was,
that upon the strong land inclining to be wet in winter,
he dared not plough with oxen, because they trod it so
much, that it was a very difficult matter to get it
again to pieces in the spring, when it became dry and
hard, so that he ploughed such land only with horses.
Oxen, singly working, were drove in the furrows, so
that the land was no more trod by ploughing in this
manner, than if horses had been used. Another advantage Mr. Davies experienced was in the motion,
which was quicker in the new method.

Mr. Clutton adopted the same practice at Cuckfield,
and he found that five used in this way, were equal to
eight in the other ; and besides this extraordinary difference, the work was performed much easier to the
animal ; and in wet seasons they ploughed in collars,
when they could not in yokes.

The Duke of Richmond is a warm advocate on the
same mode. His Grace affirms, that the pace is much
faster in harness, though the number is less, and that
by the oxen bending their necks over the bow of the
yoke, the windpipe is affected. It is certain, that four
in harness equal six in yokes. Mr. Pinnix uses all
his working oxen, at Upmarden, in collars ; and he observes, that they will plough more, by the third part
of an acre, than when yoked, and work much easier.
He found they would not, without difficulty, work
when coupled in yokes, after they had been for some
time used to the collar.

Mr.

Mr. Pennington, a very intelligent farmer, steward
to Lord Ashburnham, strongly impressed with the
idea of the superior advantages which would result
from the introduction of harness instead of yokes, then
universally used in his neighbourhood, purchased
harness for six oxen, and worked in this manner some
six and seven-year old cattle, which he had purchased
in the country at an age when, having attained their
growth, they are commonly either sold or fattened.
They were soon reconciled to harness, but were much
more sluggish than younger oxen, and though many
were not necessary to draw a load, that load moved
but slowly; and when they were required fully to exert their strength, they could not do it without extraordinary food, both in quantity and quality, which
their work only could pay, there being no hopes that
an advance in their growth would contribute towards
it.   After working some time in harness, he resolved to
fatten the old ones, and, in the mean time, having purchased many three and four-year olds, he worked those
in yokes, as, upon close and attentive observation, he
saw that hard work would stop their growth, and that,
without any inconvenience, they could use as much
power in yokes as it would be prudent and beneficial
to permit them to employ.   He perceived that the
trouble and expense of harness, of course, would have
been thrown away, even though these oxen might have
been capable of drawing a greater weight in harness,
of which he has now some doubts.   In summer he
found the harness an incumbrance, the ox requiring
all the relief and liberty that can be given in hot
weather, and that the yoke left as much as is possible
for any animal to have whilst labouring; and he thinks
it neither unnatural nor improper, to place the point
                                                        of

of draught upon the neck of the ox, just before his shoulders, that point seeming adapted by Nature to bear the pressure. He never had an ox galled by his labour; and he finds that an ox is much seldomer galled by the yoke, than the horse by the collar, which is however adapted to the form of the latter, as under a yoke he could not work one hour.

Mr. Pennington conceives the system of working, only to be profitable whilst the growth of the ox nearly pays for the keeping, and that it cannot do when the ox is hard worked. He thinks that, in the nature of the ox, there are qualities opposite to quick or severe labour; for when the ox is driven beyond his strength or wind, he is rendered unfit for work for a great length of time, and even frequently falls a sacrifice to the exertions of a single hour. When he is brought low in flesh, no art or food will speedily put him into condition. He thinks also, that as the horse is otherwise formed, he will bear the extreme of heat and cold, most frequently without injury, and if brought low by labour, will in a short time, with attention and proper food, recover his flesh. Hence, in all severe or quick labour, horses are undoubtedly to be preferred, and oxen are only profitably employed in easy regular business, without any perceptible inconvenience. This has induced him to lay aside harness entirely. If it is desirable, on account of the wet state of the ground, to plough with oxen single, some farmers frequently use a particular kind of yoke for this purpose. When Mr. Pennington first came into Sussex, he thought it preposterous and unnecessary to use such a number of oxen in ploughs, harrows, carts, and waggons, and imagined that it proceeded from their want of power in yokes; but he has discovered that the practice arises

out

out of, or is a part of a system proper in using oxen, which is very far from requiring the application of their full strength during the time they are at work.

In order to decide the respective merit of yoked and harnessed oxen, Mr. Bishop, of Westburton, and Mr. Salter, of Fittleworth, for a wager, agreed to plough an acre of land; Mr. Salter to use six oxen in double yokes, and Mr. Bishop four oxen in collars : again, Mr. Salter to plough with four oxen in single yokes, against Mr. Bishop's four in collars; the team which ploughed in the best manner, and in the shortest space of time, to be the winner. It was about the latter end of September, or beginning of October, that these trials were made. In the first, the six in yoke beat the four easy. Little exertion was used at the latter end of the match. The second was a near thing, and only three minutes difference in an acre.

The four oxen in collars ploughed the acre in four hours and seven minutes; Mr. Salter's four in yokes, in four hours and ten minutes; however, to plough it in that time, we may readily suppose it was not capitally executed.

As far as this experiment went, it proved the equality of the teams. Lord Egremont has worked his cattle each way, and in road and field work, upon a large scale; and his experience fully confirms the general opinion, that the old established mode is superior to the new method, and that any number in yoke are equal to an equal number in collar*.

5. *Pro-*

Breach Cattle.—Horned cattle are sometimes very troublesome when they get breach, particularly bulls, and frequently do considerable damage in tearing hedges, untying and breaking gates, bars, &c. It is customary in some parts of Kent, to fix a sort of axle-tree across the
horns,

5. *Proportion of Draught Oxen to Arable Land.*

The number of oxen used in husbandry in Sussex is considerable. This cattle system, as at present arranged, requires a large supply for the labour of the farm. The grazing part of the business may turn to profit and advantage, as it is argued that the winding up is the most lucrative; consequently, that an ox must not be impeded in the thriving stage of his growth by severe labour, since the increase in his value, from his birth to his death, is the strongest reason to be tender of him in his working state.

There is doubtless much experimental reasoning in this argument: still the number for work might be very considerably lessened; and it would be a singular advantage, not to speak of the saving of a driver, to reduce their teams to four oxen, of such an age as should not receive any damage from hard labour. This might be effectually performed by keeping their beasts two years longer; and even one year would be attended with considerable effect. Although it may be said, that keeping the ox two more years, would not increase his value, still the expense of their tillage would be less-

---

horns, with small wheels at the extremities, so that when a bullock would attempt to toss a gate, &c. the wheels fly upwards, and the creature receives a smart blow on the nose. A very few attempts of this kind serve completely to sicken him, and cure his breachness: so when they attempt to gore one another, it seldom amounts to more than inoffensively thrusting the snout in the other's flank. In this case it might be of use in the dairy. Metal knobs on the tips of the horns, which are become now in general use, are highly improper, as they may serve to attract the electrical fluid: horn, as being a non-conductor, would be a good substitute, well bound round the middle with pitched twine, to prevent their splitting.--*Mr. Trayton.*

ened,

ened, and the land not liable to such injuries as it
receives at present.

The proportion of draught oxen to arable land, varies with the size of the farm and the number of horses :
and here is one of the great benefits of large farms over
small ones, in the cheaper style in which they are
tilled ; for it is found to be very necessary to have one
horse and one ox-team for a farm of one hundred acres
of arable, that is, eight oxen and four horses ; but if
we extend the hundred acres to five times the quantity,
the number of draught cattle lessens in proportion.

The proportion which the cattle bear to the land,
may be seen by attending to the under-mentioned particulars :

Mr. John Ellman allows twelve oxen and nine
horses, constant working, to be the proper proportion
for 200 acres in tillage. The real proportion is higher
than this rate : but let it be noted, that Mr. Ellman
speaks from a knowledge of facts, and a reliance upon
experiment, and founds his calculations upon his own
long practice in an extensive concern.

Eight oxen, and three or four horses, are used for a
hundred acres of tillage land about Lewes ; the land
heavy and adhesive, deep and rich.

About Cuckfield, 130 acres require two ox-teams and
one horse-team ; strong clay loam. Mr. Mayo, upon
the same quantity of arable land, or 133 acres, employs at Battel sixteen oxen and six horses ; soil, a
lighter loam, friable, and moist bottom.

A South Down farm, rented at 500l. per ann. 24
oxen and 12 horses ; soil, thin chalk rubble upon a
chalk rock : in other parts of it, where the staple is
deeper, covered with a layer of flints.

When Mr. Davies farmed with oxen at Bedingham,
                                        upon

upon 200 acres in tillage, from fourteen to sixteen oxen and nine horses; chalk, rubble, and deep loam.

Mr. Gell, of Applesham, upon 500 acres, 29 oxen and 13 horses; light chalk, rubble flinty.

At Kindford, in the heart of the Weald, 130 acres arable, four oxen and six horses.

From these, selected from a great variety of estimates, the conclusion is evident, that the common allotment for an hundred acres is an ox and a horse-team; from 150 to 200 acres, 12 to 16 oxen, besides horses; upon 500 acres tilt land, six oxen and two horses and a half for each hundred.

DISORDERS.

Hoving.—South Down receipt for hoved bullocks is, a quart of lintseed-oil, which vomits them directly, and never known to fail*.

SECT.

---

* I know of no particular disorders that our cattle are afflicted with, and am a farrier. If any of my cattle get into a low weak state, I generally recommend nursing, which in most cases is much better than a doctor: have often seen the beast much weakened, and the stomach relaxed, by throwing in a quantity of medicine injudiciously, and the animal lost, when with good nursing, in all probability, it might have been otherwise: here I allude principally to cattle that are brought into a low weak state by over-working, when put to grass in the spring, and particularly so in such a year as the present, when the grass grows very quick, which often brings on the flux, or what is here called *scouring*; the best way to prevent which is, to continue giving a small quantity of hay for some time after turning to grass, and not to keep them too many hours at a time from water, which is often done here in summer, to the prejudice of the beast, by working too copiously at a time, which will tend to increase this complaint. When I see it coming on, I keep the ox as much as possible on hay and bran, and let him have water often in very small quantities.—*J. E.*

There is a disorder incidental to young stock in this county, which is called

SECT. II.—SHEEP.

Sussex is almost the only county that is at the same time in the possession of a breed of cattle, as well as sheep, both of which are of very great comparative excellence, that may be deemed peculiarly her own. Only one other can challenge her pre-eminence, and question the superiority of her stock. The Hereford breed of beasts is, upon the whole, equal to the Sussex; but the merit of the Ryland flocks, though in point of fleece superior, are in other respects perhaps a less perfect breed.

---

called being *struck*. It most frequently happens in the best pastures, and is caused by a too great nutrition of the juices, arising from feeding on succulent herbage, buds of trees, or rather of shrubs in coppices or hedge-rows, together with an over indulgence of ruminating, lying down, whereby they acquire a sluggish habit, that renders the blood torpid, and they die suddenly, as in the apoplexy. This happens chiefly among weaning calves, and yearlings: to prevent which, they are commonly bled and purged, before turned out to winter pasture; but the best way is to turn out the weanlings and yearlings in large enclosures of coarse sharp-bladed grass, and mix among them colts of a year or two's growth; thereby their mischievous and playful gambols will harass the calves thoroughly, and by keeping them in action, will keep them in health. The truth of this is well known among many graziers, but is not so generally practised as it ought.

There is another disease calves are subject to, called the *husk*, in which the lungs are inflamed, and perforated with myriads of small worms or maggots; but as these animalcula are seldom seen but in embryo, the animals commonly dying before they receive their full existence, the notion of such being engendered has been doubted by many; but I knew a very experienced farrier, who was the only person ever known to cure the disease, affirm that he had observed it in all its stages, and was well assured of the fact. This method of cure was by inflating the lungs with *nitrous air* [*]; but what the process was I never could learn.—*Mr. Trayton.*

[*] *This remedy, I suppose, would be instant death.—A. Y.*

The

Ja.^s Lambert del.^t

Scale equal 9 to 1 Ground.

SOUTH DOWN RAM, bred by Mr Ellman of GLYND, SUSSEX.

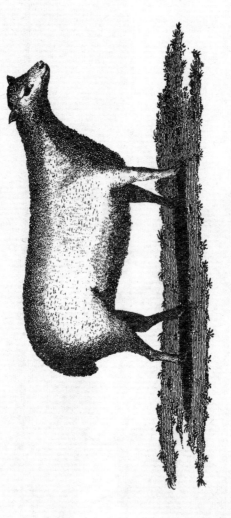

SOUTH DOWN EWE.

Bred by Mr. Ellman of Glynd Sussex.

Ja.ᵗ Lambert del.ᵗ

Neele sculp.ᵗ Strand.

The merit of the South Down breed of sheep is universally acknowledged, and the demand is so unlimited as not to be supplied; they have of late been extending themselves over the eastern, and more particularly over the western sides of the kingdom, with a rapidity hitherto without a parallel in the annals of our husbandry.

Indeed this has been acknowledged in those very counties where other breeds have, time out of mind, been established, and the consequence has been that of adopting the South Down stock; and we witness with no slight sensations of satisfaction and pleasure, the great improvement of the age. The emigration of the superfluity of the South Down breed, is rather an extraordinary circumstance in the detail of the live-stock, of this district; and it must be acknowledged as a strong argument in favour of the breed, that after several years, the demand has increased.

Some late experiments which have been very accurately made with this breed by the Earl of Egremont, who was certainly no otherwise interested than as being desirous of discovering the scale of their merit, have turned out highly satisfactory. Norfolk sheep are supposed to be amongst the inferior breeds in this island; therefore comparison with them is unnecessary.

Wiltshire, Dorset, Somerset, &c. though better than the preceding, are far inferior to the South Downs.

The Dishley breed are in some counties the favourites: and here we may hazard a remark in this early stage of our inquiry, as being founded on facts which will be hereafter explained, that this celebrated breed, which has been the theme of so much admiration, should have fallen into disgrace in the county now under review.

review. They have been admitted upon the most libe-
beral footing; and if merit had been in proportion to
price, it might have been expected that they would
have set competition at defiance.

If we set ourselves to examine the demand for the
Dishley stock, which has in some districts been so very
prevalent, and the additional surface of ground over
which they gr ze, in consequence of the improved
method of letting out tups ; whilst, on the other hand,
the South Down breed has overspread a wide extent,
by the exportation of them in droves, free from any
secret transactions betwixt individuals; it will be
allowed, that the real and active demand is great indeed in favour of Sussex.

It is not with any view of depreciating a breed of
sheep, which undoubtedly possesses merit, that these
observations are made; but it will afterwards appear
that they have been weighed in the scales, and found
wanting. They have their respective merits and peculiar advantages ; but no man of understanding will be
persuaded to believe (since no facts have yet been produced in support of the assertion), that they will thrive
upon a soil of five shillings just as well, in proportion
to land five times that rent. On the other hand, no
complete and satisfactory experiments have yet appeared before the public, sufficiently convincing to be
relied upon as authentic, that a small and close-coated
sheep is so well calculated for a marsh, as the breeds
which prevail in those districts, or that they thrive in
them to equal profit, as upon their native pasture.
We have in Sussex a surface of both these kinds ; and
facts speak a different language. The art of man has
new-modelled the creation; but it has not been effected
by traversing the laws of Nature : trusting to her directions

rections for the basis of improvement, she infallibly points out the line of demarcation, and the proper conduct for the experimentalist.

In considering the merit or demerit of any breed, the *food eaten* is the primary object, and the inquiry to weigh the wool and mutton against the food which produces it. Scales and weights must be appealed to in every step of the business : these are the only arbitrators in a national and political light; the food eaten is an object of immense import, and it has till very lately been most astonishingly overlooked. The Dishley breed *may* take up a greater proportion of land to support them, than perhaps any other breed we are acquainted with in the kingdom ; of land too that is rented at from 20s. to 40s. per acre. Of what importance is the fact, that individuals of this stock have been slaughtered of seven inches and a half of solid fat cut straight upon the ribs, if it requires an acre of land, rented at three times as high, to bring them to this pitch, as it serves to bring another stock to the butcher, in the ordinary course of fattening them ? Are these sheep spread thinly over one of the richest tracts of land in England, that the beauty of their form may attract universal admiration ?

We know in Sussex, that the South Downs are stocked at the rate of one sheep and a half to every acre: moreover, taking into this account all sorts of land over which sheep at any time of the year may be supposed to go, besides finding support for vast quantities of cattle; for oxen do the labour of the farm ; yet this land does not average above 12s. per acre.

Granting it to be an admitted fact, that even a thou-

SUSSEX.   sand

sand guineas had been given as the hiring price of a tup, without any secret understanding between the parties, and that the individual so purchasing had been equally benefitted; is the nation therefore a gainer?

It is the greatest possible quantity of mutton and wool from the least quantity of food, that constitutes excellence in live-stock. And is this to be met with in Leicestershire, connected with mutton and wool, hardiness, activity, folding, flavour, quality of wool, &c. *?

It

---

* In the month of May, 1791, myself and a farming friend, in whose judgment I had more confidence than in my own (and which confidence is founded upon his early and extensive experience, more especially in live-stock), went on to the South Downs, purposely to view the live-stock of that district, and carried with us an order from Lord Sheffield to his steward, to give every assistance in his power to assist our inquiries; which order was cheerfully complied with; and in consequence, his Lordship's flock was particularly examined, as also several other flocks upon the Down, though less accurately; and I must observe, that my expectations were much disappointed: the sheep upon the Down, struck me as ranking with the smallest breeds in England: those in his Lordship's pasture were better grown; but the best wether sheep we could find, struck mine and my friend's ideas as not exceeding 20 lb. per quarter. The wool is doubtless good clothing-wool: we carried home several specimens, which I shewed to a wool-stapler in large business, who agreed with us, that it was not superior to the average quality of the wool of the commons of Shropshire and Staffordshire, and much inferior to the Hereford, Ross, or the Shropshire Morf. The sheep may be fairly pronounced, a small, compact, and well made breed, doubtless healthy, as being preserved from internal complaints by the soundness of its native walk, and protected from external ones by being completely covered in a close, compact, and warm fleece. The breed cannot, I think, be admitted as distincly original, or peculiar to this county; for the Staffordshire Cannock Heaths, bred upon a waste of 25,000 acres, appears to me of the same origin; are very generally grey-faced; without horns; with fine wool; and if

well

It will conduce to clearness, to arrange the information upon this head under these divisions:

    I.   Breed.
    II. Management.
    III. Profit.

I. Breed.

---

well selected from a sound part of the waste, then taken into pasture, and put to proper rams, the produce is a heavier carcass than any I ever saw from the South Down stock, with a good disposition to fatten, though I admit inferior in make, beauty, and compactness of carcass, and heavier in bone. These last defects I attribute to want of former and early attention in the breeders; some of whom are, however, now making progress in this species of improvement.

I can never find the South Down sheep cut much figure in Smithfield, where I often searched for them, but not found them in any proportion of number to the Lincoln, Leicester, Wiltshire, Norfolk, and several mixed breeds, notwithstanding their local advantage of proximity of situation.

Again, I think the South Down sheep too light in carcass, to be deemed proper stock for good enclosed pasturage. With this circumstance is united another, to which my experience points out no exception: light-carcassed sheep are always disposed to break out of their pastures, or to commit depredations on the wrong side of the hurdles in your turnips. In this respect, no other stock I know, is so pleasing as the Leicestershire; wherever they are left they are found, whether upon a plentiful or a scanty allowance.

I would by no means wish to be understood, to have any intention of depreciating the character of a breed of sheep that have undoubtedly great merit; all I mean is, to call in question their superlative superiority over some other breeds that produce clothing-wool; (the Leicester and Cotswold breeds bearing combing-wool, are out of the question). The heaviest carcasses producing fine wool, are doubtless the Wiltshire. I have known wether-sheep of this breed fattened to 40 lb. per quarter, and sold at 3*l.* 10*s.* each; yet I can say from experience, that they have not clothing enough for the severest winters, of even the midland counties, in cold, wet and inclement seasons; and upon cold soils, their naked bellies, and thin open coats, exposing them to starvation, even to death; which I believe was actually the case with some very sound ones in my possession a few years since.

I do

I. *Breed.*

Under this head, the following subdivisions are necessary :

1. Form.
2. Colour.
3. Hardiness.
4. Proportion of stock to ground.
5. Prices.
6. Principles of breeding.
7. Castrating.

1. *Form.*

I do not pretend to say that the Cannock-heath breed, though somewhat similar, and superior in weight, when pastured, to any I saw in or from the South Downs, are yet brought to so correct a form; but even in this respect, I know many gentlemen and farmers who have bred from the Hereford, Ross, and the Shropshire Morf, that would by no means yield, but claim the pre-eminence. They have heavier carcasses, and I believe wool of a superior quality ; have small light crooked horns, small clean legs, the face and legs white, or a little freckled ; light in bone ; and are brought, in some hands, not only to a superior weight, but to a degree of compactness and perfection in carcass, which perhaps no other breed has excelled. These, in our neighbourhood, are called the Ryeland breed, and of which you will doubtless receive farther information from different quarters, especially the counties of Salop, Worcester, and Hereford.

The South Down sheep are, however, fairly introduced into Staffordshire, by the most noble the Marquis of Donnegal, who has a great many of that breed amongst other breeds, in his park at Fisherwick, and to which breed I know his steward to be very partial.

I shall not fail, when I visit that quarter, to make inquiries, and a faithful report concerning them.—*William Pitt, Penderford, Staffordshire.*

The main objection brought forward, in these observations of Mr. William Pitt, to the South Down sheep, is founded in a notion which has been waved by men of experiment, and given up upon grounds which are deemed no longer tenable—" want of weight." The South Down is among the smallest breeds, not exceeding 20 lb. per quarter; light carcass. Wiltshire, Cannock, heavier, up to 40 lb. All this may be

1. *Form.*

The true South Down sheep are polled, and when very well bred, have a small head, and clear neck, which

---

be very true, yet not affecting the merit of South Down sheep one iota. The West Country sheep are abandoned in Wiltshire by flock-masters, who say they can keep half as many again of the one as they used to do of the other breed. They are given up by a Wiltshire farmer, one of the greatest in England, in favour of the South Downs, by a flock-farmer, who clips five thousand every year, and who finds that three of the latter are kept upon the same ground in better condition, than served to feed only two of the former (Mr. Dyke); so that if the Wiltshire sheep average four or five pounds per quarter more than the South Downs, and not more wool, it is unequal to the circumstance of three thriving where two only did before: weight therefore signifies nothing, and is entitled to little attention. The greatest argument that was ever brought forward in favour of size (and it is very far from being admitted), is when it requires a less quantity of food to gain any given weight of flesh (say 500 or 1000 lb.) from one animal, than it requires to produce an equal weight in two of the same breed. Experiments already detailed, go to prove that this appears to be the case in large and small breeds of cattle; but with respect to sheep, facts seem to prove the reverse.

The South Downs support a much greater weight of flesh than we see in any other part of England upon land of the same value. Although it is a small breed, not averaging more than eighteen or twenty pounds per quarter, yet we have various instances of their weighing much heavier, up to thirty-nine pounds per quarter.

The question about the origin of them is of little consequence; but the fact, that there are some sheep upon a Staffordshire waste, that somewhat resemble the South Down in the colour of the leg, in being polled, and having carding-wool on their backs, proves nothing against the originality of them.

In regard to the quality of the wool, the samples which Mr. Pitt carried home with him, might perhaps be selected indiscriminately, and possibly from an inferior sort. This remark is the more applicable, because we are given to understand that no other flock was viewed with the same attention that Lord Sheffield's was. No doubt this is a fine stock;

which are very essential points; the length, indeed, of the neck, is a matter in dispute among the breeders.

Mr.

---

stock; but a person who comes into Sussex to collect information, and make his observations upon sheep, will not content himself with the examination of sheep in the Weald; but he will traverse that part of the county from whence they originally come, and not leave the county with the examination of one flock. As to the superiority of Ross or Morf wool over South Down, the practice of *trinding* the fleece in these counties, goes against him. If the stapler buys with any separation of the coarser sorts, these two kinds of carding-wool will be nearly upon a par. As to the specimens of wool which Mr. Pitt selected, the circumstance of superiority depends upon the manner in which it was done. A lock of wool taken from this or that part of a fleece, to set against another lock drawn from a fleece of another breed, is no comparison whatever. Fleece must be compared with fleece, or rather, tod against tod; for as there are eight or nine different sorts of wool in the same fleece, it requires some accuracy and judgment indeed, to pronounce upon the quality of either from such indifferent grounds. The Staffordshire stapler gave his opinion from the specimen shewn him. Was it drawn from the shoulder, the barrel, or the breech?

There is one remaining objection which Mr. Pitt makes to the breed, which is so diametrically the reverse, in point of fact, that it is not a little surprising how such a charge could have been urged—*the wildness of the breed, breaking pasture, committing depredations.* Some sheep have a great propensity to roving; the South Down are unexceptionably among the quietest and most orderly, and hardly to be excelled, even by the Bakewell breed. So far from a rambling disposition, the reverse has been repeatedly remarked. They have been at Bradfield for some years, and almost every field has its respective lot; and ground more heavily stocked is not to be found; yet no complaints are ever heard from any intermixture, though no other than very low and close-splashed hedges separate the lots.—*A. Y.*

" How far this sort (the South Down) will answer, time and experience must determine. It has already so far gained ground, that although they were only introduced into Wiltshire (by Mr. Mighell, of Kennet) in 1789, the number kept in the county is already increased to fifteen thousand, and is daily increasing. Those who keep them say, that they live so much hardier, and feed so much closer, that they can keep

three

Mr. Ellman, who certainly has brought his flock to a
high degree of perfection, thinks the length of the
neck no demerit; and other breeders, who look for
fine wool more than form of carcass, think it a merit,
as the surface produces more wool, and that of a fine
quality. Others, on the contrary, prefer a short neck,
because it is thought that lambs that are spear-necked,
are free of wool, and not so well able to bear severe
weather; and long necks are inclined to long carcasses.
Thus the form is sacrificed to a very inferior consideration. That the neck should be bred as light as possible, is at once seen in the large proportion of offal
(bone) in that part of the sheep, perhaps more than in
any other part, in proportion to flesh : the chap is
fine and small, and the bone light; points which are
very pleasing to the eye, and, as they imply a light
offal, certainly of importance. The points in which
this breed commonly fail is, that it is low and narrow

---

three hundred *well* on the same land that would only keep two hundred
Wiltshire sheep: that they are more docile, will feed more contentedly,
and stay more quietly in the fold."—Mr. Pitt, what say you to this?—
that though they are able, by keeping this kind of stock, to breed more
lambs, the ewes are such good nurses, that the lambs will be of equal
individual value with the Wiltshire lambs: that the wool, by the improved quality, as well as by the increased quantity, will almost double
the profit they have hitherto had from Wiltshire sheep; and that, by the
increased number they keep, they will be better able to dung their
arable land; and they see no disadvantage in them, but that the old
ewes, when sold off for fattening, will not yield so much individually
as the Wiltshire ewes. But then, " that they shall have three to sell instead of two, and that the wethers, when fattened, always sell for a
halfpenny, and near a penny a pound dearer in Smithfield, than horned
sheep."—They are full of wool, and that wool commonly very fine:
the weight of their fleeces is nearly as much as those of the Wiltshire
sheep, and the value is at least 6*d.* per pound more.—*Report of Wiltshire,*
by Mr. Thomas Davies.

in,

in its fore-quarters, and proportionally light; they
stand full two inches lower in their fore-end than in
their hind-quarters; a point which has been particularly objected to. Mr. Ellman cannot be brought to
satisfy himself that it is any defect: but one of the
excellencies of the Dishley stock is, the perfect beauty
of their form; a back in every part straight and even;
no rising back-bone, but the whole equally level.

The South Downs are thicker in their hind than in
their fore-quarters; and when fat, the hind-quarters
are frequently two or three pounds heavier than the
fore. Mr. Ellman considers this a great merit in the
breed, as the butchers have a ready sale for the former,
at an advance of 1*d.* per pound over the other; in
which case he entirely agrees with Mr. Bakewell, that
the criterion of breeding is flesh, not bone; and the
true point, to throw the greatest weight upon the most
valuable quarters of the carcass. The jaw clean and
thin, and covered with wool, as it has been remarked,
that sheep free of wool about the jaw, are apt to lose
it under the belly; a great defect, and what the breed
is sometimes subject to: the belly cannot be too much
covered. In a cold lambing-time, it will happen that
the lambs perish for want of wool to preserve them
warm; however, this defect the best flocks are free
from. Mr. Ellman weans one-third more lambs than
the number of his ewes; so that the South Down lambs
coming bare of wool, is chiefly seen in flocks unimproved.

Wool on the poll is not approved (they call it owl-
headed), nor any tuft on the cheeks. The shoulders
are wide; open breast, and deep; fore and hind-legs
stand wide; they are round and straight in the
barrel; broad upon the loin and hips; shut well in
the

the twist, which is a projection of flesh in the inner part of the thigh, that gives a fullness when viewed behind, and makes a South Down leg of mutton remarkably round and short, more so than in most other breeds.

In the form of the sheep, great room is open for improvement. It is only in a few of the best flocks, that much attention has been paid to the carcass. The quality of the wool has been the first object, and points of greater consequence neglected.

The improvement of the South Down sheep in the last twelve years, though considerable, is not to be attributed to the general spirit which prevailed in the county, so much as to the skill and intelligence of a few active individuals, who first set improvement afloat.

Such are the ideas which have been pretty much afloat since this breed has been improved. It remains for further experience to ascertain, whether they are not in the way to be carried too far; a point suspected by some persons;—whether the hardiness of the old breed will not gradually be lost in the modern improvements. I have no experience myself to determine this; but it is a consideration deserving the attention of those who are deeply interested in South Down sheep.

2. *Colour.*

The South Down farmers breed their sheep with faces and legs of a colour, just as suits their fancy. One likes black, another sandy, a third speckled, and one and all exclaim against white. This man concludes, that legs and faces with an inclination to white, are infallible signs of tenderness, and do not stand against the severity of the weather with the same hardiness as the darker breed; and they allege, that
these

these sorts will fall off in their flesh. A second will set the first right, and pronounce that, in a lot of wethers, those that are soonest and most fat, are white-faced; that they prove remarkable good milkers; but that white is an indication of a tender breed. Another is of opinion that, by breeding the lambs too black, the wool is injured, and likewise apt to be tainted with black, and spotted, especially about the neck, and not saleable. A fourth breeds with legs and faces as black as it is possible; and he too is convinced that the healthiness is in proportion to blackness; whilst another says, that if the South Down sheep were suffered to run in a wild state, they would in a very few years become absolutely black. All these are the opinions of eminent breeders: in order to reconcile them, others breed for speckled faces; and it is the prevailing colour.

It is merely mentioned, with a view of pointing out the various opinions which prevail. The stupidity of shepherds we do not wonder at; but that they should be able to impose these prejudices on their masters, is more surprising.

Let it be observed, that in the flocks in Sussex, grey, speckled, or mottled faces and legs, generally prevail; but any sheep with white faces or legs, though in other respects an unexceptionable animal, would not be esteemed.

3. *Hardiness.*

The merit of every species of stock depends in no inconsiderable degree upon hardiness; and it is intimately connected with the shape and make of the animal, and strongly interwoven with those points of the form which occasion it.

                                          Therefore

Therefore the observation is founded in fact, that a mould with the truest proportion, will stand the vicissitude of bad weather and hard food much better, than a thin carcass and a ridged and curved backbone : South Down breeders admit this ; but how far the quality and state of the wool contributes or not to hardiness, is yet undetermined amongst the breeders of long and short wool. The Dishley gentlemen contend, that their own breed will stand storms and hard seasons better than all others, and starve them out. And as to closeness and compactness in the fleece being any indication of hardiness, they deny it ; and suppose that people have been much deceived in this respect : that short woolled sheep should stand the winter better than other breeds ; and the reason alleged is, that close wool necessarily stands out in every direction, and consequently receives the wet to the skin, and lodges there ; whereas the open-woolled ones may be compared to the thatch upon a house, or stack which conveys the wet off as it comes on ; and the Scotch sheep are brought as an evidence, who are all flag-woolled, and yet inhabit the wettest and coldest part of the island.

On the other hand, it deserves the attention of breeders, since it is a fact no less remarkable than true, confirmed by repeated observation from various quarters of the county, that the finest fleeced sheep, with the closest pile and thickest wool, have by far a much kinder disposition to fatten, and are from one to two months sooner ready for market, than coarse woolled sheep ; and in proportion to the fineness of the wool, is the disposition to thrive, and the quality of the mutton. Indeed it is confirmed by some, who have tried the long-woolled sheep, for they were found to be more unkind in
their

their disposition to get fat. Rain and dew, drops from a close coat which is well protected by its density, whilst the long fleeces absorb the wet ; and as the wool of this breed is apt to separate on the middle of the back, it is contended that it imbibes moisture, and makes an opening for the rain to penetrate.

If we compare together in a flock of South Down sheep, those of shape and good wool, with those that have not these good points, it is admitted by South Down flock-masters, that in the same flock the fine woolled sheep will be the hardier breed. If we examine them when feeding in the same field, those that are well formed and covered in a fine and close fleece will have much the advantage, in a pinching season, of others that are worse made ; and sheep having a disposition to do kindly, and be in good order and fat, is a circumstance next to none. In all these respects the South Down breed is unexceptionable : their healthiness and freedom from losses will appear by the number set against accidents, and which is far inferior to what is commonly found in many other breeds. Mr. Ellman's annual losses have been for some years about one in a hundred (exclusive of lambing time) and other flocks are nearly upon a par with him. The natural soundness of the Down, unquestionably renders the breed that feeds upon these chalk hills peculiarly free from internal distempers ; and their hard and close coats are an excellent preservative from external ones : in these respects they are well defended against accidents ; boisterous winds blow over these high hills in winter and spring, with a violence that more level countries are free from : exposed to the fury of the elements at the extremity of the island, which do incredible damage to the houses, corn, &c. without a fence

or

or tree of any kind, for vegetation is cut by the fury of the winds. It must be a hardy breed to weather such storms.

4. *Proportion of Stock to Ground.*

In describing the South Down breed of sheep, it is this, of all other circumstances, which ranks foremost in order, and merits the most attention : the truest shaped sheep, and the finest fleece united on the same sheep, would be very imperfect, if it required as much food to feed a score of them as it took for half as many again of others without these marks of merit.—Wool, fold, tallow, flavour, hardiness, separately considered, are entitled to no attention; if not connected with the food eaten, merit is no longer merit. Of what avail is a heavy South Down fleece, or that bears the fold well, or that thrives well; that the flavour is exquisite, offal light, or hardiness unequalled, if, in order to acquire this perfection in the breed, it is endued with a voracious appetite, and the consumption of food not taken into the account. The fact is, that Sussex experiments have been but very few, so much as hardly touched upon balancing the comparative merit of sheep ; yet it is the most important and interesting of any.

Between East Bourne and Steyning, which is thirty-three miles, the Downs are about six miles wide, and in this tract there are about 200,000 ewes kept : the whole tract of the Downs in their full extent, is stocked with sheep, and the amazing number they keep, is one of the most singular circumstances in the husbandry of England. The six miles of breadth include such part of the vales adjoining as are occupied by the Down farmers, which adds a belt of flat land to the whole tract, but of very inconsiderable breadth ; and there are

are many farms on the Downs which have no vale land. This Down tract therefore, including to each farm a tract of good vale, is reckoned upon an average of the whole sixty miles, to keep one sheep and a half per acre. This is, soil considered, the highest stocking which is known in this kingdom, and ought *prima facie* to give us a good opinion of the breed, wherever it might be, that can be kept in such numbers, on a given space of country. Some parishes are rather lower stocked; but others higher: in that of Glynde 900 acres have 1000 sheep. Mr. Morris has one ewe per acre in winter, and two and one third in summer to an acre; and another farm of 2000 acres has 3000 in summer and 1500 in winter. In the above tract of thirty-three miles by six, on which 200,000 ewes are kept, there are 126,720 acres; this therefore is about one ewe and a half per acre.

As an explanation of this great stock, it should be observed, that it is a very general custom to put some, at times all their *tegs*, or lambs of last yeaning, to be wintered in the Weald, distributed thirty, forty, or fifty to each of the small farmers there, for which they pay so much per head.

It is a part of the subject which depends entirely on the quantity of food eaten, and it is very necessary and requisite to know the usual allowance of food, artificial as well as Down, in several distinct flocks, from which it will appear, that the food eaten is comparatively small in this part of the kingdom; and it must impress other counties with a very high idea of South Down sheep.

A tenantry flock belonging to Denton parish, consisting of six hundred breeding ewes, has no other provision but the native Down (seven or eight loads of hay excepted)

excepted) for the whole year : no green food. This flock
lives upon the hill the greatest part of the year, very
nearly indeed the whole of it : at lambing time it is
taken away ; and it is observed, that no where is finer
wool found ; and is an instance in favour of the quality of the wool depending upon the sort of food, and
as strong a one to shew, how small a quantity of food
serves to winter this flock.

In Adfriston parish, six acres of coleseed and eight
or nine ton of hay, are used for 450 sheep, with the Down.
A large farm at East Bourne, of cole, turnip, tares,
and ray, only forty acres with fifteen ton of hay for
the winter and spring provision of 1400 sheep, the
tegs being sent into the Weald : the Down 450 acres.
Another considerable farm in the same neighbourhood,
for 1000 sheep has fifteen acres of turnips, 10 of cole,
30 of ray, and 500 of Down, for its annual provision,
and from one ton to one ton and a half per week in winter. At Bedingham, 300 ewes are kept for a month
upon two acres and a half of turnips, and 30 cwt. of
hay and the Down. Winter-food here to keep 363
breeding ewes for four months, turnip 15 ; hay 10 ton ;
four acres of cabbage and four of cole will last till the
middle of April : in May twelve acres of rye and
tares are added : the Down 204 acres. A large farm
near Lewes, 1627 acres, of which *Down* 800, arable
500, meadow and pasture 327, has 2200 sheep, besides
157 head of draught oxen, horses, cows, and young
stock : winter and summer food—turnip 20 acres,
sainfoin 30, clover 50, rape 16, tares 50 ; one acre and
twenty-nine perches of a mixture of tares and ray have
been sufficient for 400 ewes for one week. General
Murray's farm, total 1637 acres, stocked with 4425
sheep, and 200 head of horned cattle, at the same time
that

that 680 acres of it are arable ; that is, there are above two sheep and a half over the whole, besides cattle and horses. Taking the upland farm alone, and independent of the marshes, 1150 acres are stocked by above 2200, or two sheep to an acre, besides 140 horses and horned cattle, yet 680 arable. Such proportions ought to give us a high idea of the breed that will admit such stocking upon land, none of which is more than 10s. per acre.

The native Down is stocked in proportion to the quality. Glynde and Ringmer Down, measuring 1100 acres, now maintains 5000 sheep and lambs for six months in the summer, and 2500 in the winter, exclusive of artificial provision.

Upon the whole of those accounts, a superiority is immediately discovered over other breeds, in the small proportion of food allotted for the maintenance of such numerous flocks. It is to the excellency of the breed, in union with the happy state of the Downs, to which this circumstance is to be attributed ; and partly to the beneficial arrangement of arable and pasture. In all seasons, recourse is had to the Downs for food ; and it is admirably well calculated for the purpose. If the proportion of stock to ground is extended over all the South Downs and the contiguous land, so as to comprehend a tract of 150,000 acres, the stock of sheep upon this surface, from authentic accounts, is estimated at 270,000 in summer, and 220,000 in winter ; a rate of stocking which is not to be exceeded in any other part of England, marsh land alone excepted.

5. *Prices.*

5. *Prices.*

This is another point which ought to impress upon the world an high idea of the merit of this breed of sheep: the advance in the prices of the flock proves it in the most satisfactory manner, and marks the improvement it has received.

The superiority of one flock over another, may be gathered from the difference of value in the sale of the produce: thus estimated, the success of some few breeders have been felt and acknowledged: the difference in the price is the quantum of improvement: and the constant unbroken rise of late years in the prices of sheep and lambs, denotes certainly the merit, and probably the demand for the breed. And no where shall we see such accounts of the profits of flocks, that will bear to be compared with the prices on the Downs. Such an incessant demand has existed for the breed, that the advance in the value has excited much emulation: price has done the whole.

In Leicestershire, no live-stock is highly valued which is not high priced. Except that county, there are no exertions which are not exceeded in Sussex. Ten guineas has been a high price for horned rams: the common price two and three guineas.

This is the price in the East of England, as well as in the West. It is not that the breeds of these counties is incapable of improvement; the most unthrifty sheep are open to amelioration; but the truth is, that in the face of others, they have not so much merit.

Until lately, ten guineas was the highest price that was heard of in Sussex for the sale of any ram. Now Mr. Ellman letts many of his three-year old rams for

SUSSEX.]   fifty;

fifty; inferior ones for thirty, twenty, ten guineas; and
he has lett at one hundred.

6. *Principles of Breeding.*

The management of sheep here, as in most other
flock counties, is to sell the wether lambs, and the
refuse of the ewe lambs, after keeping a sufficient
number to support the flock, and which do not take
the ram till after the second shearing, that is, till
they are a year and a half old, and a proportionate
number of old ewes: but in this respect, the South
Down farmers differ very much from some other
counties, where the ewes are kept till they are broken
mouthed, and some till they have not a tooth in their
heads: this in Sussex is considered as the worst of
management.

They universally get rid of them at five years old;
but the best flock-master at four. Mr. Ellman attributes the contrary conduct, and apparently with
much reason, to the sheep-masters listening to the suggestions of idle shepherds, who have so mush less
trouble with an old ewe, than with a young one, that
they are partial to keeping them as long as possible.
Mr. Ellman is so much convinced of this, that he
thinks it would answer better to sell the ewes at three
years old, that is to say, when they are at their highest
value. To this it might be objected, that what seemed
a rational motive for keeping ewes much longer, was
the fact generally admitted, that old ewes generally
bring finer lambs than young ones; it not being uncommon in Norfolk and Suffolk, to see a lot of very
fine lambs, from crones that have hardly a tooth left
in their heads. But Mr. Ellman thinks, that although

an

an old ewe would bring a large lamb, yet such lamb will not make a large or fine sheep ; nor will it fatten so well as the produce of young sheep. He has made the same observation on cows, mares, sows, and bitches.

Mr. Ellman's flock consists of about five hundred breeding ewes, each ewe (barrens and refuse excepted) produces three lambs, lambing at two, three, and four years old, and when four years and a half old, he sells them off, to go into other flocks. The general practice has been, to sell them to the graziers, in the Weald of Sussex and Kent, who fatten both the lambs and the ewes the following summer; but Mr. Ellman has for some years, found a better market in the great demand in other parts for his sheep, and he expects that this will continue to be the case, till the South Down sheep are generally known. He usually saves for store about two hundred and twenty ewe lambs, which gives him an opportunity to refuse about fifty-one each year. His ewe lambs at Michaelmas are sent out to keep in the Weald, amongst the small farmers, till the following Lady-day, when he takes them home, and flocks them, or folds them at night, till they are a year and a half old, when they are put with his breeding ewes. He always takes sixty of his best ewes, and puts them to his rams of the best shape, and finest wool, and saves the rams from them*. But the usual way is to give about fifty ewes to one ram, and to put all the

rams

---

* What I mean by the best wool, is a thick curdly wool, with depth of staple, and even toped; such wool as will best defend the sheep in bad weather, and will not admit the water to penetrate, as it does a thin, light, loose wool. I have found from many years experience,

rams into a flock at a time, which he very properly
condemns, for several reasons. After having taken out
sixty ewes, he then puts three of his next best rams
into the flock, and about five or six days after, he
adds two more, and continues to add two every four
or five days, till the whole are put in, by which
means his best rams have the most ewes. He begins
to put them to his flock about the 25th of October,
and lets them continue with the ewes about five weeks,
from first to last.

They are folded at night, throughout the year, except for a month or five weeks after lambing, which is
the latter end of March, or beginning of April, and
the lambs are well covered with wool when born. If
the ewes are well kept, one third of his flock will
bring twins . The lambs are weaned at twelve or fourteen

---

perience, that sheep in the same flock, of the former descriptions, will
keep themselves in better flesh, than those of the latter.

When I change my breed, which I think it absolutely necessary to
do, I get some neighbour to let me take out fifty of his best ewes (of
the former description) and put my best ram with them, and I save
ram lambs from them. By following the above practice, and drafting
thirty or forty refuse ewes every year, I have got my flock tolerably
good, both for shape and wool. The farmers on the South Downs, a
few years since, were taught by the wool buyers to believe, that it was
not possible to increase the quantity of wool, without decreasing the
quality, an opinion which was not grounded in truth. For by adhering to the rule above-mentioned, I believe I grow the heaviest wool
between Brighton and East-bourne, and sell for the highest price on the
South Downs.

I do not put more than eleven rams to five hundred and sixty ewes,
so, by saving twenty ram lambs every year, have an opportunity of
refusing eight or nine.—*John Ellman.*

* My ewes usually produce but one lamb each; but if well kept, and
full of flesh, when the rams are put to them, many will bring two,
some three, very few four. I have known instances of an ewe producing

teen weeks old. Mr. Ellman never puts ram lambs
to his flock. His cousin, at Shoreham, some years
puts no other: the former carefully avoids his ewe
lambs taking the ram; but this is no general rule.
Neither ram nor ewe lambs should be allowed to
copulate.

In the principles of breeding upon those hills, ex-
changing the rams every third, fourth, or fifth year,
is the practice almost unanimously agreed upon. It
is found to be most essentially necessary in preserving
the health, size, and bone of the flocks; although it
is contrary to the maxims of some great men, who are
advocates for *breeding in and in* continually, when
there is a good sort. Under the article *crossing* in
cattle, it will be seen that Mr. Alfrey was long in the
possession of some of the finest beasts in the county of
Sussex: by breeding in and in for more than twenty
years, they were totally ruined in constitution and
habit, and died four or five of them in a year. By the
same treatment, his flock was reduced to the same
situation, but by changing his rams, the improvement was wonderful. It was the same with his dogs.
The breeders on the Downs are one and all of them of
the same sentiments.

Crossing the South Down breed with other sorts,
has been very sparingly practised. Spanish rams
have been introduced into some few flocks. Lord
Sheffield, first introduced the Spanish breed into the
county: the wool of his Lordship's flock was considerably improved by it. The few breeders on the
South Downs who have tried it, found two capital

---

ducing five. It is seldom that more than two are saved. The lambs
are wonderfully covered with wool when dropped, and the coarsest
woolled ewes, bring their lambs with the greatest quantity of wool at
the fall.—*George Allfrey.*

defects,

defects, not to be compensated by any improvement in the wool, tender constitution, and bad shape.

Mr. Morris, of Glynde, has gone into the Spanish cross (the only one which has been tried by breeders who have valued themselves on their flocks) more than any other breeder on the Downs, that I have met with. The ram from which it proceeded was half Spanish, half Ryland; but this nominal Spanish was from France, and by no means of the true Segovian breed. Mr. Morris is out of the breed now, from being convinced that they are not so hardy as the South Downs, by their not bearing equally well the sharp winds which blow on the hills, with a violence that flatter countries are free from. Quere, if this tenderness be Ryland or Spanish? Mr. Morris however improved his wool very considerably by his cross.

Mr. Ellman observes, that he knows of no crossing, generally speaking, though two of his neighbours tried a cross with the Spanish, but found them delicate, and not well shaped; and the South Downs have done them away as fast as possible, and returned to their original breed.

7. *Castrating.*

The best time for this operation is, eight, ten, or twelve days old. Mr. Ellman cuts off the tails of his lambs at the time of castration : thus a considerable quantity of blood is lost, which he considers as preventing the part from the gangrene.

II. *Management.*

1. Food.
2. Watering.
3. Fattening.
4. Distempers.

This

This may be divided into summer, winter, and spring. The *summer* provisions for a flock of sheep, besides the native Down, is tares, coleseed, and artificial grasses; though many of the flock-masters in the central parts of the hills, are not so well situated, respecting the food, as those whose farms are adjoining the rich land at the roots of the Downs; so that the chief summer food which the farmer relies upon, is what the native Down is able to produce. This is a very short, sweet, and aromatic herbage, peculiar to these hills, and by far the best which they can have, provided they are able to fill their bellies before folding time. It is the herbage of the Downs which renders the flavour of the mutton so exquisitely fine, the flesh so firm, and the wool so excellent. Artificial grass, clover, ray-grass, besides rape, tares, turnips, and all other succulent food, are considered as enemies to the production of fine wool. The richness and luxuriancy of the food, is thought to contribute to render the wool more coarse, but abundant, in the same proportion that the fine quality is injured.

This circumstance may be remarked by an examination of the flocks between Lewes, East-bourne, and Newhaven, where the finest wool is produced in the county, yet the food they feed upon is no other than the Down in summer and winter, except a little hay distributed in hard weather on the hill. This fact coincides with observations made upon other flocks in other parts of the county, which feed upon little other food than the Down and hay, and they have the finest fleeces. *Si tibi lanicium curæ fuge pabula tocta*, was laid down as a maxim two thousand years ago, and it is no less founded in reason, than confirmed by practice.

But

But it will be said, why then does Mr. Ellman sell at the highest price, when he feeds his sheep very much upon artificial food, as his Down of only 150 acres, is not in any proportion to the size of his flock ; but art and attention will perform much.

It is his exertions in the improvement of his flock, that have enabled him to sell at the highest value on the hills; and his Down, though not large, goes a very great way, when we account for the quality of his wool*.

As some of the flock-masters have little other food than the Down to summer their sheep, others have little of it ; and consequently their sheep in the summer are at a very considerable expense, on rape, tares, and grasses, none of which are so beneficial to wool as the natural herbage.

I cannot fail of impressing any person with a high idea of the breed of sheep, and the value of the food, to view them grazing in the summer upon the South Downs. The number of the flocks seen at the same time in a small tract of land, instantly strikes any man of reflection, who examines into the state of the sheep-walk, that they must be a very profitable sheep, comparing together the weight of flesh and the food eaten.

After harvest the flocks are turned into the stubbles, and at Michaelmas many are half fat, which they ose as quickly in the spring, as they gained it in autumn, as winter food is not sown in that quantity which it ought to be.   Tail seeds are not unfrequently

---

* My Down, or sheep walk, is but small, and my enclosed land extremely wet, so that I cannot stock with store sheep; am therefore obliged to depend on artificial food.—*John Ellman*,

sown with the corn in the spring, for the flocks after harvest: but it has been known, that they have sometimes been very violently purged, by turning them into wheat stubbles, and the flux has been fatal to many.

1. *Food.*

Winter Food.—Turnips.—The introduction of this root into the English husbandry has been a vast improvement. Norfolk is quoted by every man on this occasion. The great exertions which have been made upon the Downs within the last twenty or twenty-five years, have been equally great, and they have chiefly related to sheep. Turnips are sown by almost every flock-farmer, in some quantity, as food for sheep; though indeed the cultivation is far inferior in breadth of land (and management) to what is thought necessary in Norfolk or Suffolk: a few years back, no such thing as turnips was seen in Sussex; at least in any quantity.

The flocks are penned over the field, usually some time before Christmas : turnips will last well for four months, and even longer, if alternate frosts and thaws do not rot the crop at the breaking up of the winter, or when a warm or open spring drives them to seed. With proper management, they are the best dependence; if in some degree restricted as to the quantity eaten, South Down sheep are never subject to any ill effects from them. Several farmers, who had lost many of their sheep upon turnips, by giving them hay, or even pea, bean, or wheat straw, have obviated the effects of the watery nature of them : a very small quantity of dry food, is found to correct the properties of the fluid.

Sainfoin

Sainfoin hay with the turnips, is the best provision in winter. The *redwater* often attacks the sheep feeding upon turnips; the effect of wet seasons. Hay prevents it from breaking out.

Mr. Ellman never loses any of his sheep by this disorder, from his attending to the above.

Mr. Ellman of Shoreham, generally gives his sheep hay, in hoar frosty mornings. He finds that it preserves them from the *gall*.

The turnips are more generally drawn two days before folding them, by which means they do not burst, which is sometimes the case when not drawn. It has been objected to turnips, that they occasion the ewes to *slink* their lambs, but by previously drawing them, no inconvenience of that kind is experienced*.

Turnips are sometimes stacked, but not so often as they ought. Mr. Milward always feeds with dry turnips, on which occasion he always stacks, and in such a manner, as to prevent the frosts from injuring them.

Potatoes—Have been tried as food for sheep, and found upon experiment to answer; and perhaps superior to turnips, as being a more regular and certain dependence. The farmer who relies upon his tur-

---

* I remarked in a field of Mr. Car, at Bedingham, a practice which deserves noting. He was eating off turnips for sowing wheat: I observed, all within the fold were drawn a day or two before the sheep were allowed to enter, in order that the turnips might wither and evaporate their water. I demanded the motive; they said, that when the sheep ate them in the common manner, they not only disagreed with them, but even some were lost by it. I think the practice very rational, and combines with a great number of other observations on different foods. and different subjects.—*(Annals of Agriculture,* vol. xv. p. 433.)

nip

nip crop, will, in some seasons, run great hazard and
danger from the frost. A crop which depends for its
preservation, on the mildness and regularity of the
season, is not to be considered as a certain one, if we
recollect how any sudden change from frost to thaw,
frequently occasions the destruction of the whole crop
in a few days. The consequence of such accidents
at the most critical season of the year, is easily fore-
seen. But this is a part only of the loss sustained.
The great difficulty in raising any crop at all, and
not seldom the utter impossibility of insuring a full
and fair one, is another heavy deduction from the
value of the crop. In a dry seed-time, it never
comes to perfection : in showery weather, the young
plant is devoured by the fly, and the ground three
times sown with little chance of success. Mildews, and
various other accidents, render turnips by no means a
certain dependence; they are liable to destruction at
that season of the year when they are most wanted : for
after all other hazards, a hard frost and sudden thaw
destroy them at once.

Not one of those objections holds good against potatoes : no accident, but what may be easily guarded
against. The frost is no longer any formidable enemy,
when the store is deposited in well-formed pits. General Murray fed 5000 sheep with potatoes and hay :
1651 of his breeding ewes ate 51 bushels every day,
giving a quart to each; and that, for 120 days, is
6120 bushels. A Norfolk flock-farmer provides for
720 sheep, 80 acres of turnips, 16 ton of hay, 20 acres
of rye : let us compare the provision.

If 720 sheep require 80 acres of turnips, 2240, the
upland flock at General Murray's, require 248 acres
of turnips; but they have only 50.

If

If 720 sheep require 16 tons of hay, or 10 acres, 2240 should require 49; instead of which they have 120, which is 71 surplus, or, at one load and an half per acre, 48 acres.

If 720 sheep require 20 acres of rye, 2240 should require 62; instead of which, they have none at all.

Winter Food of 2240 *Sheep, as provided for in Norfolk.*		*Winter Food of* 2240 *Sheep, as provided for in Sussex.*	
	Acres.		Acres.
Turnips,	248	Turnips,	50
Hay,	10	Hay,	80
Rye,	62	Rye,	0
Potatoes,	0	Potatoes,	20
	320		150

Now let us value these crops, so as to apply fairly to Sussex and Norfolk, equally rejecting each table of expenses: the following rates will not be far from the truth.

	£.	s.	d.
Turnips,	2	0	0
Hay,	5	0	0
Rye,	0	10	0
Potatoes,	4	0	0

Expenses.

Expenses. Norfolk.		Expenses. Sussex.	
	£.		£.
Turnips,	496	Turnips,	100
Hay,	20	Hay,	160
Rye,	31	Rye,	0
Potatoes,	0	Potatoes,	80
	£.547		£.340

Which is a difference of 63 per cent.

Now, is this vast difference to be attributed to potatoes being a cheaper food than turnips; or to the distinction between the one flock being Norfolks and the other South Downs? That a very considerable portion of this superiority is to be attributed to the breed, there is not the least doubt; for the general turn and colour of all the intelligence has given us, on every occasion, reason to think this: but as to feeding sheep with potatoes, it is, though ascertained on General Murray's farm, on the largest scale, a more doubtful circumstance; and for this reason they are allowanced, or limited in their consumption, which is not the case with turnips: these, on the Norfolk farm, are fed on the land, and consequently, in the greatest plenty. Another-contrast, however, is not to be forgotten. Turnips are subject to frosts, to fly, to mildew, to various accidents: potatoes are a regular certain crop, and subject to few accidents. The General was using his potatoes while we were with him, and found them safe and secure, notwithstanding the severity of the frost; and another gentleman we were with afterwards, had one of his pits uncovered, and hardly a rotten potatoe

tatoe was to be seen in a hundred bushels. When this circumstance is well considered in the pinch of a severe season, every one will agree, that the vast experiment made by the late General Murray, in the introduction of this root as a winter and spring provision for sheep, is truly important.

The Rev. Thos. Fuller, of Heathfield, used potatoes in the same way *. Mr. Fuller's experiment is upon fat sheep, and General Murray's upon a lean flock,

----

* My general method has been, to let the sheep of the true South Down breed have the after-grass, and about the middle of the month of November, to take them into the yard, with a shed or lodge adjoining, and confine them till they are ready for the market, at the end of February, or the first week in March.

The potatoes are cut into two or more slices, as may be deemed necessary, and put into troughs, which are fixed under the shelter. On an average, I have observed that a sheep will eat one gallon a day. I have generally purchased my lambs at 12s. 13s. and 14s. and have always sold them at good prices, as I have seldom got less than 4s. per stone. They have ever proved well upon examination; and at the age of two years, have amounted to nine, ten, and eleven stone: the internal fat of one, which I had the curiosity to weigh, was equal to 15 lb. It will not admit of a doubt, but that if fair trial is made, the potatoe system will prove the most expeditious in fattening sheep. I have made the experiment with sheep of the same age, and of the same flock; on corn of different kinds, and even with oil-cake; and have found the potatoes will do the business in the shortest time. As a farther proof of what I have said in favour of potatoes, I have remarked, that on placing two sheep from the common stock, in a yard with those that had enjoyed the best after-grass, I had found, at the usual time of selling them, little or no difference at all in point of fatness, and have sold them at the same price. I presume, therefore, you will allow me to say, that I think this method of preparing mutton for the table is the most expeditious, and the most profitable plan to be pursued in accomplishing the end designed.

You will please to observe, that I give the sheep a little hay, morning and evening; and, if the yard is properly attended to, you may easily conjecture what a mass of manure, both in point of quality as well as quantity, may thus be procured.—*Thos. Fuller.*

which

which accounts for some difference in the quantity eaten
per diem, but it by no means accounts for the difference of four to one. The obvious conclusion is, either
that the General did not allow potatoes sufficient, or
that his flock had other food unnoticed.

Cabbages—Have been applied to the feeding of sheep
in Sussex, and, where the practice has been adopted,
with great and uniform success. It is, perhaps, of all
other sorts of food, that which demands the greatest
attention, since it is by means of this food that great
improvements might be expected in the Weald, as the
soil is perfectly well adapted to the production of cabbages, in any quantity, as food for sheep. Objections
raised from its requiring a richer soil, are too trifling
to refute. There cannot be any doubt but that by cabbages, tares, and potatoes, with turnips, where the
soil is suited, with the great command of dry food and
pasture in the Weald, far greater than what the Down
farmers enjoy, they might be enabled to maintain, acre
for acre, thrice over the number they do at present
with a well-regulated management. If those absolutely useless exertions (the only efforts worth notice)
in liming a fallow field at the expense of 5*l.* per acre,
to gain five sacks, were exerted in raising crop of sheep
by means of cabbage, potatoe, tares, upon their
arable land, and laying down to permanent pasture
a soil *only* adapted to grass, some such an arrangement would be rather more satisfactory than the present system of husbandry.

Spring Food.—Artificial Grasses.—Ray-grass is one
of the earliest that is cultivated, and in much request
for a flock when the turnips, &c. are gone, and it comes
at

at a time when it is much wanted. It is to be lamented, that this grass, as well as some others, are not cultivated in a greater degree for sheep-feed, as a succedaneum for clover, especially upon those soils where clover fails. The culture of some of the best of the grasses, would lay open an almost inexhaustible mean of improvement: such a variety are at hand, some of them known to be more productive than any that have been yet cultivated, and that possess the three requisites of quantity, quality and earliness. The *Alopecurus pratensis*, and *Dactylis glommerata*, are admirable plants, and much is it to be regretted that they are not more frequently introduced into our artificial lays. There cannot be better grasses than the *Poa trivialis* and *pratensis;* and upon calcareous soils, sainfoin and burnet (*Hedysarum onobrychis*, *Poterium sanguisorba*) are indigenous. The *Dactylis glommerata* is rough and coarse, if let to grow old, and very early, but hardy and productive. Here are then about six or eight grasses, from which might be selected specimens for clay, mixed, and light soils.

Tares, Rye.—All these are sown as spring provision for the flock. Arable lands, tolerably clean and in heart, or rendered sufficiently so with manure, are ploughed in September and October, and sown with winter tares, rye, or cole, according to the nature of the soil, as tares upon the stronger, rye upon the lighter, and cole chiefly on the calcareous hills. These crops come sufficiently early to be fed off in April or May, when the turnips are finished, and are hurdled off in the same manner. After they are taken off, the land is again ploughed, and spring tares are then

sown,

sown, which are to be fed at the end of autumn, when the land is in admirable order for the ensuing crop of wheat, if the autumn is favourable, or for barley and seeds in the spring. This double crop of tares is worthy a journey of many miles to see it, and the more such husbandry is analyzed, the better it will appear. The mixture of tares and rye answers better for soiling than for sheep-feed; for the horses are soiled at a time when the tares are young, and have no great strength in them; and the rye is a very dry food, which counteracts the moisture of the tares: but for sheep-feed it is not equally good; for the rye and the tares being sown in September, the former, upon good land, will be fit for folding by the middle of April, and the tares by the middle of May upon the same soil: if the rye is preserved till the tares are ready, the rye will hardly be touched, or trod down —one of the two must suffer.

Let us, however, consider this husbandry.

Instead of an unproductive and expensive fallow, the skilful and active farmer raises two crops of tares, to answer the great end of fallowing (clearing and meliorating) equally well. The ploughing is at a season of the year when the ground can easily be worked; and in the western part of Sussex, with a light plough, and two horses, and a man who holds and guides it (a great saving of labour), he secures food for his stock at the most critical period of the year, and enriches the ground with the manure arising from the fold, or stock fed upon it. Mr. Thos. Ellman sows ray-grass for two years; it is twice folded, then broke up, and two bushels of tares and a gallon of cole are sown in May or June, fed in August and September. By such means, one acre and a few perches are sufficient for

SUSSEX.]   400

400 ewes a week. The value of the food, at 2*d.* each
for a week, is 3*l.* 6*s.* 8*d.* ; the fold, 1*l.* 5*s.*; together,
4*l.* 11*s.* 8*d.* the value of the crop for seed and fold.
Where shall we go to find management better than
this? To break up a layer in order to sow with wheat,
the common system, would be a useless and barren fallow, made at 4 or 5*l.* expense. But setting aside this
practice, here is a crop of tares, expenses more than
paid, and the land in hearty order for the succeeding
crop ; whilst his neighbour gains his crop, which is
not a better one, at the expense of 5*l.**

Stubble Turnips.—After harvest, the stubble is
ploughed, and turnips sown, which come round for
late spring feed ; but some harvests are too late for this
excellent practice : other green crops, however, render it equally good ; nor can it be sufficiently commended, for it is in the true spirit of good husbandry.

About Petworth it is a common practice, either to
sow stubble turnips, or rye and tares, upon the
wheat, barley, or oatersh. The whole practice of
throwing in one crop upon the back of another, is a
feature too good to be passed over.

Rouen.—One of the most capital arrangements for
the support of a flock that was ever thought of. Valuable as all the preceding crops certainly are, they are
inferior to this ; yet we see but little attention paid to
a species of food so well adapted for ewes and lambs.
It is nothing more, than making a reserve in a time of
plenty for the hour of want.

---

* Annals of Agriculture.—*Editor.*

Mr.

Mr. Ellman reserves his best pasture for them. The Earl of Egremont is strongly in this practice, which will be described in another place.

Sheep feeding Wheat—Is practised in different parts of the county. It is alleged in favour of the custom, that the wheats rise the stiffer and more abundant. The fact appears to be, that it is not done so much to benefit the wheat, as through mere necessity; since it is allowed, that as other food is scarce, this becomes necessary. Even the best farmers are frequently compelled, by having no other provision, to feed their wheat. The practice is pretty general. On light lands, it may be right to fold sheep, in order to close the ground about the roots of the plants; and when it is thin set, feeding dry land will give a better stock; but it is more frequently done through necessity.

Winter Barley—Has been sown for sheep-feed, but the practice confined to a few individuals.

2. *Watering.*

This is very necessary in the management of a flock on the South Downs; and as there is no other water than what is to be collected in reservoirs, artificial ponds are constructed to retain the rain-water; these are generally circular, and very gently sloping to the centre: the bed very strongly rammed down, to prevent any loss by soaking through to the chalk. As the surface of the South Downs is found to be waving, every farm presents an opportunity of collecting any quantity of water; though in very dry weather, many

of

of the flocks having no water of their own, are driven to their neighbours' ponds, sometimes at a considerable distance.

Mr. Sneyd says, " Previous to the mode now pretty generally adopted, of forming ponds or basons in various parts of every hill farm, in long droughts the sheep have been driven some distance to water. It is much to be lamented, that the exertions lately used to furnish a supply of water, have not been attended with more general success ; as many farmers in the neighbourhood of East-bourne have been at great trouble and expense in forming these ponds, which is done by lining them with chalk, puddled and trod down till it makes a kind of plaister floor, and they generally hold water well enough for some time; but are apt to become leaky, and a hard frost spoils them. There is a pond on the top of Friston-hill, which I never knew dry : it was formed many years ago, and, I am well informed, has the bottom paved with very small flints. Ponds which have no run of water into them answer best. I made one which, from receiving a large run of water, is perpetually choked up; while another I made at the same time, and which receives no water but what rains perpendicularly into it, has answered better, and never wanted clearing."

In Italy, the flocks are regularly watered morning and evening. Inde ubi, &c.

Ad puteos aut alta greges ad stagna jubeto currentem ilignis potare canalibus undam, &c.

Tum tenuis dare rursus aquas, &c.

Adduxere sitim tempora.—So Columella, 4 v. vii.

3. *Fat*-

3. *Fattening.*

This important point of Sussex management resolves itself into the following subdivisions :

1. Age,
2. Food,
3. Thriving disposition,
4. Live and dead weight,
5. Flesh,
6. Tallow,
7. Offal,
8. Pelt,
9. Distempers,
10. Interesting experiments.

1. *Age.*

The South Down wethers are generally turned off to fatten from one to two years old. It is considered as bad policy to keep for profit more than two years and a half; and indeed it is usually allowed, that they pay better at one year and a half old, than at any other age. Few, or rather no experiments have been set on foot to ascertain the precise time when they fatten to most advantage; but it appears, that the profit lessens as the age increases; and it is pretty generally acknowledged, that the quickest return is the most profitable, and accordingly, the sheep are turned off at an early age. Moderately fat at a year and a half old, a wether pays much better than if he is much fatter at double the age *.

2, *Food.*

---

* South Down wethers arrive at perfection at five or six years old; ewes at five and wethers at six. They will continue improving as long
as

2. *Food.*

Turnip is the usual food : and it is well worth no-
ticing, as late experiments tend to confirm the remark, that to fat sheep upon this food, after summer
pasturing them, they will fall off very considerably in
flesh : so far from having gained any flesh, they decrease, so that there is little profit by keeping sheep
through the winter.

The Duke of Bedford's experiments, inserted in the
twenty-third volume of Annals of Agriculture, proves
this to be the case. From whence it appears that
four breeds, South Down, Dishley, Coteswold, and
Wiltshire, all lost upon turnips.

The loss of money from keeping fat sheep through
the winter is considerable, and affords a lesson well
worth remembering : to get fatting sheep so forward,
as to sell them between the first of August and the
first of October. The Michaelmas markets are sometimes not high, but the difference of price will by no
means pay for winter food. Apparently the winter
food is thrown away.

That this circumstance is not at all peculiar to the

---

as their teeth remain sound, which generally decline after the sixth
year.—*Geo. Allfrey.*

Mr. Ellman says, " To discover the age of mutton, is to observe the
colour of the breast-bone when a sheep is dressed ; that is, where the
breast-bone is separated, which, in a lamb, or before it is one year old,
will be quite red ; from one to two years old, the upper and lower bone
will be changing to white, and a small circle of white will appear round
the edge of the other bones, and the middle part of the breast-bone will
yet continue red ; at three years old, a very small streak of red will be
seen in the middle of the four middle bones, and the others will be
white ; and at four years, all the breast-bone will be of a white or
gristly colour."

South

South Downs, appears clearly by the late Mr. Macro's
most accurate experiments on Norfolk sheep, in the
Annals, where he details the winter food (cabbage,
turnip, and hay) of some, and none gained any
weight of consequence, but most of them lost. One
must take for granted, that farmers and graziers, on a
great scale, know this fact: when they keep a great
number of sheep upon turnips, do they wait for a market only? if so, they wait at an enormous expense. It
should seem that the profitable consumption of turnips and cabbages by sheep, is by the breeding stock
and hoggits, which demand keeping only, and not
fattening.

Such is the language of the experiments hitherto
published; but when we compare it with a very general practice, there is a great disagreement; for too
many farmers are in the constant practice of winter-
fatting sheep, to permit us to conclude that they all
lose money by it. It should therefore be considered as
a question by no means sufficiently ascertained. Probably much will be found to depend on breed; for
the Norfolk and the Wiltshire, bad as they may be
in other respects, have been found to pay well in winter feeding.

3. *Thriving Disposition.*

The merit of the South Down breed is beyond a
doubt, if we consider the food eaten, which is at once
ascertained from the number kept.

A sheep and a half per acre, including all sorts of
land, is very high stocking, and rarely to be met
with. In fattening, the remark is equally applicable
to the breed. A thriving sheep is seen in what it pays
for the food it eats: and this point is in union with
another,

another, which has not that attention paid it which it merits. Good South Downs, fine shaped, and fine woolled, will sooner be ready for the butcher, than others of the breed ill shaped and coarser woolled. This is the language of experiment, which Lord Egremont, the two Ellmans, Mr. Allfrey, and other breeders and graziers have determined.

4. *Live and Dead Weight.*

In the 20th vol. of the Annals, Mr. Ellman has weighed frequently the live and dead weight of his three year old South Down wethers, bred by him and slaughtered at Glynde. As many objections have been raised against the breed, from want of weight, it will be seen, that thorough-bred wethers will fat at three years, to 30lb. per quarter: but, what is of greater consequence than weight, the proportion of the dead to the live weight is very great.

Live weight, ... 192 lb.

	lb.	oz.
Blood,	6	0
Entrails,	11	0
Caul,	16	4
Gut fat,	5	0
Head and pluck,	8	12
Pelt*,	15	12

---

* Washed the pelt and clipped 5lb. of wool, when dry:

	lb.	oz.
Skin weighed,	5	8
Lost by washing,	5	4
	10	12

This great loss may be accounted for by the pelt being thrown under the sheep to receive the blood, &c. while dressing. The skin was not dry when weighed.

Carcass

Carcass next morning, ... 125 lb.
                                                   ___
Carcass, ... 125
Offal, ... 67
                                                   ___
                                                   192
                                                   ___

If 192 gives 125, what will 20 give?—Answer, 13.
Slaughtered the 21st, and cut up the 24th of December:

	lb.	oz.
First fore quarter,	29	0
Second,	28	12
First hind,	33	8
Second,	32	0
Lost,	1	12
	125	0

Had one side cut into joints and weighed.

	lb.	oz.
Haunch,	23	0
Loin,	10	4
Neck,	12	0
Shoulder,	11	12
Breast,	4	8
Lost,	0	12
	62	4

The above weighing does credit to the South Down
sheep: the quarters were divided in the usual way,
leaving one short rib to the hind, and twelve to the
fore. The hind quarters of this wether were heavier
than the fore, which Mr. Ellman very justly considers
as a merit in the breed, as the former sell at 1*d*. a
pound more than the latter. In the 15th vol. of Annals,

nals, is an account of the live and dead weight of three South down wethers, slaughtered at Lord Sheffield's. They are an average specimen of the breed.

Weight alive, ... 133 lb.

Dead next day, ... 73
Tallow, ... 10
Blood, ... 4
Entrails, ... 14
Skin and feet, ... 16
Head and pluck, ... 9

126

Proportion, half and one-tenth.
One of General Murray's :

Alive, ... 129 lb.
Dead, ... 62
Tallow, ... 6
Not half.

At four o'clock in the afternoon, two fat wethers were weighed alive, directly from their food :

First, ... 126 lb.
Second, ... 110

236

After twenty-two hours fasting again weighed :
First, ... 117 lb.
Second, ... 102
They lost 17lb. they were then killed :
First, ... $58\frac{1}{2}$ lb.
Second, ... $53\frac{1}{2}$

112

Tallow of the two, $13\frac{1}{2}$ lb.

From

From some experiments which have lately been made at Petworth, by Lord Egremont, and which will be presently detailed, much valuable information will be added, tending to elucidate the subject, and give us a clearer knowledge of the proportion between the live and dead weight of different breeds of sheep.

5. *Flesh.*

South Down wether mutton, in point of delicacy and flavour, is thought equal to almost any that is killed ; and in summer, as preferable to some other fine flavoured breeds, especially to Norfolk mutton. This circumstance is attributed to the closeness of the grain, or the specific gravity being greater, rendering it more impermeable to the air than coarser and looser fleshed mutton, which is of course more subject to putridity*.

The older the mutton, the finer the flavour ; though this is a circumstance not thought of by the grazier. Those who are connoisseurs in the flavour of mutton, will find, that a spayed ewe kept five years before she is fattened, is superior to any wether mutton. The Duke of Grafton sent a haunch of it (a cross between

---

* I was informed at Lewes, that Mr. Gates, a butcher and grazier at Steyning, of considerable experience, had given it as a fact, that Hampshire sheep when killed, stiffen sooner, and keep twenty-four hours longer than South Downs; yet that the South Downs are of all other sorts, the finest grain, and indeed the best of mutton. I called on him at Steyning with Mr. Gell, and he confirmed it, as a fact with which he was well acquainted. It seems rather to militate against the undoubted fact given in this work, of the South Down and Norfolk mutton, made by Mr. Vyse, butcher at Eton college ; but the latter is the result of such large experience, that it will admit of no doubt ; it however militates merely against the mode of accounting for the fact, by attributing the quality of keeping to fineness of grain. A loose open texture of flesh, seemed to be more adapted for admitting air, and if so, ought to putrify the sooner.—*A. Y.*

Norfolk

Norfolk and South Down) to Lord Egremont; and the admirers of mutton confessed it was truly excellent.

6. *Tallow*.

It is by no means a settled point upon the South Downs, how far a sheep which gathers its fat upon the intestines, is or is not preferable to another which collects it upon the back and the neck. The Leicestershire graziers contending as much for the latter as the former, is considered as a test of merit in Norfolk and various other counties. But when it is considered that it requires a certain portion of food to create a given quantity of fat, the question is, which is the best part to collect it upon—within, or without? As long as the fat of the latter will sell at more than one-third of the other, it would seem that there cannot be a doubt, which of the two is preferable; and upon the principle of food eaten to produce the tallow or fat, that which tallows least is the best breed. The tallow, with the major part of the fifth quarter, is all the butcher's profit, who would no doubt encourage that breed which tallows best and yields most offal.

The South Down sheep are not great tallowers, compared with some other sorts; but what they loose in tallow, they make up in a disposition to fatten. The tallow of a wether in common management, will generally average from an eighth to a tenth part of its dead weight. In Mr. Ellman's fat wether, one-seventh part of the dead weight was inside fat (caul and loose fat). In another which he killed last winter, one sixth was inside fat. In others that have been slaughtered, the variation has been from a seventh to a tenth. The quantity

quantity of inside fat depends much upon the age, and time of fattening. It gathers itself much more in old sheep than in young ones.

A circumstance with respect to fat meat, is worthy of being mentioned, because it shews how much further very fat mutton will go, than that which is not equally so. At Petersfield (a great thoroughfare), the inn-keepers of that place agree with the butchers to give them 1d. per pound above the common price of mutton, provided it be very fat. It is the same with beef.

7. *Offal.*

The lightness of the offal (head, horn, feet, entrails, pluck, blood, pelt), characterizes a good sheep. Dishley wethers well fattened, it is said, are in the proportion of an ounce of bone to a pound of flesh.

The offal of Mr. Ellman's fat wether, was but a fifth part and a fraction of the live weight.

	lb.	oz.
Alive,	192	0
Offal,	42	0
Carcass,	125	0
Fat,	21	4
Lost by killing,	3	12
	192	0

8. *Pelt.*

Sheep pelts are usually sold to the fellmongers in the neighbourhood, by contract for the year, at different prices: viz. from shearing time to Michaelmas, at 12d.; to Shrove-tide, at 2s.; and from Shrove to shearing, at 3s. These are lower than usual.

9. *Distem-*

9. *Distempers.*

The distempers which the South Down sheep are subject to, are these :

    1. Redwater.
    2. Gall.
    3. Dropsy of the brain.
    4. Rot.
    5. Flux.
    6. Slipping the lamb.
    7. Hoving.
    8. Drunk.
    9. Feeding on charloc, poppy, &c.

1. *Redwater.*

Upon being first turned into turnips they are sometimes subject to this complaint, which is caused by their eating too large a portion of turnips in wet seasons. It also originates in the sheep being let out of the fold when the ground is covered with hoar frost, and often from feeding in the oatershes about Michaelmas, if the young oats are strong. It is soon obviated by allowing a small quantity of hay to counteract the wateriness of the turnip. Half a pound, or even a less quantity per day for each, is enough. It is thought that clover stubble and folded land, produce it in wet weather.

2. *Gall.*

Occasioned by feeding on turnips, and other green food of the like nature. Sometimes they have been bled for a cure, in the vein immediately under the eye. It is a purging which generally continues till

                              they

they die. Feeding upon land lately folded, seeds, rape, turnips, in wet weather, occasions it.

3. *Dropsy of the Brain, or Paterish Dunt,*

The principal malady to which the South Down sheep are liable; it is in other districts called the *sturdy,* or *duntheaded;* in Sussex being *paterish.* Trepanning has been recommended, but without effect. The most advisable mode is, to slaughter them immediately as the disease seizes them. A paterish sheep appears to be deprived entirely of its senses, and is continually turning round instead of going forward. The disorder is caused by a bladder or bag of water that surrounds the brain, in which is a hydatid, but there is no cure for it. Every farmer is more or less subject to annual losses in his flock by this incurable distemper : for it is, without doubt, one of ,the most destructive maladies that attacks the flock.

4. *Rot.*

Of this there are three sorts, the plain, the gravel, and the flesh : the two last are deemed incurable. Some few attacked with the former have been saved. A physician in Sussex once tried half a score by way of experiment : three doses of preparatory inoculation-powders for this rot : five of the worst died soon after the third dose ; the remainder lived two years after, but never grew much better. The rot was never known to be caught upon the South Downs. When the farmers suffer in that way, it is sheep that have been put out to keep in the Weald, or turned into the marsh to fat. A marsh which is occasionally overflowed with salt-water, was never known to rot sheep, but is a most admirable method to keep them sound and
healthy ;

healthy; and if any thing can cure the rot, it is such land. Mr. Ellman observes upon it, that if, after a frost, even so early as October, sheep are turned into these meadows and brooks, which are at other times so liable to rot them, they will not at this time suffer at all; as the animalculæ which the insects deposit in the summer among the herbage, are destroyed by the frost. The flounder found in the liver of the animal, is taken up with its food. August, September, October, and November (provided there is no frost), are the most favourable months to bring the rot; but after a single night of any sharpish frost, it is over for that year.

5. *Flux.*

The above are the principal disorders of South Down sheep. Others of less note are, the flux, a purging occasioned by feeding in wheat stubble.

6. *Slipping the Lamb.*

Mr. Gilbert, of East-bourne, some few years ago, lost 80 or 90 by this complaint. It was attributed to the feeding them upon rape about Christmas; yet he had fed them upon it before, without any bad effect. The sheep had been hard kept. The same thing has happened among other farmers; but it is remarkable, that a neighbour fed his rape over the hedge at the same time, without any inconvenience of the kind.

7. *Hoving,*

Or bursting with eating luxuriant plants, clover, rape, &c. Mr. Ellman remarks, that they are never subject to it when the food is wet from rain or dew; an erroneous idea, very common. He always chooses to turn into such crops at such a time; but when quite

dry,

dry, and the leaf at all withered from a hot sun, the danger is considerable. The remedy : half a pint of lintseed-oil to each sheep, given with a horn, which vomits them directly, and never known to fail.

8. *Drunk.*

Mr. Davies, of Bedingham, had, one year, eight acres of buck-wheat, which his shepherd fed with the flock for two hours when in full bloom : all were drunk; the glands of three were swelled quite to the eyes. On hogs it had the same effect. Bleeding made the sheep worse. However, none were lost.

9. *Feeding on Charlock.*

Great injuries have been felt by lambs feeding upon charlock, amongst the turnips, cole, and sometimes on the fallows. Old sheep are not subject to it, only lambs.

It is the fault of the present age, that we have no public institution, conducted by men of real science, for improvements in this branch of the farmer's art. Nothing essentially beneficial in curing the diseases of live stock, has appeared from any establishments yet founded, except in respect of horses.

III. *Profit.*

Under this head may be classed the following articles :

1. Expenses.
2. Produce.
3. Fold.
4. Wool.

SUSSEX.]   1. *Expenses.*

1. *Expenses.*

This will be best explained by stating the expenses of a flock of 560 South Down sheep, upon an average of seven years, drawn up by Mr. Ellman :

To 20 acres sown with clover and tre-foil, 8 lb. each sort per acre, at 5d. per pound together, ...	£.6	13	4
To sowing the above, at 3d. per acre, ...	0	5	0
To one year's rent, and parish rates, ...	14	0	0
To 15 acres sown with ray-grass amongst wheat, one bushel and a half per acre, 2s. 6d.	2	16	3
To sowing ditto, at 4d. per acre, ...	0	5	0
To one harrowing, at 4d. per acre, ...	0	5	0

No rent and rates to the ray-grass; the whole year's rent, &c. I charge to the succeeding crop of turnips. I let the ray-grass remain in the ground only one year, from the time it is sown; I plough it up for turnips the last week in May, or the beginning of June, at which time the roots, not having much hold in the ground, shake out very easily with the harrows. I give four ploughings for turnips.

To 15 acres sown with rye in the wheat-stubbles, as soon as the wheat is reaped; seed two bushels per acre, at 3s. 6d.	£.5	5	0
To one ploughing, at 8s. per acre, ...	6	0	0
To sowing of ditto, at 4d.	0	5	0
To three harrowings, at 4d.	0	15	0

The rent and rates charged to the succeeding crop of turnips.

turnips. I never sow any rye for seed; sow it for the purpose of sheep-feed in the spring, as it comes early, and produces a great deal of feed for my couples.

To 20 acres sown with winter-tares on the oat and barley-stubbles; seed two bushels per acre, at 5s. 6d.	£.11	0	0
To one ploughing, at 8s.	8	0	0
To sowing of ditto, at 4d.	0	6	8
To three harrowings, at 4d.	1	0	0
To three-fourths of a year's rent and rates for ten acres of the above, being fallowed up (at Midsummer) for wheat, when the tares are fed off: as the land lies one-fourth of the year under fallow, I think it right to charge the whole year's rent and rates to the sheep.			
To three-fourths of a year's rent and rates, at 22s.	8	5	0
The other part of the tare ground, I charge the whole year's rent and rates to the sheep, being sown with rape for their use, after the tares are fed off, ...	11	0	0
To rape-seed, 10 gallons, at 1s. 3d.	0	12	6
To one ploughing, at 8s.	4	0	0
To sowing, at 3d.	0	2	6
To four harrowings, at 4d.	0	13	4
To twice rolling, at 6d.	0	10	0
To 20 acres sown with spring-tares on barley or oat-stubbles, two bushels and a half per acre, at 5s. per bushel, ...	12	10	0
To one ploughing ditto, at 8s.	8	0	0
To sowing ditto, at 4d.	0	6	8
To three harrowings, at 4d.	1	0	0
To one years rent and rates, at 14s.	14	0	0

After

After the above tares are fed or mowed (if I do not want the feed, I mow part of it for hay, to set up for my sheep), I sow rape; sometimes I let it stand till the spring, and sow the ground with barley; but in common, feed it off at Michaelmas, and sow wheat.

	£		
To rape-seed, one gallon per acre, at 1s. 3d.	1	5	0
To one ploughing, at 8s.	8	0	0
To sowing, at 3d.	0	5	0
To three harrowings, at 4d.	1	0	0
To one rolling, at 6d.	0	10	0

Rent, &c. charged to the tares.

	£		
To 30 acres sown with turnips; seed, one pint and a half per acre, at 4d. per pint, ...	0	15	0
To three ploughings for 15 acres, at 7s. ..	15	15	0
To four ditto; the other 15 acres at 7s.	21	0	0
To 10 harrowings, one with the other, ...	5	0	0
To three rollings, at 6d.	2	5	0
To sowing of ditto, at 3d.	0	7	6
To 20 acres of turnips, hoed twice, at 9s. per acre, ...	9	0	0
To 10 acres, once hoed, at 6s.	3	0	0
To rent and rates, at 14s. per acre, ...	21	0	0

I observe in the Duke of Grafton's account of his flock, are charged only two extra ploughings and three harrowings; and in Mr. Macro's account, two extra ploughings, and two harrowings only, are charged to the turnip crop.—*Quere*. Is the above, the whole of the ploughing and harrowing given for the turnips?

It is not the whole, but that which is given extraordinary for turnips, beyond what would be given if it were a fallow.

The

The Duke of Grafton calculates in this way, **because**
Mr. Macro had done so, that the comparison of different flocks might be just. The method however is
certainly objectionable, because if not sown with turnips, there is certainly no necessity that it should be
fallowed ; it might be sown with tares, rape, potatoes,
&c.

In such calculations, the food given to stock should
be charged either at what it would sell for on the spot,
or at the actual expense of it to the farmer.

To 30 acres of grass in the lawn, rent and rates at 20s. per acre : I charge only three-fourths of the above to the sheep, as my cows run there in the winter,	£.22	10	0
To eight weeks' keep in my meadow and pasture-land, in and after lambing-time, for 560 couples, at 3d.	56	0	0
To herbage of 120 acres of stubbles, after harvest, set at 6d. per acre,	3	0	0
To mowing, haying, and carriage, of 20 loads of clover, or tare-hay, at 7s. 6d.	7	10	0
To thatching, and straw for ditto,	1	10	0
To rent and rates of 150 acres of sheep-down, at 3s.	22	10	0
To 40 new wattles each year, at 2s. 6d.	5	0	0
To repairing the old ones,	1	10	0
To carriage of the wattles about the farm,	2	0	0
To shepherd's wages,	30	0	0
To boy's ditto,	3	18	0
To an assistant in lambing-time, four weeks, at 10s.	2	0	0

To

To washing, shearing, and winding of wool, 1440 ewes, tegs and lambs, at per score, ...	£7	4	0
To carriage of the wool to market, ...	0	10	0
To expense of keep for 320 lambs put out in winter, from Michaelmas to Lady-day; 10 lost lambs, not paid for, at 3s.	46	10	0
To expenses driving out to keep, and bringing home, ...	1	0	0
To expenses at fairs, for wattles, &c.	0	15	0
To the use of 11 rams, at 1l. 1s. each, ...	11	11	0

I am rather at a loss to know how to make out a fair account with respect to my rams, as they do not feed on the lands which I charge to my flock, only five weeks in the year; that is, the times my ewes go to. I breed my rams for sale, which has turned to advantage for some years past. My practice in breeding my rams is, to take 50 of my best ewes out of my flock (those of the best shape and best wool), and put my best ram with them. What I mean by the best wool is, a thick curly wool, with depth of staple, and even topped; such wool as will best defend the sheep in bad weather: from being very thick, and even topped, it will not admit the water to penetrate to it, as it does a thin, light, loose wool. I have found from many years' experience, that sheep (in the same flock) of the former description, will keep themselves in better flesh than those of the latter. When I change my breed, which I think it absolutely necessary to do, I get some neighbour to let me take out 50 of his best ewes (of the former description), and put my best ram with them, and I save ram lambs from them.

By

By following the above practice, and drafting out 30 or 40 refuse ewes each year, I have got my flock tolerably good, both for shape and wool. The farmers on the South Downs, a few years since, were taught by the wool-buyers to believe, that it was not possible to increase the quantity of wool, without decreasing the quality; an opinion which was not grounded in truth; for by adhering to the rule before-mentioned, I believe I grow the heaviest wool between Brighthelmstone and East-bourne, and sell for the highest price of any wool on the South Downs.

I do not put more than eleven rams to 560 ewes; so, by saving 20 ram lambs each year, have an opportunity of refusing eight or nine: the refuse lambs I sell from one to two guineas a sheep. As I do not keep my rams in the flock, as I mentioned before, I have not brought the profit of them to this account.

To tithe of 1440 ewes, tegs and lambs,   £.30 0 0

To interest of Stock :

560 ewes, at 20s. £.560
326 lambs, at 13s.   208
Wattles, ...   30

Interest of .... 798, at five per cent.   39 18 0

Total expenses, ... £.501 14 9

2. *Produce.*

Total flock ewes, ... 560
Losses, ... 6
Profitable stock, ... 554

Twins made up for losses and barren ewes, as the number of ewe and ram lambs are nearly equal.

Say

Say ewe lambs, ... 280

Wether ditto, ... 260

Ram ditto, ... 20

                            560

I take out for stock,

Ewe lambs, ... 220

Wether ditto, ... 100

Average price of sale ewes and lambs for seven years:

140 wether lambs, sold at 13s. 2d. per lamb, £.92   3   4

20 refuse ditto, at 8s. 6d. 8   10   0

50 ewe lambs, sold at 11s. 8d. per lamb, .. 29   3   4

10 refuse ditto, at 8s. 4   0   0

20 ram lambs kept for my own use; I va- } 20   0   0
lue them at 20s. ....}

I keep only three ages of the ewes in my flock, viz. two tooths, four tooths, and six tooths. At four years old, I sell them off, adding 210 ewe-tegs to 554 flock: total ewes, 764.

For sale, 170 old ewes, sold at 18s. 6d.   £.157   5   0

34 refuse ewes of two tooths, four tooths, } 30   12   0
and six tooths, sold at 18s. ....}

100 best wether-lambs, kept for stock. I ⎤
put them out to keep in the winter |
(from Michaelmas to Lady-day), at |
3s. per lamb: they are kept in the |
flock from Lady-day to Michaelmas   ⎬ 92   3   0
following, and then turned off for fat- |
tening; allowing three for losses, 97 |
only are turned off; value them at 19s. |
per sheep ....⎦

                                        The

The weight and price of my wool for the last seven years.

	Fleeces per tod of 32 lb.	Price per tod.		
1782	14¼	£.37	0	0
1783	14¼	34	6	0
1784	13¾	38	6	0
1785	15	38	6	0
1786	13¾	36	6	0
1787	14	40	0	0
1788	12½	41	0	0

Average, at 14 fleeces to the tod, 38s. per tod.

To quantity of sheep shorn:

Flock ewes,	554
Tegs,	310
	864 fleeces.

Average, at 14 to the tod, gives 61 tod 23 lb. at 38s.	£.117	5	1¼
560 lambs shorn, weight of the wool, 8 oz. per lamb, gives 280 lb. at 6d. per pound,	7	0	0
Folding 50 ewes of arable land, at 20s.	50	0	0
Ditto 10 acres of down in the winter, when the arable land is wet, 15s.	7	10	0
One month in lambing, folded on litter in the sheep-yard, exclusive of cold nights in the winter; set this standing fold at	7	0	0

Total produce,	£.622	11	9¼
Total expense,	501	14	9
Profit,	£.120	17	0¼

I find,

I find, in looking over the accounts of the Duke of Grafton and Mr. Macro, that no rent and rates for the turnip land is charged in either account. The Duke of Grafton has set the tithe of his flock at 17*l.* 13*s.* 6*d.* I have set the tithe of mine, as you desired, at 30*l.* the same as Mr. Macro; though I think no tithe ought to have been brought to the account, as I hire my lands tithe-free, and pay a higher rent accordingly; and if no rent or tithe for the turnip lands was charged, it would make the balance in favour of the sheep-master, 174*l.* 17*s.* 0¼*d.*

This account, I flatter myself, will be thought a fair one, as I have endeavoured all through the account, to divest myself of every partiality in favour of my own breed of sheep.

It may be thought my losses in my flock are set too low; but to the best of my knowledge, and from the private account of my flock where the losses are entered, believe it to be a fair statement, as my sheep in general are very healthy.

3. *Fold.*

1. Space.
2. Value.
3. Stock.
4. Advantages.
5. Standing fold.

Undoubtedly, one of the most valuable practices ever established on the South Downs, and the universal attention paid to it, shews how well adapted the breed is to support bare keep and distant folding; for the position of great numbers of farms, in this respect, is such, as to put the flocks to the severest trials.

The

The practice upon the Downs, it appears, is, to fold upon the arable lands : in the winter, upon such as are intended for pease, oats, or turnips. At this season, two folds are thought necessary ; one on the Downs, where the sheep are penned in rainy nights, when the arable lands are too wet for them to set on. The early part of the summer, they fold on such lands as are intended for turnips ; after which, upon lands which are in rotation for wheat. It is not a common practice to fold upon pasture land, although Mr. Ellman frequently does it soon after lambing time. Folding begins soon after lambing, when the lambs are about a fortnight old, and continue folding, except in very wet weather, till the ewes begin to lamb again ; and it may be said that, during the lambing season, they are penned either in the standing fold or in the pastures. But this is Mr. Ellman's mode of management, and not the usual practice of the county, since some of the flock-masters allow their sheep to lay out of the fold on the Downs for three or four months during winter.

1. *Space.*

Mr. Ellman states, that a flock of 500 sheep will pen 28 square perch each night, which is 50 acres in a year ; allowing them to be left out of the fold two months in the year, which is a fair estimate for the best farmers.

2. *Value.*

This is in proportion as the farmer considers the profit of the fold. It varies from 35s. to 42s. per acre, which for 500, is from 87l. 10s. to 100 guineas for the 50 acres, which, if we take the average at 94l. for the flock, the annual value of the fold will be, per head, 3s. 9d. and a small fraction: at 100 guineas, it is
4s.

4s. 2½d. per head. Of what great consequence the fold
is to the farmer, when the value of it is found to be so
high!

3. Stock.

All the sheep, excepting the fat stock, are regularly
folded: these are never folded; and this is one of the
reasons why the Dishley sheep are never folded, as they
are inclined to fatten, which the fold has a great tendency to reduce.

4. Advantages.

The benefits which accrue to the farmer from the
fold, are sufficiently striking, and will be immediately
perceived, if we consider the flock as a moving dunghill, manuring the land without any expense. If the
sheep are well kept, it is equal to a coat of muck for the
first crop, but is not so durable, particularly on lands
where it is often repeated. It has been affirmed, that
folding on chalky lands, makes the wool harsh, and
not mill well, not being so soft and silky as other wool,
but the colour very fine. Such remarks as these should
be treated as they deserve, when it is considered that
the finest wool in England is grown on chalk hills.

Although it is a feature in the business of folding
altogether unknown in Sussex, yet if some trials could
be made by dividing the flock, it would seem that there
are great advantages which would flow from it.

Mr. Boys, in Kent, with a flock of 1200 South
Downs, is so convinced of the benefit of it, as to be
surprised that it should be called into question for a
moment. He does it entirely for the fold; and there
can be no doubt but many more sheep might be kept
upon the same space which it requires to feed any given
number, according to the present practice. The waste
of

of food is not inconsiderable, when a numerous flock is turned off at once over a large piece of ground; but in an open country like the Sussex Downs, this practice would, generally speaking, be hardly possible to carry into execution.

5. *Standing Fold.*

In Mr. Ellman's management of his flock, there is a circumstance which should be more universally attended to; not indeed that he is singular in it. He has two or three yards well sheltered, for the sheep to lie down in at night, in very rainy and stormy weather. One contains, including the sheds, 355 square yards. The sheds around it are about four yards wide, and the whole thoroughly well littered. These yards are extremely warm, and preserve many lambs in bad weather; around the whole circumference is a rack for giving hay. The late General Murray's standing folds were equally well contrived, enclosing an area of fifty-seven yards in length, and twenty broad, containing 1140 square yards; above 700 ewes were folded in it at night, and for that number it is more than a yard and a half for each sheep. All around it was a shed, nine or ten feet wide, and also across the middle, which latter was open on both sides. A rack for hay placed against the wall, which was boarded, surrounded the whole; and another, which was double, to be eaten out of on both sides, stood along the central shed; under the rack was a small manger, in which the food was given.

4. *Wool.*

Great exertions have been made of late years, amongst the South Down farmers, to improve the fleece, both in quantity and quality: the extraordinary demand which
has

has been created for the woollen goods of England in foreign countries, since the termination of the American war, has had its effect upon the grower; and the improvement of the wool has been the consequence of an increased demand for the commodity, notwithstanding the monopoly of the raw material. But in that eagerness, so prevalent for improving the fleece, it has happened that the flock-master has sacrificed points of greater value, in order to the production of fine wool. The shape of the carcass has not had that attention paid it in Sussex as in Leicestershire; nor has that desire been manifested which is necessary, if it is expected to combine a fine fleece with a fine form.

Mr. Ellman's is generally admitted to be the first flock, whose exertions have been unremitted : Mr. Thomas Ellman's flock is not far behind his relation of Glynde. They have each of them united those valuable properties which so essentially contribute to the perfection of the breed; *wool* and *shape*. It was before considered as impossible, to bring the shape of the animal to any degree of perfection, without sacrificing the quality of the wool; which idea originated in the fact, that the finest wool is found only on a sheep-walk : and where the carcass is not considered of so much consequence, and therefore not improved. If the county is examined between Lewes, East-bourne, and Brighton, but especially about Bourne, in this district the finest wool is grown; but in proportion to the fineness of it is its lightness, requiring 20 or 21 to a tod : and the shape of the carcass is out of proportion. What is the cause of a better fleece here than elsewhere? It is certainly not the effect of any peculiar good management, for it may be said to arise from the poverty of the land, the situation of whose farm is such as to admit

few

few opportunities of feeding with any other food than a scanty sheep-walk. There are some flocks about East-bourne, and in central situations, that having no land at the foot of the hills, have no opportunity of feeding upon artificial food; no winter and spring provision but the native Down. Here the wool is, without doubt, excellent; the shortness and bareness of the feed gives it a fine quality, but no weight: and it appears that the succulent food, as turnips, cole, rye, tares, and artificial grasses, throw out a coarse and luxuriant staple, but diminish the value of it for carding.

But these circumstances, as in the case of Mr. Ell-man, are to be counteracted.

Fine wool, therefore, may be called natural to a South Down sheep; and in proportion to the improvement of the hills will the quality of the wool be diminished, if due attention be not paid to it, and without more active exertions than any that have yet appeared. If one man at Glynde, or elsewhere, sells the wool of his flock at the highest money value, and yet trusts in winter to tur-nips; in spring, to rye-grass, clover, and rye; in sum-mer, to tares and cole-seed; and John Ellman's pasture so wet that he is unable to keep store-sheep in it; yet his down or sheep-walk but 150 acres; what else but inactivity prevents others from pursuing the same course?

The Downs west of Arundel river are very much covered with rubbish, such as beech-wood, chiefly scrubs; and with furze, &c. so that the natural herbage is not equal to those districts further east-ward. The farmers on the western side of the county have got into the notion, that no rams, or even ewes, but those that come from the other side

side of the county, are good, but especially in their wool. The fact is, that they do not exert themselves equally with the others, and then make a discovery that their sheep are inferior. Such idle opinions taking possession of the minds of a large class of men, certainly is not the way to improve the breed; but the natural good sense of farmers will surely not allow such opinions to be propagated any longer.

The inquiry into the question of wool, admits the following sections :

1. Washing.
2. Shearing.
3. Weight of fleece.
4. Value.
5. Quality.
6. Proportion of weight and value.
7. Number.

1. *Washing.*

The mode of washing, as practised by Mr. Ellman, is detailed in the following information. A stream is always to be preferred. Mr. Ellman has generally four men in the water to wash, and pens made in the water, pointing against the stream, so that the thick or foul water keeps draining from the sheep, and particularly where they are. His pens in the water are, the first where the sheep remain about three or four minutes, to soak the wool. In this pen he generally puts about twenty at a time; from which they are put forward to another pen, where the men stand to wash, which is performed by pressing the wool between their hands, after which the sheep swim out against the stream for about 15 or 20 feet, which cleanses the outside of the wool.

wool. Fourpence per score for flock sheep and lambs, and 6*d.* for fat sheep, are the prices given by Mr. Ellman for washing.

2. *Shearing.*

Midsummer is the shearing time for the flocks; earlier for the fat sheep. Clipping the lambs, has been considered in some places injurious; as an operation which hurts the growth of the lamb. In Sussex no such effect is perceived. The profit is very trifling; it about pays the expenses, or rather more, but it tends to improve the wool, and cause it to throw out a more luxuriant staple.

Mr. Ellman has a practice, which he thinks answers to him: it is, to clip off the coarsest of the wool on the thigh, and dock a month before washing and shearing, which he sells as locks; the quantity is about 4 oz. per sheep; it keeps them clean and cool in hot weather.

Fifty sheep are sheared by each man daily, at 2*s.* 6*d.*; or 1*s.* per score, and board.

Mr. Ellman stores his wool in upper chambers, as the moisture it would produce on a ground floor, if it remains there any length of time, is injurious to it*.

Twice shearing in the same year, was once tried as an experiment by Mr. Kemp, at Coneyborough; the first clip was six weeks before the usual time, the second in September. Clothing the sheep has been attempted, but it failed†.

3. *Weight.*

---

* It can be no injury to the wool, to shear the sheep as soon as it is dry, as the washing takes nothing but the dirt and filth from the wool: it loses very little of the greasy substance, or yelk.—*Mr. Allfrey.*

† Clothing the sheep must, I imagine, be prejudicial both to the sheep and the wool. Without sun and air, rains or dews, the wool would not

grow

3. *Weight.*

The weight of the fleece is various, and depends much on the food : about East-bourne it is light; upon rich food it is, of course, heavy—two pounds and an half is the average. Mr. Ellman has indeed clipped more than five pounds from several of his own breed. The improvement of the Glynde flock may be seen in the weight of it.

		lb.	*oz.*
1770 to 1774, averaged at	....	2	12
1775 to 1779, ditto	....	2	2
1780 to 1785, ditto	....	2	10
1786 to 1790, ditto	....	2	6
1791 to 1795, ditto	....	2	12
1795 to 1799, ditto	....	3	0
1800 to 1806, ditto	....	2	8½

When it is considered, that the price of the wool has been constantly rising, and that it has carried the highest price upon the Downs, it will be found an experimental refutation of a notion not uncommon, that you cannot increase the weight of the fleece, without adding to the coarseness of the staple. The contrary has been the fact here, most evidently; and it proves clearly, that there is no necessity for deteriorating wool

---

grow to more than two-thirds of its usual length (if so much), and the hair would be weak and rotten, and not possess sufficient strength for the different operations of carding, spinning, &c. As to the sheep itself, it is more than probable that its constitution would be much weakened, by having the external air kept from its body, and that the flesh would acquire a rancid unpleasant flavour, as we find that a small degree of heat before a sheep is shorn, has a surprising effect on the taste of the flesh, and that it does not recover its usual flavour in less than three weeks or a month after the fleece is taken off.—*Mr. Allfrey.*

by

by improvements of the soil; an d that the evidence which has lately been given by certain woollen manufacturers, at the bar of the House of Commons, assigning to enclosures, and consequently turnips, and artificial grasses, the effect of damaging wool, is a merely speculative idea, unsupported by facts. And if, for want of due attention, the tillage of Downs has had this effect, it has been a consequence not necessarily flowing from the measure, but arising solely from the inattention, and other circumstances, personal and local.

In twelve years, 1769 to 1780, 315,238 fleeces were registered at East-bourne Custom-house, weighing 22,135 tod; the general average, 2 lb. 4 oz.; and of lambs 88,855, weighing 41,642 lb. or 7 oz. each.

In ten years, 1783 to 1792, 385,532 fleeces, weighing 27,439 tod, were entered at Brighton; average; 2¼ lb.; and 217,446 lambs, weighing 83,112; average, 5¼ oz. Farther westward, the fleeces are heavier, but coarser.

Lord Egremont's two Spanish fleeces, sheared from rams sent to Petworth by his Majesty, weighed, the first, 5 lb.; the second, 6 lb.; and the wool beautifully fine.

4. Value.

The monopoly of wool by the manufacturers, has had such an effect in depressing it below its real value, that it makes it difficult to form any fair calculation about it. We know that our woollen fabric has flourished very highly, yet the price of wool has sunk. Before the war (1792), the finest South Down wool brought 3l. a tod, and even so high as 3l. 4s. Mr. Ellman sold at that price; Mr. Ellman, of Shoreham, at 2l. 17s.; the Duke of Richmond, at the other end of the county;
at

at 2*l.* 16*s.*; Lord Sheffield's Spanish cross higher than any.

Lord Egremont's flock for 1795, was—Hereford, 50*s.* per tod; South Downs, 42*s.*; Leicester, 24*s.*

The buyers give less for wool that is grown upon land north of the Downs; they have no reason for it, but it affects the price.

The following is the account of John Ellman's flock:

		Weight.		*Price per Tod.*			*Weight of the Fleece.*	
		tod.	*lb.*	*s.*	*d.*		*lb.*	*oz.*
1770	....	54	16	....	31	0	....	2	1
1771	....	56	23	....	32	0	....	2	$3\frac{1}{2}$
1772	....	50	9	....	31	0	....	1	15
1773	....	61	16	....	28	0	....	2	3
1774	....	70	0	....	29	0	....	2	3
1775	....	69	17	....	31	0	....	2	2
1776	....	66	0	....	31	0	....	1	15
1777	....	61	19	....	29	0	....	2	6
1778	....	62	4	....	26	0	....	2	3
1779	....	67	14	....	24	6	....	1	15
1780	....	87	4	....	29	6	....	2	$2\frac{1}{2}$
1781	....	87	14	....	37	3	....	2	$2\frac{1}{2}$
1782	....	67	3	....	37	0	....	2	$3\frac{1}{2}$
1783	....	72	24	....	34	6	....	2	$3\frac{1}{2}$
1784	....	72	18	....	38	6	....	2	5
1785	....	74	20	....	38	6	....	2	2
1786	....	91	21	....	36	6	....	2	5
1787	....	55	10	....	40	0	....	2	4
1788	....	80	9	....	41	6	....	2	$8\frac{1}{2}$
1789	....	91	6	....	40	6	....	2	$11\frac{2}{3}$
1790	....	76	19	....	43	0	....	2	$6\frac{1}{2}$
1791	....	83	8	....	47	0	....	2	12

1792

	Weight.		Price per Tod.		Weight of the Fleece.	
	tod.	lb.	s.	d.	lb.	oz.
1792	81	0	64	0	2	10
1793	—	....	39	0	2	15
1794	—	....	43	0	2	13
1795	—	....	53	0	2	13
1796	—	....	53	0	3	0
1797	—	....	48	0	3	0
1798	—	....	50	0	2	15
1799	—	....	64	0	2	9
1800	—	....	58	0	2	13
1801	—	....	57	0	2	13
1802	—	....	62	0	2	14
1803	—	....	61	0	2	13
1804	—	....	64	0	2	14
1805	—	....	86	0	3	0
1806	—	....	68	0	3	2

The progress of improvement may be seen in the following table :

1793. At this period no polled breed existed, west of Shoreham-bridge to the borders of Hampshire; all the flocks consisted either of Dorsets or Hampshire. Wool was now 19s. per tod : lambs', 6d. per pound.

1774, 20s. 6d.—lambs', 7d. per pound.

1775, 23s.—lambs', 7d. per pound.

1776, 24s. ; wool on the coast, 23s.—lambs', 7d. per pound.

1777, 21s. 6d. ; coarse, 21s.—lambs', 6d. per pound.

1778, 18s. ; coarse, 17s.—lambs', 5d. per pound.

1779. About Shoreham, the quality was improving ; rams from the South Downs were turned into some of the horned flocks; which gradually increased every

every year; 16s. 6d.; coarse, 16s.—lambs', 5d. per pound.

1780, 19s.; coarse; 18s. 6d.—lambs', 6d. per pound.

1781, 22s.; coarse, 20s.—lambs', 5d. per pound.

1782, 22s.; coarse, 20s.—lambs', 5d. per pound.

The wool from Lord Pelham's flock sold this year for 33s.; at Arundel, 21s.; about Shoreham-bridge, 23s.—lambs', 4d. per pound.

1783. The quality and demand increased. Between Arundel and Shoreham, South Down wool sold for 25s. per tod; horned flocks, 21s.; coarse, 20s.—lambs', 5d. per pound.

1784. The same wool as last year now sold for 29s.; horned, 25s. and 24s.—lambs', 5d. per pound :—about Sompting and Findon, 32s.; about Brighton, 37s.

1785. Horned, 26s.; fine wool, 33s.; coarse, 24s. —lambs', 6d. per pound.

1786. Horned, 24s.; fine, 28s.; coarse, 23s.— lambs', 6d. per pound.

1787. Horned flocks, 27s.; fine, 32s.; coarse, 26s.

1788. Horned, 30s.; fine, 34s.

1789. Horned, 30s.; fine wool, 34s.

1770. Fine, 32s.: the wool that was horned, now converted to South Down, from 28s. to 30s.

1791. Fine wool, 37s.; some few horned flocks left, 30s. to 34s.; about Mitchel-grove, Stake, Westburton, Westmarsh, to Arundel, 37s. to 39s.

1792. Fine wool, 48s. to 54s.; coarse, 40s.—lambs', 10d. per pound, in general.

5. *Quality.*

The South Down fleece is composed of a very fine sort of carding-wool, next in quality to the Ryeland.

The

The superiority of which, however, is not so clear, as it appears that they divide the fleece, and separate that which grows upon the thigh, fore-legs and belly, to the amount of a third of the whole, which they sell at an inferior price. This, if true, explains the remarkable high prices at which they have been said to sell*.

Sussex wool is soft and fine, and from three to five inches in length of staple, when stretched out. The finest of it is largely mixed up with the Spanish in the manufacture of broad cloth; the rest is wrought into a coarser kind.

Sussex wool will make a good cloth in light and full blues, and whites, and some other very sound colours; but in olives, snuffs, &c. will not mill to a firm substance of cloth. "We never were in the county of Sussex, but are told the wool of that county varies very much, according to the kind of soil the sheep graze on. Sussex wool being the freest from black hairs of any English wool we are acquainted with, must, on that

* Our wool is little inferior to the Hereford, if they were to sell the whole of the fleece without sorting; a practice not known on the South Downs.—*John Ellman.*

Mr. Campbell, in a letter to Lord Egremont, observes upon this practice: "As Herefordshire is the only county that I know, which continues the practice of *trinding* (or winding the wool in tops, ready sorted, in some degree, for fine drapers), I thought it likely your Lordship might have seen, or have knowledge, of what the practice is; and as many false suggestions and surmises have been published in regard to the Hereford fleece, I thought it might be agreeable to you to have a true state and sample of the business in your possession; have therefore taken the liberty to send you by Drew, a trinded top of wool, being one of the fleeces shorn from one of the ewes you have from Mr. Pantall, with the locks left out of it at trinding, and a card annexed of certification. But I rather think, on recollection, that I omitted to say what weight the different prices were for; if so, please to insert per stone of twelve pounds and a half."

account,

account, be properest for light-coloured kerseymeres;
and for dark-coloured kerseymeres, the same wool is
suitable for them, as for other plain wove cloths of the
same dark colours."—*Extract of a Letter from a
Woollen Manufacturer in Yorkshire.*

Our English wool may thus be arranged, beginning
with the best :

1. Hereford, Shropshire, Upland, Welsh.
2. South Down.
3. Norfolk.
4. Wiltshire.
5. Cambridgeshire.
6. Dorset.
7. Romney.
8. Lincoln.
9. North country.

Quantity of wool has been the chief object of attention with the farmer; and the carcass generally thought
the second point to be considered.   In the west of Sussex, Mr. Pinnix at Upmarden has greatly improved
his flock.   At the outset of his improvement, he was
very particular in his ewes and ewe lambs which he
bought : of the latter he bought a third more than he
wanted for the succeeding year; and when they were
two-toothed, he kept only such as he liked best to breed
his flock from : at the same time he was very careful
and attentive in the choice of good rams.   In 1793, it
was only five years since he began to improve his flock;
yet he had then by great attention, reared in the centre
of the Dorsetshire breed, as fine a flock of South
Downs, both in regard to weight of the wool, and fineness and shape of carcass, as almost any in the county.
His leading object was to give the greatest attention

to

to his rams, and to draw off all that were defective, either in wool or shape, or in any other point. Mr. Pinnix very properly considered the weight of the fleece as a great point to be gained, observing that a fleece of 2 lb. at 20*d.* is greatly exceeded in profit by the coarser but weightier breed. His flock averages at $3\frac{1}{4}$ lb.

It is a fact of the first consequence, that so far from fine wool being incompatible with a fine form, it is clearly ascertained and established, that late improvements have united them together. How far the quality depends upon the food, has already been considered. Upon a sheep-walk the wool will be fine and light; upon pastures coarse and heavy : this is the natural tendency, but breed will counteract it.

There is nothing in which Mr Ellman is more praiseworthy, than in the attention he has given to improve the breed : there is hardly a greater object than this, let the breed be what it may : if very bad, it admits an infinite improvement; and if very good, it may be made better; this is sufficiently obvious, by comparing his flock with the common ones of the Downs: he has by a long and continued attention to the rams he saves, and to refusing his ewes, brought them to be more equal, and remarkably fine woolled. This breed, like others, are apt to be coarse in the breech, which he has very much corrected; and has them less and less so every year. This attention has brought his breed into such request, that he sells his rams at 50 guineas each ; while the common price is from five to six or seven.

In his ideas of breeding, deduced from long observation, he is of opinion, that what he calls the *stain* of the breed, is for some years difficult to remove. A ram and ewe both with fine wool on the breech, will in an ordinary

nary

nary flock, very probably produce a coarse breeched sheep; but if they have been well bred for several generations, then there is good reason (and not else) to rely on the progeny.

To this attention to the breed it is that Mr. Ellman attributes the change in the wool in another respect: it was always usual to consider the value of the wool as nearly in proportion to the number that made a tod; the lighter the fleece, the better the wool. Mr. Ellman's fleeces are the heaviest on the Downs, and yet he gets the highest price: this is an object of prodigious consequence.

That increasing the quantity, does not of necessity hurt the quality, appears from the wool-staplers giving as much for that of fat wethers as of lean sheep: it must, however, be allowed, that they assert it is not intrinsically worth so much, because, when sorted, it yet cannot be made so equal; they agree that it works better in the mill.

Some experiments by Lord Sheffield upon the introduction of the New Leicester and Spanish breeds into the county, deserve insertion.

" About seventeen years ago, I crossed the best South Down ewes with one of Mr. Bakewell's rams, which improved the breed, by giving weight in the fore-parts, and more than doubled the quantity of wool. At first I got extraordinary prices for the lambs; 3s. or 4s. each, more than common; but the South Down breed being excellent, and much improved, I found it gained upon the mixed breed, and they are become very much like the South Downs in appearance; and the weight of the fleece is reduced from five pounds to three pounds. If the mixed breed had continued to suit the fairs of this county, and I had continued to cross with the

Bakewell

Bakewell breed, I still think it would have answered
very well for the Kentish men who attend our fairs, and
for Romney-marsh; but I believe the whole breed of
Bakewell would answer better for the common consumption of London, or of a manufacturing country. Five
pounds of the wool of the mixed breed, although very
indifferent, pays better than two pounds of the fine
wool, which is about the average of the finest South
Down fleeces; and so would seven or eight pounds of
the whole breed of Bakewell, although it is very coarse.

" But on the subject of fine wool, I have much more
to say than I can attempt at present. As an excellent
experiment is in its progress, prompted by the first person in the empire, and encouraged in the manner that
it may be expected from that quarter, I have little doubt
of its proving, that as fine wool may be raised in Great
Britain, and Ireland, as is brought hither from Spain.
You have heard that we do not import the best: I have
a considerable number of the three-quarter, the half,
and the quarter breed of Spain. The ewes were the
best woolled of the South Down and Hereford breed.
In respect to carcass, the three-quarter breed does not
answer, but the wool is of a good staple, of a very fine
quality, and the fleece is at least a third heavier than
the finest fleeces of England.

" Many of the half breed, and all the quarter breed,
are well shaped, and nearly as handsome as the best
South Down, and their fleeces are considerably heavier,
and very much superior in quality to the finest fleeces
of this kingdom; but such is the extravagance of the
monopoly of wool in this kingdom, and its bad effects,
that there is no prospect of an adequate price being obtained for fine wool raised in Great Britain.

" The consequence at last must be, that the wool-
growers

growers will neglect the fine, as they have done in general, and will cultivate the coarse and heavy, which pays much better. But this is a large subject, and I shall only add, that in respect to the above subject, all that can be learned, may be acquired from the President of the Royal Society, who has forwarded the experiment in question with his usual liberality and good sense, and with an accuracy peculiar to himself.

" I have specimens of wool of a Spanish breed brought from Spain twenty years ago, which seems to have preserved in France and England its original quality of fineness. It has been the fashion to suppose a fine quality is derived from the soil. I am satisfied it is only to be expected from peculiar kinds of sheep. A rich soil will increase the weight of a Spanish fleece, but it will make it hairy or harsh, like the coarse wools of England. The difference of climate alone is to be apprehended : a certain degree of warmth and of perspiration, may be of service to produce that softness, and silkiness, which is to be found in the best Spanish wool.

" The instance of Shetland wool seems to remove that difficulty. But I have more to learn relative to the sheep of those islands : some very curious information is likely to be brought forward soon concerning their kindly breed."

So few are the experiments of any consequence, which have been registered upon the comparative merit of various breeds of sheep, that we are, even to the present day, very much in the dark in all that bears relation to the proportion of food to mutton, offal, and tallow, live and dead weight, &c. &c. With a view of ascertaining these essential points, so absolutely requisite to a perfect knowledge of the subject, in the month of August 1795, the Earl of Egremont ordered
                                                                his

his wether lambs which had been lambed the preceding spring, to be put by themselves, into a paddock adjoining the Home-park. His Lordship had

29 South Down ram lambs, of which the twelve best were saved for rams, and seventeen cut for wethers.

25 New Leicesters, of which the six best were saved for rams, and nineteen for wethers.

12 Half-bred New Leicester and South Down wethers, from the same get as the preceding.

7 Romneys, out of ewes of Mr. Wall's, which came in lamb by a ram of his own, and, according to the custom of the Marsh, where they depend entirely upon grass, without any turnips or other artificial food, were not born till May, so that they were nearly two months younger than all the others.

The 55 wether lambs were turned into the paddock in August 1795, and were brought up and examined June 25, 1796, when it appeared that twelve of the South Downs, and all of the half-bred South Down and New Leicester, were marketable; but none of the New Leicester and Romney were in any condition for sale. Ten of the twelve South Downs were sent off to Smithfield a few days after, and fetched 34s. each; and ten out of the twelve half-bred Leicesters and South Downs were sent to the same market a week before, and brought 33s. each. It is necessary to observe, that the half-bred were apparently the better sheep, but they went to a bad market and prices low.

The remaining two of the half-bred, which were in equal condition for Smithfield, were kept back, in order to form a part of the following experiment.

But here it is necessary to pause, because the experiment is already decisive of one fact—that at this age of sixteen months (from March 1795 to July 1796)

those

those two breeds were so much more advanced than the others, that they might be profitably cleared from the land, and a fresh stock sent. It will remain for the future progress of the trial, to ascertain whether such fresh stock would not pay better than continuing the old; and for this purpose we may calculate that the sheep now sold at Smithfield at 34s. with the addition of 3s. for wool, pays for 64 weeks 7d. per pound from their birth. This is a very considerable profit, and if it should turn out, that keeping them much longer is not attended with an advantage somewhat proportioned, it will clearly prove the superior benefit of that breed which may be got rid of at so early an age. And I cannot help further observing, that not one of the Leicester being in any condition to be drawn off in this lot for Smithfield, is most strangely contradictory to assertions without end, that fatting at an early age is almost peculiarly a characteristic of that breed. Now let us proceed with the trial.

	lb.	Ten Weeks Gain.	Gain per 100 lb.
Sept. 7. South Downs,	273	33	13
Leicesters,	258	46	21
Half-bred,	294	34	13
Romneys,	270	34	14

The result is not very different from what might have been expected; for as the Romneys and Leicesters were very much behind the South Downs and half-breds, ten weeks before, it was natural to suppose, that when they did begin to thrive, they would do it more rapidly.

December

December 1, weighed again :

	lb.	Loss in 12 Weeks.	Loss per 100 lb.	
			lb.	oz.
South Downs,	264	9	3	0
Leicesters,	251	7	2	0
Half-bred,	282	12	4	0
Romneys,	269	1	0	6

It is very material, in all experiments of this sort, to note the losses, for it makes a trial double, as it not only shews when the sheep thrive, which do best, but it marks equally when they go backward, which breed is most able to withstand those circumstances which operate against all. The difference is not very material in the above scale. In that lot which did the worst, the loss amounts to about 1d. per week; but it is unfavourable to every lot, that in a period including the best part of the autumn (as sheep ought to thrive deep into November, unless the weather is very bad) none of them should have gained, which they ought to have done considerably. It should however be observed, that their pasture, though good in quality, was bare.

Upon finding this result, Lord Egremont ordered them to be starved for 24 hours, and after such starving, to be turned out for 24 hours, proposing, by thus weighing them, to ascertain the quantity of food eaten, and the quantity voided; his Lordship rightly conceiving, that if, upon a repetition of such experiments, there existed any remarkable superiority, or any material difference between the respective breeds, it might throw some light upon the general inquiry.

South

	Loss by Starving.		Loss per 100 lb.
South Downs, ...	8	....	3 0 oz.
Leicesters, ...	11	....	4 0
Half-bred, ...	17	....	4 6
Romneys, ...	5	....	0 14

They were then turned out, and were twice weighed after twenty-four hours eating each time.

	Gain in the 1st 24 hours.	Gain in the 2d 24 hours.	Total.	Gain per 100 lb.
				lb. oz.
South Downs,	1	6	7	2 10
Leicesters,	6	6	12	4 13
Half-bred,	10	9	19	6 12
Romneys,	—	5	5	1 13

Hence it appears, that the half-bred lost most and gained most; that the Romney lost least and ate least; that the Leicesters lost more than the South Downs, and ate more. Such trials must be repeated many times, before conclusions are ventured to be drawn. How the Romneys, in the first twenty-four hours, could gain nothing, is not to be accounted for, as the weighing was carefully executed.

March 30, 1797. Weighed again, and as this weighing will mark the loss sustained by the severest part of the winter, it deserves particular attention. They were at grass during the whole season.

		Loss in Four Months.		Loss per 100 lb.
	lb.	lb.		lb.
South Downs,	253	11	....	4
Leicesters, ...	214	37	....	14
Half-bred,	253	29	....	10
Romneys, ...	254	15	....	5

It

It is remarkable, that the Leicesters have suffered the most, from which we may very fairly conclude, as far as one trial goes, that the great peculiarity of that breed is by no means what has been contended for, a capacity of supporting itself on little food ; but that, on the contrary, they demand a very plentiful nourishment, and will bear the want of it worse than any of the other breeds. The next in demerit are the half-bred ; the SouthDowns are the best of all.

June 19. Weighed again.

		Gain in 12 Weeks. lb.	Gain per 100 lb. lb.
South Downs,	299	46	18
Leicesters, ...	275	61	28
Half-bred, ...	310	57	22
Romneys, ...	317	63	24

As the period from the 30th of March to the 19th of June, takes in the whole flush of the spring growth of grass, it necessarily forms another interesting period of the experiment. And here the result is remarkable, and a strong confirmation of the preceding observations on the Leicesters; for when in favourable circumstances relative to food, as in the present case, from season, they exceed all the rest. The Romneys, however, approach nearer to them ; and as these had lost in pinching circumstances, much less, their superiority upon these two weighings, seems to be clearly ascertained ; and this will appear the plainer, by comparing the weight of December 1, with June 19.

	Dec. 1.	June 19.	Gain.	Gain per 100 lb.
			lb.	lb.
South Downs,	264	299	35	13
Leicesters,	251	275	24	9
Half-bred,	282	310	28	9
Romneys,	269	317	48	17

The merit of the Romneys, in this stage of the trial, is conspicuous. The South Downs are the next, and the Leicesters and the half-bred are equal.

September 7. Weighed again.

		Gain per 100 lb.
		lb.
South Downs,	316	5
Leicesters,	312	11
Half-bred,	340	8
Romneys,	337	6

Here we observe the Leicesters continue to take the lead throughout the summer. So long as the food is plentiful, they beat all the others : and this part of the experiment goes to prove a most important point, which has indeed been long suspected, that in good situations, no breed is so profitable. Next to these, are the half-bred.

July 4. The remaining five of the Romneys were sent to Smithfield, and brought 48s. ; and August 7, ten of the remaining Leicesters went at 48s. and seven at 40s. So that the profit for two years and two months' food, that is, from May 1795, added to the value of the wool, is 5d. and a fraction per week for the Romneys, and from 4d. to $4\frac{1}{2}d$. for the Leicesters, from the time of their birth. By referring back to the former part of the experiment, it will be seen that the South Downs

and

and half-bred, in 64 weeks age, brought 7*d*. per week profit; and that the Romneys and Leicesters, kept till they were near twice the age of the others (108 the first, and 117 weeks the latter), only gave a profit of 4*d*. to 5*d*. per week. This is a most interesting circumstance, and it manifestly tends to ascertain how much better it would be to the grazier to get rid of these sheep at a younger age, and re-stock his land with those which are most saleable at the earliest.

November 21 to December 25.   Weighed again.

		Gain.	Gain per 100 lb.
		lb.	lb.
South Downs,	320	4	1
Leicesters,	326	14	4
Half-bred,	346	6	1
Romneys,	331	lost 6	lost 1

This perhaps is the most striking period of the experiment. If we turn to the last weighing, it will be seen that the Leicesters had outstripped all the rest, and this superiority is still maintained.

Under the article of fine wool, it will not be deemed foreign to the subject just to observe, that Spanish sheep, in their greasy state, average 6 lb. of wool; and it loses half in the washing. The greatest part of the produce of the fine wool is exported to work up the fabrics of more industrious nations.

Mr. Newland, of Chichester, the great importer of Spanish wool, estimates the produce of Spain exported at 40,000 bags (240 lb. each); 24,000 of it comes to England. Formerly, France took off a much greater quantity than the above: in 1781, 11,000 bags to Rouen only. From the intelligence of Mr. Newland, it appears that the finest fleeces are the growth of old Castile,

Castile, and the province of Leon, which travel to the southward; he attributes the fine staple of wool to that equality of temperature occasioned by the annual emigration of the flocks.   The best flock, and that which is remarkable for containing the finest pile, in any part of that kingdom, is in the possession of the family of Negretta.   The Marquis d'Infantado's flock produces an annual receipt of 14,000*l.* a year, but the other a larger.

7. *Number of Sheep in Sussex.*

Great improvements have been lately brought forward by ploughing up the Downs, and thereby increasing not only the corn and cattle, but likewise the number of our sheep.   It is an erroneous idea to suppose, that the sheep have decreased by ploughing up the native Down, since we have the most decisive proof to the contrary, the testimony of the farmers themselves upon oath.   The Down has been ploughed about Brighton, and in various other places, and the sheep multiplied in consequence.   In ten years, from 1784 to 1793, the flocks about Brighton increased 13,395. About East-bourne, the same; and in no part of the county have they decreased.   To what cause is this assignable? palpably to that which has been brought forward as proof of the contrary—ploughing the Downs.

We have a most correct account of the number of sheep in the county, including the horned flocks on the western, and the Romney on the eastern side, by referring to the Custom-house entries.

Chichester

	No. of Sheep.	Weight. Tod.	Lambs.	Tod.	lb
Chichester	60,983	4537	133,81	444	0
Arundel	30,942	2805	9852	219	5
Shoreham	28,215	2280	10,782	189	11
Brighton	43,258	3182	24,866	336	12
Newhaven	45,605	3247	18,566	278	21
East-bourne	30,638	2207	8274	124	13
Hastings	13,118	1098	4555	95	28
Bexhill	11,785	1351	2271	41	14
Winchelsea	9627	1392	3816	42	2
Rye	67,544	8801	21,212	952	13
	341,745	30,900	113,605	2714	23
	113,605			30,900	
Sheep & lambs	455,350		Tod of wool	33,614	

This includes all the sheep that are registered, and
nearly all that are kept in the county: in this account, however, some should be deducted from the Rye
register, as belonging to Romney-marsh, and some few
as entered from Hampshire; but as the entries hardly
embrace any part of the Weald, where some sheep are
kept, the extra number not registered will make the balance nearly even: the whole county then contains
455,350 sheep; twice this is 91,070: and Mr. William
Gardner's estimate of the number of acres, delivered to the
writer, was 933,360, or very nearly half a sheep per
acre, including all sorts of land—but not more than one-
fourth part of Sussex is stocked with sheep. If the odd
55,350 sheep be taken as belonging to the Weald, the
remaining 400,000 are fed upon 250,000 acres, which
is considerably more than a sheep and a half to each
acre. Some of this land is stocked at a much heavier
rate

rate than other parts. The marshes carry two and
three per acre. The Downs one and a half: so that
taken altogether, it is (soil considered) as high stocking
as can be found in any part of the kingdom.

There are other breeds of sheep in Sussex besides the
South Downs; but as these are the well known breed of
the county, and are a native original breed peculiar to
it, there is the less occasion to be particular as to the
others; as the Kent sheep have had such justice done
them by Mr. Boys, and the others will no doubt be properly attended to.

Others are, besides Romneys, Hampshires, Dorsets,
Wiltshires, and Somersets: the Earl of Egremont has
introduced the Hereford, of which breed there is a very
noble flock in Petworth-park. His Lordship has likewise introduced the Notts, from the neighbourhood of
Tiverton in Devonshire, a stock very much resembling
the New Leicester. The ewes rise to 20 lb. per quarter; wethers as high as 30 lb.: they are generally white
faced, but larger boned, longer in the legs and body,
and not so broad in the back: 18 lb. of wool has been
shorn from a ram of this breed: the carcass coarser
than Dorset, and the wool 2d. per pound cheaper.
Lord Egremont has in his park three large flocks of the
Hereford, South Down and Dishley; and it is a rather
curious circumstance, that these three flocks keep themselves perfectly distinct and separate from each other,
although each has as much opportunity of intermixing
with the other as they have with themselves.

Besides these, his Lordship has imported, through the
medium of his Excellency Count Orloff, the Kalmuc,
and Astrackan breed: their chief peculiarity is, that
in the place of a tail they have a very large projection
of fat, or rather a kind of marrow of most exquisite
delicacy:

delicacy: their fleece is a short but not coarse wool, but with hair growing through it. Lord Egremont has also the shaul goat of Thibet: from the fleece of which the finest and most valuable manufactures of the East Indies are made. In 1796 it was in its perfection: they are not shorn, but the wool is combed off: about a pound to a fleece: the hatters give a guinea a pound for it*.

SECT.

---

* If it is possible to give you any information about wool, that would tend to put the breeders of sheep in this county on thinking of the advantage which might be derived by attention, without its being construed into sounding my own praise, I should be highly gratified. I will just state to you what improvement might be made by a little attention, in the article of wool only, without mentioning the carcass. This is what you collected from the Custom-houses at East-bourne, Newhaven, Brighton and Shoreham, the number of fleeces and the weight entered at those places, and which includes what is here called the South Downs and part of the Weald, a space of about twenty-six miles in length adjoining the sea coast, and ten miles from the sea. The South Downs do not extend so far from the sea: but by the Wool Bill, all the wool shorn within ten miles of the sea coast must be entered at the several Custom-houses. The South Downs, or what is here generally understood to come under that name, extends from East-bourne on the east, to Shoreham river on the west. And by the late Survey, by Yeakell and Gordners, is about twenty-six miles in length, and a little more than five miles in width. I have taken it at six miles from the sea, as most of the farmers on the North side of the Downs dip nearly a mile into the Weald. On this district the South Down flocks are kept, although some small flocks are kept off the Downs and within ten miles of the sea, and which are entered at the several Custom-houses before mentioned; but taking the Downs at twenty-six miles long and six wide, contains 99,840 statute acres; on which there are kept in summer, about 180,000 sheep and lambs, and 120,000 in winter; which is (deducting lambs put out to keep in winter) something less than two sheep per acre in summer, and one and one-fifth per acre in winter. You will be pleased to observe, that, full one-half of this district is arable land, and the other grass land. But here I am going from what I was about to notice, i. e. the advantage to be gained by attention in improving the breed. Taking the whole

SECT. III.—HORSES.

THE horses employed in the husbandry of the county, have nothing in them which deserves particular notice. Consi-

---

whole quantity of wool shorn between East-bourne and Shoreham river, and within ten miles of the sea, according to the entries in the years 1788, 1789, 1790, 1791, and 1792, average at 134,041 fleeces per year, weight 9490 tods 13 lb.; from which I will deduct my own flock, 1007 fleeces per year, which average at 82 tods 8 lb. or a small fraction short of 2 lb. 10 oz. per fleece, the account will stand thus:

	Fleeces.		Tods.	lb.
Total entry per year,	134,041,	weighed	9490	20, of 32lb.
From which I deduct my flock,	1007		82	8 —
Which leaves	133,032		9408	12

Average at 2lb. 10 oz. per fleece.

      2   4   per fleece, a difference of 6 oz. per fleece for 133,034 fleeces, is 1559 tods of wool per annum, which I will set at 3s. per tod less than I sold for, for five years.

1788	- - - - - - - -	£. 2 1 6
1789	- - - - - - - -	2 0 6
1790	- - - - - - - -	2 3 0
1791	- - - - - - - -	2 7 0
1792	- - - - - - - -	3 4 0
	Average - - - - -	£. 2 7 4

1559 tods at 47s. 4d. is 3325l. 17s. 4d. per annum, or 16,629l. 16s. 8d. for the five years of which this account is taken. I will observe, that I set the price of my wool at 3s. per tod above the average of this district; which I am inclined to think is under done.

By this statement, you will see what a prodigious loss of property to the sheep-masters in this small district, for want of attention to the article of wool only, the which, if I was to add the difference of price, would

Considerable numbers are annually bought up at the fairs and markets, which come from other places. As the business of the farm is more profitably conducted with oxen, the only material inquiry is the comparison.

Those who have worked oxen, and are well acquainted with their powers of draught, know that they are equal to horses in the tillage of a farm: therefore the fair way to put the case between them, would be to compare them singly, horse against ox; but in order to make an allowance every way equal, it shall be admitted, that eight oxen can only plough as much land as four horses.

---

would be nearly equal to the above statement, which is confined to weight only; to which let us add the carcass, and here I flatter myself is the greater loss by far, as I will be bold to say, that the price I sold my ewes and lambs for, the last ten years, is more than 10s. per lamb above the average of the whole Down; which, if added to quality and weight of wool, would even surprise you.

My reason for the above statement, is to show at one view, the great advantage to be gained by attention; and with hopes of seeing my neighbours take a share in so important a matter to this country.

With respect to this sheep account, you will see, my dear Sir, that I had two objects in view; one for the purpose of shewing the quantity of sheep kept on the South Downs, and the number per acre, and which you will see I have done by deducting about one-tenth for what is kept off the Downs in the Weald, and within ten miles of the sea; and this I believe is stating the matter fairly. The other is, to shew at one view, the advantage which my neighbours might partake of by a little attention in the article of wool only.—*John Ellman.*

Eight

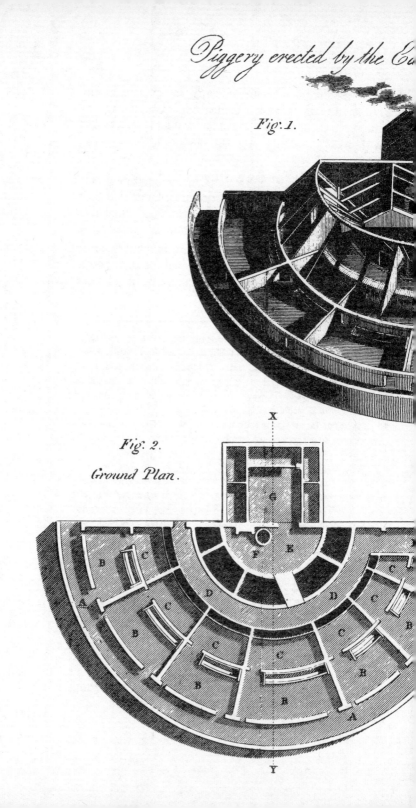

Piggery erected by the Ea

Fig: 1.

Fig: 2.

Ground Plan.

remont at Petworth.

*Perspective View
from an elevated situation*

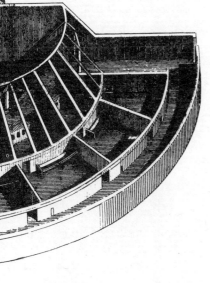

Fig. 3.

Section thro' **XY.**

A B C D d F G

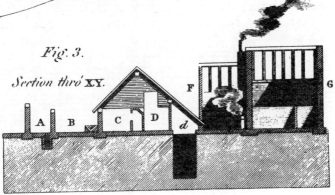

AAAA. *Passage of entrance to the Sties.*
BBBB. *Open Pounds with the Troughs.*
  CCC. *Lodging places with fatting hutches.*
    DD. *Internal passage to ye whole.* **d**. *Cisterns for food.*
      E. *Inner Court.*
      F. *Boiling house.*
      G. *Granary.*

10   20   30   40 feet.

Scale for Fig's 2 & 3.

Engraved by Neele.

Eight oxen, at 12*l*.	£.96	0	0
Yokes and chains for six, ...	4	4	0
Six summer months' work, at 2*s*. per week,	20	16	0
Ditto winter, at 2*s*. 6*d*.	26	0	0
	£.147	0	0
If they rest two or three months, they }	8	0	0
may afford a profit of 8*l*.			
	£.139	0	0
Four horses, at 25*s*.	£.100	0	0
Harness, at 45*s*.	9	0	0
Oats, 2 lb. per week, ...	52	0	0
Hay and herbage, at 6*s*. per week, ...	15	12	0
Farrier, wear and tear, ...	4	0	0
	£.180	12	0
An ox-team will plough one acre per day the whole year through, at 42*s*. per week; six acres, at 7*s*. .... }	£.105	0	0
Horse-team the same, ...	105	0	0*
In favour of oxen, ...	£.41	12	0

This

The common allowance to a farm-horse in winter, or while they are kept in the stable (which is seldom less than 30 weeks), from the beginning of October to the beginning of May, per week, two bushels of oats, at 2*s*. 6*d*. - - }	£.0	5	0
We give little hay, mostly pease-haulm, or straw, which I value, with the hay, - - - - -}	0	2	0
Per week, whilst kept in the stable, - - -	£.0	7	0
The 22 weeks in summer, or whilst he is out of the stable, we give about one bushel of oats per week, -}	£.0	2	6
If turned out to grass, and which is usually done in Sussex with the working oxen, I value the pasture per week,}	0	2	6
We generally give some hay when first taken up in the morning, and before turning out in the evening, and which I set at per week, - - - -}	0	1	6
Per week in summer, when turned out to pasture,	£.0	6	6

If

This too, calculating the work according to the common custom, that of yoking three and four times over
the number it requires to plough the same ground with
horses : but if a pair of Sussex oxen in harness, are able
to turn up an acre of stiff loam in Suffolk to a proper
depth with an iron plough, with as much ease as any
pair of horses, and this too for nine or ten months in
twelve, the same work may certainly be performed in
other counties, where the soil is not peculiarly tenacious; and the average of England is not of this description.

Expenses of eight horses. .... £. 361 4 0
Ditto eight oxen, ... 147 0 0

In favour of oxen, ... £. 214 4 0

If this balance be thought high, some deduction may
be made, by allowing the oxen two or three months'
rest in the year ; and perhaps something further should
be allowed for additional provender; but whatever the
deduction, the balance is greatly in favour of the ox-
team ; and when it is considered that the older the ox,
the greater his value, and *vice versa* with the horse, it
then becomes apparent enough : the one, at his death,

If soiled with tares or clover, a horse will consume about
  one square perch of tares, or two square perches
  of clover, of a middling crop, per day; the tares   £ 0 3 6
  value at 6d. per perch, and the clover at 3d. or per
  week,   -   -   -   -   -   -   -

One bushel of oats per week,   -   -   -   -   0 2 6

When soiled, cost per week,   -   -   -   -   £. 0 6 0

John Ellman.

gives

gives beef for mankind, the other is horse-flesh for dogs.

Is not this a circumstance, if every other was wanting, sufficiently engaging, when we reflect too, that the immense quantity of beef which would be brought to market, would then sink the price, and allow the poor cottager to partake of a diet more hearty and nourishing than his present humble fare ? Such a change in the circumstances of that class of people, is too striking to need any comment.

Sussex possesses a breed of beasts equal to any in England : they are in general use ; but the system is not carried to that extent which it might, for every little farmer has his horse team. The substitution of oxen instead of horses, is extremely to be desired ; but in the county under review, where the tillage of the farm has for ages been done by means of them in a great measure, the extension of the benefit, to the utter exclusion of horses, would not create that difficulty in the execution of the system, which it would necessarily have in those counties where the practice is unknown*.

SECT.

---

* I am no advocate in favour either of horses or oxen ; but for that mode of business that can be done with the most ease and expedition, and with the least expense.

I have on my farm some as strong and heavy land as any in the kingdom, and some land as light ; and three horses, with the allowance of two bushels of oats per week each horse, are able to plough an acre in the heaviest and strongest land, if it has ever been broke up into tillage ; and will plough it up to any depth from four to eight inches deep ; breadth according to fancy, if it is ploughed at a proper season of the year, as such land requires ; and when it has been ploughed a second time over to the full depth of the first ploughing, the two horses, in the spring and summer months, will plough one acre and a half per day in that land that has been before twice ploughed ; so there is a spare horse

for

SECT. IV.—HOGS.

The hogs of Sussex are either descended from the large Berkshire spotted breed, or from a cross between that

---

for harrowing seed in, if sown broad-cast, or for any other use the farmer may require. I plough with the horses both double and single, whichever answer the business best; and average the heavy and light soils together on my farm. A three horse team will average to plough the year through seven statute acres per week; which at 7s. per acre, is 49s. per week; and have a spare horse for eight weeks out of the team. My plough with cast iron mould-board (I have them of different strength, according to the work and the land to plough in; and I can put a single wheel to any of them, if the land requires it, and can occasionally put a pair of double wheels to any of them), are drawn by a chain fixed to the axis of the wheels, which chain is fastened to the end of the plough, so that a good ploughman will plough his ground all truly up, and to one depth, let the land lie ever so uneven; which is not the case with most of the wheel-ploughs.

And three horses, at the following expense of keep, &c. will average to plough seven acres per week the year through; and will allow to take one horse out of the team for eight weeks, either for rest, or any other business the farm may require.

	£	s	d
Three horses, at 25l. each,	75	0	0
Harness, at 4l. 4s. each,	12	12	0
Oats, at six bushels per week, for six months,	19	10	0
Oats, at three do. per do. for do.	9	15	0
Hay for six months, at 1l. 1s.	27	6	0
Grass and green crop for six months, at 15s. per week,	19	10	0
Wear and tear of two ploughs, a year,	3	3	0
Ditto horse, year,	1	5	0
Horse shoeing,	1	11	6
Farrier,	0	15	0
Cost and outgoings of horse team,	170	7	6
Ditto of ox team,	147	0	0
Balance in favour of ox team on first cost,	£.23	7	6

that and a smaller black, or white breed.   The Berkshire hog rises from 50 to 80 stone, and some of this
                                                    sort

To the income of the horse-team for one year, 49s. per week, £.127	8	0
To profit on two young horses, 1l. per year each, - -	2	0	0
	£.129	8	0
To the income of the ox-team for nine months, 30s. per week, 54	0	0
To profit on the oxen, - - - - - - -	8	0	0
Take - - - - - - £.62	0	0
From - - - - - -	129	8	0
Remains - - - - - -	£.67	8	0
To balance in favour of the ox-team's out-goings, - -	23	7	6
Balance in favour of the horse-team for one year, -	£.44	0	6

The above statement is what a horse-team will do on any farm in England, where they have proper implements, and properly applied.—*Mr.*
Harper, Bank-hall, Liverpool.

This statement is not founded on the general system of the county, but
upon calculations brought forward as the result of Mr. Harper's private
practice, and is what a horse-team will do upon any farm in England,
where they have proper implements, and properly applied : the annual
balance in favour of the horse-team is 44l. 6s. estimating a team of oxen
at eight, and the other at three ; but as these eight will make four teams
as well as one, and that too upon strong loam, his ideal balance sinks at
once.   Yet this is practice, and what a pair of oxen will do upon most
farms in England, where they have proper implements, &c.

Two Sussex oxen in harness will plough daily an acre of strong land
with as much ease as two horses.   I say they will do it, because several
have done it for years in Suffolk, and the fact confirmed.

The food created for the consumption of a numerous and increasing
population, is surely of *some* consequence, especially when we contem-
plate the origin of scarcities, and consider the land set apart for the
maintenance of an unproductive live stock.   The food eaten by the ox
is not wasted to raise up an animal for the dog-kennel. The multiplication
of these has never, like horses, been the cause of scarcity, but every
blade of grass consumed is so much beef produced.   The country does
not import near a million of quarters of corn, to feed our oxen with
oats.   No man has encouraged the rearing of oxen in preference to horses
with such spirit as Lord Egremont.

                                                    Three

sort have been killed which exceeded 100 stone: the Cantam breed weighs fat from 15 to 25 stone; the cross much approved. The whole breed are thought tender; an erroneous idea.

A very large stag hog of the Berkshire breed was lately fattened by Mr. Dale, miller at Petworth, and brought in a waggon to be weighed in the market-place.

	cwt.	qrs
The hog and the waggon at 112 lb. per cwt. weighed	28	0
Waggon,	21	2
Hog,	6	2

Weight of the hog, 6 cwt. 2 qrs. which, at 112 lb. per cwt. is 728 lb. or 91 stone.

His offal was:

	st.	lb.
Blood,	1	6
Guts,	4	0
Flay,	6	7½
Caul, sweetbread, &c.	0	6
Heart and lights,	0	7
Liver,	1	2
Crow,	0	7½
Loose fat,	1	4
Hair,	1	6
Pisser,	0	2
	20	0

Three per cent. upon the annual rent, is returned at the spring audit to any of Lord Egremont's tenants who shall, during the year, have done the whole work of their farms with oxen, and who shall not have used any horses for draught upon any land which they shall have occupied, of their own, or belonging to his Lordship, or to any other person.

They will not be entitled to the drawback, if they have not paid their half-yearly rents regularly upon the days of the audit.—*A. Y.*

Brought

				st.	lb.
Brought forward,				20	0
	st.	lb.			
Flay,	6	7½	}	7	3½
Caul,	0	4			
				12	4½
Weight of the carcass,				66	7
Flay and caul,				7	4
				74	3
				12	4½
				86	7½
Wasted,				4	0½
				91	0

Another hog, fatted by Mr. Dale at the same place.

			st.	lb
Weight of the hog alive,			113	0

Weight sold, viz.

	st.	lb.	st.	lb.
Carcass,	82	7		
Flay,	9	4¼		
Caul,	0	6		
Total weight sold,	93	1¼	} 95	5½
Weight of the crow,	0	6¾		
Fat taken from the entrails,	1	4¼		
Sweetbread,	0	1¼		
Weight lost in killing,			17	2½

Entrails, maw, caul, crow, fat, and sweetbread weighed, when taken out of the hog,		71 lb.
Caul,	6	} 44¾
Crow,	6¾	
Fat from the entrails,	12¼	
Sweetbread,	1¼	
Entrails and maw,	18½	
Lost in cleaning,		26¼

Liver,

Liver, ... 6½
Haslet, ... 9½
Blood, ... 20
Bladder, &c. 3½
Hair, ... 14

The hog was shut up to fat 28th August, 1797, and killed 8th March, 1798, being 192 days fatting.

The fatting was meal of pease, barley and oats, in the whole 78 bushels, which at the market prices amounted to .... £.15 13 3

The hog when killed amounted to 15 3 0

Balance against William Dale, who fatted the hog, reckoning nothing for the ori-} 0 10 3
ginal value of the hog, ...

N. B. The value of the fatting, per week, was, on the average, 11s. 5d.

Hogs are either fattened upon barley, pease, oats, or potatoes : the two first frequently mixed together : five sacks of barley and one of pease, will fatten a hog of 60 or 70 stone : one bushel of pease to four of oats, and four of barley, or three or four bushels of potatoes, with two bushels of ground oats and barley boiled, is a good mixture. An average sized hog of 50 stone, will eat two bushels in a week, and if a fair thriver, gains two stone in that time.

Lord Egremont has tried a great variety of hogs, and made many experiments, to determine the most profitable food, which is barley ; the white hog for store and grazing, is the best he has yet tried. They are killed after summer grazing in the park ; and it is a most advantageous method : no corn is given : nothing but grass. They are turned out in May, and in October and November brought to the slaughter-house, and die good porkers. This is a curious experiment, and deserves further trial.

In this experiment the hogs ranged over an extensive park. In another trial made, they were confined in a cage, exactly fitted to the size of the animal, which was augmented as the hog grew larger; and no more space allowed him, than what was sufficient for him to lie down upon his belly.

As there were some hogs that we wanted to keep over the summer, seven of the largest were put up to fat on the 25th of February; they were fatted upon barley meal, of which they had as much as they could eat. Some days after, the observation of a particular circumstance suggested the following experiment: a hog nearly of the same size as the seven, but who had not been put up with them, because they appeared to be rather larger, but without weighing them, was confined on the 4th of March, in a cage made of planks, of which one side was made to move with pegs, so as to fit exactly the size of the hog, with small holes at the bottom for the water to drain from him, and a door behind to remove the soil. The cage stood upon four feet, about one foot from the ground, and was made to confine the hog so closely, that he could only stand up to feed, and lie down upon his belly. He had only two bushels of barley-meal, and the rest of his food was boiled potatoes: they were all killed on the 13th of April, and the weights were as follow: (8 lb. to the stone.)

The

	St.	lb.
The hog in the cage	13	2
	12	2
	12	3
	11	2
The other hogs, all of the same breed	11	4
	11	4
	11	2
	12	2

The hog in the cage was weighed before he was put in: alive 11 stone 1 lb.; he was kept five weeks and five days, and then weighed alive 18 stone 8 lb.; he had two bushels of barley-meal, and about eight bushels of potatoes. He was quite sulky for the two first days, and would eat nothing.

This is a most singular result, and as the hog thus confined was so much superior to all the others, though not equally fed, it can scarcely arise from any other circumstance but the method adopted: it is extremely curious, and deserves to be farther examined in a variety of trials.

His

Lord Egremont has tried rice for hogs, and compared it with barley, in the following experiment.

January 1, 1798.	March 16.	Dead.
No. 1. 36¼lb.	155 lb.	114 lb.
2. 47½	228	172
5. 56	252	194
8. 45	200	166

	St.	lb.		
No. 1.	14	2	at 3s. 4d. per stone,	£.2 7 6
2.	21	4		3 11 8
5.	24	2		4 0 10
8.	20	6		3 9 10
				£.13 9 10
				2 17 8¼

Four hogs bought in, at ....

	St.	lb.		
	4	4½	at 2s. 6d. per stone,	£.0 11 3
				0 14 10½
				0 17 6
				0 14 0¾
				£.2 17 8¼

There remains .... £.10 12 1¾
Barley consumed, ... 8 12 2

Gain, ... £.1 19 11¾

	January 1, 1795.	March 16.	Dead.			
No. 3.	93 lb.	201 lb.	114 lb.	93 lb. (11 st. 5 lb.), at 2s. 4d.	£.1	7	1½
6.	94	191	154	94 lb. (11 st. 6 lb.), at 2s. 4d.	1	7	5
7.	186	220	167	186 lb. (23 st. 2 lb.), at 2s. 4d.	2	14	10
4.	89	182	149	89 lb. (11 st. 1 lb.), at 2s. 4d.	1	5	11½
					£.6	15	4

	St.	lb.			St.	lb.		
No. 3.	14	2, at 3s. 4d. per stone,	£.2	8	2	5	6, at 2s. 6d. per stone, is £.0	14	3½
6.	19	2	3	4	2	5	7	0 14 7¼
7.	20	7	3	9	7	11	5	1 6 6¾
4.	18	5	3	2	1	5	4½	0 13 9
			£.12	4	0			£.3 9 2½

Four hogs bought in, at ....

There remains .... £.8 14 9½
The rice consumed, ... 12 2 11

Loss, ... £.3 8 1½

In the above experiment, the rice was given dry; in the following trial it was boiled with water.

Names.	Nov. 29.	Dec. 30.	Gained alive.	Weight when killed.	Loss, dead, on each hog.
	St.	St.	lb.	St.	lb.
Blackhead, ...	142	166	24	127	39
Longhead, ...	162	200	38	154	46
Spot, ...	160	176	16	143	33
White, ...	146	164	18	136	28
Chubb, ...	148	178	30	138	40
Spotbox, ...	142	168	26	126	42
Whitebox, ...	136	178	42	136	42
Dumpling, ...	118	131	13	97	34
Slimslack, ...	128	145	17	117	28
Blackside, ...	141	162	21	128	34
	1423	1668	245	1302	366

Total weight of the hogs, Nov. 3, when put up to fat on rice, was } 1045

Gained from November 3, to December 30, } 623

Total weight when killed, } 1302, at 7*d.* per pound, £.37 19 6

Rice consumed by the above hogs, was } 3033, at 1*s.* 4*d.* per pound, 15 15 11¼

Balance, ... £.22 3 6¼
Supposing the hogs bought in at 4*d.* per pound, 17 8 4

Gain, ... £.4 15 2¼

SECT. V.—RABBITS.

THIS stock is the nuisance of a county; they flourish in proportion to the size of the wastes, and are therefore productive in Sussex. From Horsham forest and Ashdown, &c. considerable quantities are sent to London.

SECT. VI.—POULTRY.

NORTH CHAPPEL, Kindford, &c. are famous for their fowls. They are fattened here to a size and perfection unknown elsewhere. The food given them is ground oats made into gruel, mixed with hogs'-grease, sugar, pot-liquor, and milk; or ground oats, treacle, and suet; also sheep's plucks, &c.; and they are kept very warm: they are always crammed in the morning and at night. They mix the pot-liquor with a few handfuls of oatmeal; boil it: it is then taken off the fire, and the meal is wetted, so as to be made to roll into pieces of a sufficient size for cramming: the fowls are put into the coop two or three days before they begin to cram them, which is done for a fortnight, and then sold to the higlers. They will weigh, when full grown, 7 lb. each, and are sold at 4s. 6d. and 5s.; the average weight, 5lb.; but there are instances of these fowls weighing double this.

Mr. Turner, of North Chappel, a tenant of Lord Egremont's, crams 200 in a year. Many fat capons are fed in this manner; good ones always look pale, and waste away: great art and attention is requisite to cut them, and numbers are destroyed in the operation.

The

The Sussex breed are too long in their body, to cut them with much success, which is done at three quarters old. The Darking fowls, as they are called, are all raised in the Weald of Sussex; but the finest market for them is Horsham. The five-clawed breed have been considered as the best sort: this however is a great mistake, and it took its origin in some fowls with this peculiarity, that happened to be very large and fine, which laid the foundation of what have been since called *the Darking*, or *five-clawed fowl*, and considered in other parts of England as the prime stock; but such a thing is hardly known in Sussex; it is a bastard breed. The fowls at Lord Egremont's table, of the Sussex breed, have very frequently astonished the company by their size.

SECT. VII.—PIGEONS.

THE dove-house is sometimes attached to the farm-house, especially on the Downs; but they are not propagated to any extent.

SECT. VIII.—BEES.

BEES are the cottager's stock as well as the farmer's, and some profit is made by them, but too trifling to merit attention. It is only in desert countries, like the North of Europe, where wax is made an object of commerce, and an article for exportation. Dungeness light-house, on the coast of Kent, is surrounded on the land side with beech and gravel for several miles, and

it

it is so deep, that hardly any thing grows on it. By much industry, the man who lives in the light-house has enclosed a small patch of ground for a garden, and he keeps some bee-hives: the quality of his honey is excellent, and from each hive he gains 12 lb. in favourable years.

SECT. IX.—FISH.

This is an object of some consequence in Sussex. The ponds in the Weald are innumerable; and numbers of them date their origin from that part of the county having once been the seat of an extensive iron manufactory, which has now deserted the country; and the mill-ponds now raise large quantities of fish. A Mr. Fenn, of London, has long rented, and is the sole monopolizer of all the fish that are sold in Sussex. Carp is the chief stock; but tench and perch, eels and pike, are raised. A stream should always flow through the pond; and a marley soil is the best. Mr. Milward has drawn carp from his marl-pits 25 lb. a brace, and two inches of fat upon them, but then he feeds with pease. When the waters are drawn off and re-stocked, it is done with stores of a year old, which remain four years: the carp will then be 12 or 13 inches long, and, if the water is good, 14 or 15. The usual season for drawing the water, is either autumn or spring: the sale is regulated by measure, from the eye to the fork of the tail. At 12 inches, carp are worth 50s. and 3l. per hundred; at 15 inches, 6l.; at 18 inches, 8l. and 9l.: a hundred stores will stock an acre; or 35 brace, 10 or 12 inches long, are fully sufficient for a breeding pond.

pond. The first year they will be three inches long ; second year, 7 ; third year, 11 or 12; fourth year, 14 or 15. This year they breed.

Mr. Biddulph has, in Burton-park, a fine reach of water, which yields carp, tench, perch, pike, &c. in great abundance ; and as it is an extensive pond, I shall insert some account of the produce.

March 10, 1789. Number of fish taken out of the Mill-pond, of 50 acres.

Carp,	1517
Tench,	473
Pike,	806
Perch,	50
	2846

March 12. Number of fish taken out of Chilford-pond.

Carp to Crouch-stews,	200
Carp to Mill-pond,	1800
Tench to Mill-pond,	274
Tench returned from Crouch-stews to Mill-pond,	180
Carp returned from Crouch-stews to Mill-pond,	70
Pike returned from Crouch-stews to Mill-pond,	500
Total number of fish put into the Mill-pond, March 12, 1789,	2824

March 19.

Carp from Trout-pond to Crouch-stews,	300
Small tench,	80

Fish sold, March 10, 1789.

	12-inch Carp.	14-inch Carp.	16-inch Carp,
To Richard Fenn,	1062	550	20
To ditto,		247 tench	46 tench
To ditto,		148 pike—400 cwt.	

Sold

Sold to sundry people.

William Windle,	50 pike,	150 perch.
Shenfe, ...	50 carp,	16 pike, 20 tench.
Hodik, ...,	12 ——	
Earl of Neald, ...	12 ——	
Mrs. Budd, ...,	20 ——	
Lady, ...	6 ——	
Aling, ...	14 ——	
Milford, ...	100 ——	
Ditto, ...	24 ——	
Kent, ...	4 ——	

April 16, 1789. Small store carp bought of William Milford, Esq. 6300, or five gallons.

One gallon put into Lawn-pond, for stock.
Two gallons put into Chilford-pond, for ditto.
Half a gallon to the Briant-stew, for ditto.
Half a gallon into each of the other stews, for ditto.

January 28, 1790. Bought of John Serjeant, Esq. 4600 store carp, from four to six inches.

March 1, 1791. Account of fish taken by Fenn.

60 14-inch carp, at 5l. 10s. per 100, £.3	6	0	
740 12-inch ditto, at 4l. per 100, ... 29	12	0	
400 12-inch tench, at 4l. per 100, ... 16	0	0	
100 under-size ditto, at 3l. per 100,	3	0	0
500 ditto carp, ... 11	11	0	
March 3. 335 pike, 784 lb. at 6d. 19	12	0	
14. 52 12-inch carp, } at 4l. per 100, 3	0	0	
25 12-inch tench, }			

£.86 1 0

March

March 5, 1791. Account of fish put into Mill-pond.

500 pike.

422 tench, returned from Crouch-stews.

1355 carp.

2277

March 5, 1791. Account of store carp put into Chilford-pond and Trout-pond.

5287, from five to six inches long.

October 15, 1792. Store carp from New-stew to Troutpond, 217.

October 12, 1791. Store carp from Trout-pond to Mill-pond, 130.

Small pike, ... 14

Upper stew, ... 200

Lower, ... 320

February and March, 1793. Account of fish sold, and to whom.

Mr. Fenn.

735 carp and tench 12 inches, at 4*l.* per 100, £. 29   8   0

130 ditto 14 ditto, at 5*l.* 10*s.* per 100, ...   7   3   0

5 ditto, 16 ditto, at 8*l.* 8*s.* per 100, ...   0   8   5

125 tail tench at 3*l.* per 100, ...   3 15   0

840 lb. of pike, at 6*d.* per pound 21   0   0

61 14   5

Sold to sundry persons, {
1 17   6
1 14   6
1 12   0
6 15   0
1 14   6
0   9   6
0   6   0

£.76   3   5

Stock

Stock fish carried to different ponds in 1793, to the Mill-pond.

Carp from Chilford-pond, ... 1854

Ditto returned from Crouch, ... 71

Ditto ditto from Trout-pond, ... 193

Ditto ditto ditto Upper-stew, ... 144

———

2262

———

Tench from Chilford-pond, ... 33

Pike from different ponds and stews, ... 552

To Chilford-pond.

Carp, 372 from seven to eight inches long, from Trout-pond and Lower-stew.

29 from nine to ten ditto, from Crouch-stews.

4 from 14 to 16 ditto, ditto.

350 from six to seven ditto, from Lawn-pond.

Tench, 123 from four to five inches, ditto.

21 from seven to eight ditto, from Crouch-stews.

100 from three to four ditto, from ditto.

Carp, 300 from eight to nine ditto, from ditto.

1299

———

1793. Stock fish carried to different ponds.

To the Trout-pond from Lawn-pond.

Carp, 193 from six to seven inches.

Tench, 37 from four to five ditto.

Carp, 20

——

250

———

Stock

Stock fish from Mr. Baker.

Carp,   157 three inches and upwards.
Tench, 1772 from two to three ditto.
                —————
                 1929
                —————

Feb. and March, 1793. Fish sold to sundry people, 19s.

March 14, 1795. Bought of John Biddulph by Fenn.

1180 carp 12-inch, deduct 24—1156 at 4l. £.46   4   9
107 14s. (fourteens) deduct 2—105 at 5l.   5 15   6
98 pike 240 lb. at 6d. per pound, ...   6   0   0
                                              ——————————
                                               58   0   3

To cash received of sundry per- ⎰ 26 2 0
sons, ... ⎱ 6 3 0
                                          3 11 6
                                        ————————35 16   6
                                              ——————————
                                               93 16   9
                                                2   2   0
                                              ——————————
                                            £.95 18   9
                                              ——————————

March, 1795.   Fish carried to Mill-pond.

	Carp.	Tench.	Pike.
Fountain-stew, ...	68	91	0
Chilford-pond, ...	1292	104	11
Lawn-pond, ...	10	115	0
Old Stable-pond, ...	0	30	0
Stew Lodge-coppice, ...	20	0	0
Crouch-stew, ...	0	0	239
From Mr. Baker, ...	310	50	0
	1700	390	250

March,

March, 1795.   Fish for the house.

20 carp.

20 pike.

14 tench.

9 perch.

Fish to Trout-pond from Crouch-stews.

184 carp.

24 pike.

Stock fish to Chilford-pond.

Carp from stews, ... 103 .... Tench, ... 843
From Mr. Serjeant, ... 1600

                        1703   843
Fish to Fountain-stew, 820   373

Mill-pond fished March 2, 1797.

Fish taken. Carp, ... 1631

             Tench, ... 226

             Pike, ... 1095

                                        2952

Mill-pond stocked, March 4, 1797.

Pike,   300 returned.

Carp, 1700 ditto.

Tench, 270 ditto.

The

The following is the account of a pond of Sir John Shelley's, at Mitchel-grove.

Particulars of Coppice-pond, fished February 15, 1798, one acre 30 poles.

	£	s	d
125 carp, weighing 266 lb. at 6d. per pound,	5	10	10
22 ditto, weighing 30 lb. at 4d.	0	10	0
	6	0	10
Store fish sold,	0	13	0
	6	13	10
Value of fish preserved in a little pond, in case of company,	0	15	0
	7	8	10
Supposed expenses of fishing, &c.	0	8	10
*3)7	0	0
	2	6	8

besides a sufficiency to stock two ponds.

Lord Egremont has several noble ponds for breeding, and others for fattening, one immediately under another, with streams running through them. They are fished every third year; and the best reserved for the stews; but none sold.

1793. Nov. 6. Fished the Frith-pond. Brought home 172 carp (the selling price of these fish would be from 7l. to 8l. per hundred) 300 tench (8l. to 9l. per hundred) and 100 perch.

1799. March 26. Brought home 225 carp, 200 tench, 50 pike, 140 perch.

* Ponds are fished every three years.

The

The tench remain for two fishings in Lord Egremont's ponds, as they are a slow growing fish. At the last fishing of this pond, 300 store tench, and as many store carp were put into it: the stores are worth 10s. 6d. a hundred. Male tench are good for nothing, and are thrown away: hen tench only are preserved. It is rather difficult to mark the distinction between the male and female. Male perch are known by the appearance of milk upon squeezing: tench by the thick fin of the males. If the water is good, about 70 two-year old store carp, and as many tench, are a fair allowance for one acre*.

SECT.

* The monopolies in this useful branch of commerce is a very great nuisance to the public, by sending almost all the fish to the capital; so that though our coasts abound with an inexhaustible supply, the country receives but very little benefit, as the natives that reside six or seven miles from the coasts are as destitute of fish as if they lived in the interior of the kingdom. I should imagine that, by making a law to divide upon the strand whatever fish is caught, one half for the London markets, and the other for the consumption at home, it would give great encouragement to the fisheries, and exceedingly increase that useful body of men, so necessary to our defence in manning the fleets. I think it would be good policy to oblige every market-town in the county to keep open a fish-market; there would be no fear of buyers, if such an useful regulation were adopted, nor would there be any fear of a supply. Judge then what an increase of hands would be employed, to what there is now, under its present restrictions.—*Mr. Fuller.*

This Note recurs to the idea, that London devours the produce of the country, which it starves, and so raises the price of provisions. But *Quere*, Whether it is true, that in proportion as this is the case, is the flourishing state of our country? What implies the high price of any commodity, but an increased demand? Population multiplied, consumption doubled, trade, manufactures, agriculture, circulation, all increased; communication between the capital and the provinces laid open; new people, ideas, and exertions created: it is these which raise the earth's products, and are consequently the pillars of agriculture, and

the

SECT. X.—DEER.

No experiments have been made or registered respecting deer: the following are the particulars of two that were killed, among many others at Petworth, 1796.

Weight alive, ... 196 lb.

	lb.
Fore quarter,	34
Ditto,	32
Haunch,	30
Ditto,	23
Kidney-fat,	2
	108

	lb.
Chine and other offal,	16
Blood,	7
Loose fat,	$3\frac{1}{4}$
Horns,	$5\frac{1}{4}$
Head,	$8\frac{1}{2}$
Legs,	3
Liver and lights,	7
Guts and paunch not weighed, but remaining at	$29\frac{1}{4}$—88 lb.

the supports of our industry. If the inhabitants of London were scattered over England, the country would be a comparative desert. It may be assumed as a fact, that *no cheap country was ever a rich one.*

I do not by this mean, that regulations for the supply of the country are improper, but only that they ought to be so framed, as not to impede the supply of London: such, I should conceive, might easily be devised.

Another:

Another :

Live weight, ...	194¼ lb.
Butcher's weight, like a sheep, ...	129
Fifth quarter, ...	65½

This buck was far from being a very fat one, and in the fifth quarter the horns weighed seven pounds, and were at that time full of blood vessels. If the buck had not been killed till the end of August these would have been dried up, and would not have weighed more than four or five pounds, and the buck would have been much fatter. He was not weighed alive, but was shot, and weighed immediately, and the blood was received into a pail, and the weight added to the carcass.

It deserves inquiry, in what degree this is an unprofitable stock : it would not be difficult, by certain experiments, to ascertain the exact degree of their benefit or demerit.

CHAP. XIV.

RURAL ECONOMY.

———◆———

SECT. I.—LABOUR.

THERE are in most counties three descriptions of labourers—domestic servants, task-workers, and weekly labourers. The first class are the least numerous, but best provided, generally unmarried ; the last description are the most numerous and necessitous. The wages of servants vary from 7*l*. and 8*l*. to 11*l*. a year. Task-workers will earn upon a medium from 1*s*. 6*d*. to 2*s*. a day ; perhaps the average is 2*s*. or near it. The weekly labourers from 16*d*. to 18*d*.

The price of labour is above the medium of many other counties; in the neighbourhood of the sea are seen many old labourers, as the young and active find smuggling a more lucrative employ, which is very successfully pursued in Sussex. At Rye and Hastings, Bourne, &c. it is highly flourishing, whilst the health of the inhabitants is injured, the revenue defrauded, and labour extremely high. It has been computed, that the revenue in this line of country is cheated to the amount of 80,000*l*. per annum : between 3 and 400,000 gallons of gin, rum, and other spirits, are annually smuggled into this district. The principals engaged in the business have about 10*s*. 6*d*. each night : the common men a guinea a week ; and in the conveyance from the

vessel

vessel to the shore, from 2*s.* to 7*s.* per night: 12,000 gallons of spirits have been landed in a week at Dungeness, in Kent. Light goods from Flanders, into Sussex and Kent, 105,000*l.* a year, upon which the profits have been so high as 30,000*l.*

This great consumption of spirits is very pernicious to the labourers, and equally injurious to the farmers: but the cheapness of gin recommends the sale of it, and unlicensed gin-shops are without number.

The price of labour in Sussex, is in some measure according to the local situation. The standing price is lower on the western side of the county than it is in the eastern: it has advanced in half a century about thirty per cent.

The following is a table of the prices of labour in different parts of the county.

A Table

A Table of the Prices of Labour.

	Cuckfield.	Hamsey.	Kitchinam.	Salehurst.	Battel.	East-bourne.	Applesham.	Selsey.	Arundel.	Average.
	£. s. d.	£. s. d.	£. s. d.	£. s. d.	£. s. d.	£. s. d.	£. s. d.	£. s. d.	£. s. d.	£. s. d.
In winter, - -	0 1 4	0 1 6	0 1 6	0 1 6	0 1 4	0 1 6	0 1 6	0 1 4	1*s.* 4*d.* to 1*s.* 6*d.*	0 1 5
Summer, - -	0 1 6	0 2 0	0 2 0	0 1 8	0 1 6	0 2 3	0 2 0	0 1 6	0 1 6	0 1 9
Harvest, - -	0 2 0	0 2 6	0 3 0	0 2 6	0 2 3	0 3 0	0 2 6	1*s.* 9*d.* to 2*s.*	0 2 6	0 2 4¾
Reaping wheat, -	8*s.* to 9*s.*	0 8 0	0 9 0	0 8 6	7*s.* to 10*s.*	8*s.* 6*d.* to 9*s.*	8*s.* to 10*s.*	0 7 0	0 8 0	0 8 5½
Oats, - -	1*s.* 6*d.* to 2*s.*	0 1 6	1*s.* 8*d.* to 2*s.*	0 1 6	1*s.* 8*d.* to 2*s.* 2*d.*	0 1 4	1*s.* 2*d.* to 1*s.* 6*d.*	0 2 0	0 2 6	0 1 8¾
Barley, - -	1*s.* 6*d.* to 2*s.*	0 1 6	0 2 0	0 1 6	1*s.* 8*d.* to 2*s.* 2*d.*	0 1 4	1*s.* 2*d.* to 1*s.* 6*d.*	0 2 0	0 2 6	0 1 8¾
Pease, - -	0 3 0	0 3 3	0 3 6	0 3 0	0 3 0	2*s.* 9*d.* to 3*s.* 6*d.*	0 3 6	0 3 0	0 3 0	0 3 1¾
Mowing grass, -	0 2 0	0 2 6	0 2 6	0 2 3	0 2 8	2*s.* 3*d.* to 2*s.* 6*d.*	0 1 9	0 2 0	0 2 0	0 2 2¾
Clover, - - -	0 1 6	0 2 0	0 2 0	0 2 2	0 2 0	1*s.* 6*d.* to 2*s.*	0 1 6	0 1 6	0 1 6	0 1 6¾
Hoeing turnips, -	0 5 6	0 6 0	5*s.* 6*d.* to 6*s.* 6*d.*	0 5 6	0 7 0	0 6 0	4*s.* to 7*s.*	0 4 9	0 5 6	0 6 0¼
Thrashing wheat, -	0 3 0	0 2 8	0 3 0	0 2 9	0 3 0	0 2 3	2*s.* 9*d.* to 3*s.* 6*d.*	0 1 6	0 2 6	0 2 7½
Barley, - -	0 1 8	1*s.* 6*d.* to 1*s.* 8*d.*	0 1 6	0 1 6	0 1 6½	0 1 4	0 1 4	0 1 4	0 2 0	0 1 6¼
Oats, - -	0 1 0	1*s.* to 1*s.* 4*d.*	0 1 6	1*s.* to 1*s.* 4*d.*	0 0 10	0 1 4	0 1 0	0 1 0	0 1 8	0 1 2
Pease, - -	0 1 6	0 2 0	0 1 6	1*s.* 4*d.* to 1*s.* 6*d.*	0 3 0	0 2 0	0 1 6	0 1 4	0 2 6	0 1 9¾
Women in winter,	0 0 6	0 0 6	0 0 7	0 0 8	0 0 6	0 0 7	0 0 6	0 0 6	0 0 8	0 0 6½
Summer, - -	0 0 9	0 0 10	0 0 8	0 0 8	0 0 7	0 0 8	0 0 8	0 0 7	0 0 10	0 0 7¾
Harvest, - -	0 0 10	0 1 0	0 0 10	0 0 10	0 0 8	0 0 10	0 0 10	0 0 9	0 1 0	0 0 10
Yearly earnings, -	25 0 0	30 0 0	30 0 0	28 0 0	26 0 0	30 0 0	30 0 0	28 0 0	29 0 0	28 8 10½
Rent of cottage, -	3 0 0	3 3 0	2*l.* 10*s.* to 3*l.*	3 0 0	2*l.* to 3*l.*	2*l.* 10*s.* to 3*l.*	3 0 0	2*l.* 10*s.* to 3*l.*	3 0 0	2 16 1

The reaping, mowing, hoeing, by the acre; thrashing, by the quarter.

Rate of Labour, Wages, Clothing, &c. &c. in Sussex, and in Suffolk. Communicated by Lord Egremont.

The supposed sum of money earned by a labourer in one year, in Sussex.

	£	s	d
52 weeks' common labour, at 9s. per week,	23	8	0
For his harvest month, 2l. 10s.; from which deduct 9s. per week, his common wages, which is 1l. 16s.; deducted from 2l. 10s. he clears	0	14	0
Suppose saved in his board 3s. per week, which he must live on, if at home,	0	12	0
For three weeks at hay-making, for mowing he gets 2s. per day, instead of 1s. 6d.: it is more than in general is earned on that score,	0	9	0
Suppose by barking, which does not fall to the lot of one in a hundred, in three weeks he earns three guineas, out of which deduct per week for lodging, and extra living to support his hard labour, 5s. per week; 15s. from three guineas, leaves	2	8	0
	27	11	0
Suppose by bad weather, slight illness of short duration, and his loss of time, is six days in the course of the year, he loses,	0	9	0
Total,	27	2	0

This calculation amounts to 10s. 5d. per week.

A husbandman may earn in winter 9s. per week, from St. Michael to Lady-day, and in barking season from

from 12s. to 20s. per week, for three weeks, and mowing grass from 12s. to 18s. per week, for four or five
weeks, and from 40s. to 55s. for harvest month, and
the rest of the time about 9s. per week for the year.
The wife may earn in winter about 15s. or 20s., weeding
and hay-making about 20s., and gleaning corn in harvest, and raking oats and barley, about 20s. If a boy
about eight or nine years old, 3d. per day; 11 or 12
years old, 6d. per day. Girls earn but little in winter:
weeding or hay-making, 3d. or 4d. per day.

SUFFOLK.

*Communicated by Mr. Capel Lofft to Lord Egre-
mont, March, 1797.*

Estimate of what it would cost to clothe a family, viz.
a man, woman, and five children, the eldest under
twelve years old.

Man, kersey waistcoat and breeches, ... £.0	15	0
Woman, red gown and two coats, ... 0	13	1¼
First boy, waistcoat and breeches, ... 0	11	0
Second ditto, ditto, ... 0	8	0
First girl, red gown and two coats, ... 0	8	0
Second ditto, ditto, ... 0	7	0
Third ditto, ditto, ... 0	5	0
32 yards of cloth for the family, ... 1	17	4
14 pairs of stockings, ditto, ... 1	0	0
Seven pairs of shoes, ditto, ... 1	6	0
Seven hats, ditto, ... 0	12	0

£.8   2   5½

Communicated from the parish of Hepworth, in Black-
bourne hundred, Suffolk, March 6, 1797.

Mr.

Mr. Capel Lofft to Lord Egremont, 1795.

In my letter in the Annals, I had stated wages in this year, 1795, as generally in this neighbourhood 16*d*. per day. They were so early in the season, but they soon became 18*d*. and have so continued.

The statement of the earnings of the same number of persons as I have stated for 1773, at the same age of the children in the present year 1795, will be:

Man, 26 weeks, at 16*d*. per day, ... £. 10	8	0
—— 21 ditto, at 18*d*. ditto, ... 9	9	0
—— 5 ditto harvest, including malt, ... 6	0	0
Advantage by job-work, ... 1	12	0
	£. 27	9	0
Boy at 12 years of age, ... 9	12	0
Girl at 10 ; spinning, pease and wheat-} dropping, gleaning, &c. ....} 4	0	0
Wife, ... 1	10	0
	£. 42	11	0

Mr. Capel Lofft to Lord Egremont, September, 1795.

Twenty-two years ago (1773), I learn that wages and price of corn were thus in this part of Suffolk :

Price of wheat, from 24*s*. to 28*s*. the coomb of four bushels, Winchester measure.

Wages in 1773.

Winter, 1*s*. per day.

Summer, 1*s*. 2*d*. ditto.

Harvest, 3*l*. 10*s*. or 12*s*. Scarcely any malt, or allowance for entertainments during harvest.

Total

Total earnings, taking harvest at 3*l*. 12*s*. with the above wages, will be found, I believe, as under:

26 weeks, at 6*s*. per week, ... £. 7	16	0
20 ditto, at 7*s*. ditto, ... 7	0	0
Harvest*, ... 3	12	0
Job-work†, ... 1	12	0
	£.20	0	0

---

* Harvest is mostly taken by job, at about 20 acres per man, to see it into the barn.

† Supposing labourers to work most part of their time by the piece, or at job-work, which is the case here, at which they usually earn considerably more than by day-wages, this advantage may be fairly set at 1*l*. 12*s*. per year, as above stated.

If the expenditure of flour be taken at three-fourths of a bushel, and the price at that time at 24*s*. per coomb, the annual out-goings for flour only, is 11*l*. 14*s*. There must then be calculated the expense of clothing, and other necessaries for the family.

To balance this expenditure against the earnings, as they then were, the average earnings of the rest of a family, supposed to consist of six persons, must be taken into the account.

On consulting on this subject, I state them thus:

Boy of 12 years, - - - - - - £. 8	8	0
Girl of 10, spinning, gleaning, &c. - - - 4	8	0
Wife, - - - - - - 1	10	0
	£.14	6	0

Two infant children gain nothing.

Man, - - - - - - £.20	0	0
Wife and children, - - - - 14	6	0
Total earnings of the whole family, - £.34	6	0

If the annual expenditure of this family, two of them only being infants, be taken in flour at its then lowest price, and at a bushel per week, it will be 15*l*. 12*s*. Taken at 7*s*. per bushel, 18*l*. 4*s*.

This would have been the statement in this county (Suffolk) in the year 1773.—*Capel Lofft to Lord Egremont.*

SECT.

411

SECT. II.—PROVISIONS.

The high price of many of the necessaries of life is an object of great consequence: political arithmeticians and calculators have quarrelled for more than a century, whether or not the price of labour is in proportion to the price of provisions. That it should have been ever doubted is surprising.

Most clearly the wages of labouring families are inadequate to support them in that comfortable condition which they are entitled to expect: it is evident from the general increase of rates; but far more so to any man who thinks it no disgrace to visit the dwellings of the poor: their clothes, their bedding, diet, fuel, and cot; and when the interior of the cottager's house is inspected, will it be made a question whether labour and provisions are upon a par? 'Tis absurdity to question it. A labouring family, honest as they may be, and industrious as their strength and activity renders them (if numerous), it is hardly possible they can be maintained upon the present wages of labour with ease and comfort.

A Table

A Table of the Price of Provisions, &c. 1793.

	Cuckfield.	Hamsey.	Kitchinam.	Salehurst.	Battel.	East-bourne.	Applesham.	Selsey.	Average.
	£. s. d.	£. s. d.	£. s. d.	£. s. d.	£. s. d.	£. s. d.	£. s. d.	£. s. d.	£. s. d.
Flour per gallon	0 0 11	0 0 11	0 1 0	0 1 0	0 0 11½	0 0 11½	0 0 11½	0 1 2	0 0 11¾
Peck-loaf	0 1 1½	0 1 1½	0 1 0	0 1 0	0 1 2	0 1 0	0 1 0	0 1 3	0 1 1½
Cheese per pound	0 0 6	0 0 6	0 0 6	0 5¼d. to 6d.	0 0 6	0 0 6	0 0 5½	0 0 6	0 0 6
Butter per pound	0 0 9	0 0 9	0 0 9	0 0 8	0 0 9	0 0 9	0 0 8	0 0 9	0 0 8¾
Pork per pound	0 0 8	0 0 7½	0 0 9	0 0 8¼	0 0 8	0 0 7	0 0 5		0 0 7¼
Bacon per pound	0 0 7	0 0 8½	0 0 8½	0 0 8½	0 0 8½	0 0 9	0 0 6½		0 0 7
Malt per bushel	0 6 6	0 6 6	0 6 6	0 6 6	0 6 6	0 6 6	0 5 9	0 6 4	0 6 4½
Brush-faggots per load*	1 0 0	18s. to 20s.	18s. to 20s.	0 17 0	1 2 0	1 4 0	1 3 0	1 3 0	1 0 8
Potatoes per bushel			18d. to 20d.	0 2 0	0 1 0		0 1 6		0 1 6
Cord-wood				0 18 0	1 0 0		1 4 0		1 0 8

* A load is 100 faggots. A common family consumes 300, and a cord of wood (14 feet in length, three high, and three wide). Some families consume ten bushels of coal, in addition to the above, per annum.

SECT. III.—FUEL.

COAL or wood, in a few places turf is used. The woods are very extensive; yet the price has greatly increased: great quantities are made into charcoal, and still larger (of the smaller sort) burnt for lime.

It is sincerely to be lamented, that some steps are not taken by those who have it in their power, to convert the present method of warming the cottagers' houses at a large expense of fuel, by recommending or substituting a cheaper and more effectual plan, according to the excellent idea laid down by Count Rumford, in his Experimental Essays; by adopting which, a great expense of fuel might be saved, and the houses more effectually heated: and in cooking the food, by a small alteration in the construction of the stove or grate, and the fireplace, the Count's ingenuity has so contrived it in a very simple manner, that a copper that holds 50 or 60 gallons, may be kept boiling for several hours at a comparatively trifling expense of fuel, by confining the heat to its proper place, and allowing none to be wasted; and by throwing a flue in a different manner, that when it once becomes heated, which is very soon done, a very small quantity of wood or coal is necessary to keep it in that state. Coppers have been hung upon the Count's plan by Mr. Poyntz. Every grate in Lord Egremont's house is completely Rumfordized.

But in the country, where fuel enters so largely into the expenses of living in a cottage, it is to them an object of immense consequence, to make the smallest
quantity

quantity contribute in the best manner, and last as long
as it can; but the expensive mode of burning it in the
present unsystematic and unphilosophical method
adopted by almost all ranks of the people, cannot be
too universally reprobated.

It is greatly to be wished, that an improvement so
excellent might be encouraged, and its benefit extended.
To those who are unacquainted with the Count's plan,
it may be necessary to observe, that the great size of the
throats of chimnies in general, is the principal objection in the construction of them, which serves as a
passage for the warm air to escape up the chimney,
and the loss of heat is consequently great—but the
warm air is not only lost, but the room filled with cold
air from without, which is well guarded against by
diminishing the size of the throats of the chimney. A
few shillings expended in the purchase of the Count's
Experimental Essays, will add a very valuable store
of knowledge eminently useful to the purchaser; but
it is particularly incumbent on all those who are desirous of contributing to ameliorate the condition of the
lower class of life to read this work, and see at what a
trifling expense great numbers of people can be fed
upon a nourishing and wholesome diet, and how cheaply
they can be warmed*.

Saving of fuel.—I have known a leg of mutton and turnips boiled in
a wooden pail. The trick was thus performed : a six feet barrel of a
fowling-piece was inserted at the muzzle in the pail, the other end
placed against the fire; the water flowing to the breech of the barrel,
the whole was made to boil. *Quere,* might not furnaces and vessels
be heated in different rooms by the kitchen-fire only, by means of
tubes of cast iron, with a large butt at the end conveniently fitted to be
heated?—*Mr. Trayton.*

Undoubtedly all the rooms of a house might be equally heated,
by

by conducting the heat through flues, in the way it is done in the North of Europe and Germany. The equal temperature of a room, is an object of vast consequence to the health and comfort of the owner. At present, we are roasted near the fire, whilst we are almost frozen every where else.—*A. Y.*

CHAP. XV.

POLITICAL ECONOMY,

As CONNECTED WITH AGRICULTURE.

———————

SECT. I.——ROADS.

THE turnpike roads in Sussex are generally well
enough executed: the materials are excellent: whin-
stone, the Kentish rag, broken into moderate sized
pieces. Where this is not found, or not used, the
roads are not so good; though turnpikes are numerous
and tolls high: in some places in the east they are narrow and sandy. From Chichester, Arundel, Steyning,
Brighton, Bourne, the roads to the metropolis, and
the great cross road near the coast, which connects
them together, are very good.

Before Shoreham-bridge was founded, the communication to the West of England was very troublesome
and inconvenient, and at high water very dangerous;
but building this bridge (by a tontine) has essentially
contributed to the general benefit of the county; though
the tolls are scandalously high: for every four-wheel
carriage 2s. is exacted; and for every horse 3d. besides
a halfpenny for every foot passenger, and all this every
time of their passing.

This is a grievous imposition on the public.

The cross roads upon the coast are usually kept in
good order: the gravel or sea beech keeps them firm
and

and dry, but not binding : but in the Weald, the cross roads are in all probability the very worst that are to be met with in any part of the island : yet it is affirmed that they have been considerably improved.

The transport of vast loads of timber, corn, &c. through a heavy clay soil (for there is no bottom) renders them nearly impassable in winter for wheels of any description; and in dry weather the hardness of the clay is very prejudicial to the feet of the cattle. As there is no bottom for the felloes to move over, the wheels are frequently buried up to the nave, to the great damage of waggon and horses.

Good roads are an infallible sign of prosperity; but so indifferent is the state of the Weald respecting its husbandry, arising partly from the predilection which gentlemen have for their shaws and woods in a very stiff soil, that to have good roads is hardly possible. It is the free circulation of the wind upon the road, which takes off the moisture the very hour it falls, that so essentially contributes to this desirable end; and it is the want of this requisite that renders them so bad.

The forest-like appearance of this part of Sussex is such, that it cherishes every drop of rain that falls, by sheltering the roads from the wind and sun, and preventing the absorption of the waters.

The 44 miles of turnpike from Bury St. Edmunds to Huntingdon, is perhaps the very best road in England; yet hardly a hedge to be met with. All around Newmarket, but particularly to Cambridge and Bury, no road in the world surpasses it; and the expense of mending them is so trifling, that much of the materials have laid there since the road was first formed. The goodness of it is owing to two circumstances: 1st, the dry nature of the soil: 2dly, the openness of the country.

try. Now as the first is not to be had in Sussex, there is the more necessity for the second ; and when we consider that the materials are at hand, it is the more to be wished.

But those detestable screens of hedge-row must be extirpated, before any improvement is to be expected; its utility is obvious enough ; yet where shall we find any thing useful that is not violently opposed!

There is such an instance of the benefit of a turnpike-road at Horsham, as is very rarely to be met with : the present road to London was made in 1756 ; before that time it was so execrably bad, that whoever went on wheels, were forced to go round by Canterbury, which is one of the most extraordinary circumstances that the history of non-communication in this kingdom can furnish. The making the road was opposed—for what measure of common sense could ever be started that would not be opposed ! It was no sooner completed than rents rose from 7s. to 11s. per acre: nor is there a gentleman in the country who does not acknowledge and date the prosperity of the country to this road; and the people who were the greatest opposers of it, are now so convinced, that there is a general spirit of mending their cross-roads by rates.

A justly celebrated French writer, the Count de Mirabeau, has lately questioned very much in detail, the advantages of great cities to a country. Such an instance as this is surely sufficient to do away many of his objections. Before the communication with London, low rents, low prices, a confined consumption, and no improvements: open the communication, and high rents, high prices, a rapid consumption, and numerous improvements: yet from the frontier of the county by sea, there was always an open communication : the instance

instance therefore is the more striking : it is to be attributed not to the power of carrying heavy loads by land to London, but to the general impetus given to circulation and fresh activity to every branch of industry : people residing among good roads, who were never seen with bad ones, and all the animation, vigour, life, and energy of luxury, consumption, and industry, which flow with a full tide through this kingdom, wherever there is a free communication between the capital and the provinces.

——◆——

SECT. II.——CANALS.

THE advantages which England has derived from extending its inland navigation, have been prodigious ; and to agriculture it has been no less beneficial than to manufactures and commerce. When we consider that the power of a nation is in proportion to her industry, and that her industry is multiplied as markets for the products of her soil are encouraged ; whatever has a tendency to enlarge them deserves universal encouragement. Though it be true that Sussex has hardly the shadow of any thing that deserves the name of a manufacture, yet the advantages which the county has received, and is likely still farther to gain from increasing her navigation, will be very considerable. The principal productions of Sussex are :

1. Corn,
2. Timber, bark, charcoal,
3. Chalk, lime, marl,
4. Iron, marble, limestone,
5. Cattle and sheep, hides and wool.

It

It is evident that all these articles, most of them of a very heavy and bulky nature, can either be exported from the county, or transported from place to place within it, at a much less expense by water than by land; and consequently that both farmer and landlord are equally interested in the management of such useful designs.

Not only the above-mentioned articles, but every other the produce of the farm, have their consumption and value increased by a more speedy conveyance to the place of their destination. In like manner, the productions of other districts are imported with equal advantage. Very large quantities of timber, as well in its rough state as in scantling, were formerly sent from Sussex, by a tedious land carriage, to the coast. Considerable quantities travel the same course at present, but the length of the land carriage is not equal to what it used to be; though the consequence to the roads is such, that they are in many places almost impassable.

Cordwood for charcoal, and oak-bark, are exported in considerable quantities; and the value and consumption of all these articles enhanced by Lord Egremont's canal.

Lime is an article for which there has always existed a great demand: it is now carried by land-carriage from the Downs to various parts of the county; so that the farmer raises his crop of wheat in many places at an expense of five or six guineas per acre in the article of manure only. The navigation of the Rother has opened a market for this as well as other valuable articles, at a much less expense to the farmer, and increased the consumption many thousand ton.

Sussex is a corn county, and produces over and
above

above what is sufficient for the supply of her inhabitants. Much goes to Portsmouth, and the west; and some to the east; and a still greater quantity will be shipped off, whenever the communication shall be more completely opened between the Weald and the sea.

The Arun is navigable from the sea to its junction with the New Cut, 17 miles 3 furlongs; and from thence a company of merchants extended it as far as Newbridge. The first cut nearest the sea, called in the plan the *new canal*, is a mile and three quarters long, and has a tunnel of about a quarter of a mile $13\frac{1}{4}$ feet wide, and as much in height, which cost 6000*l*. The new canal is 26 feet wide at top, and four deep, having three locks. The circuitous navigation by Greatham and Pulborough, was too shallow to be navigable at all times of the year; but the tolls on the New Cut, drive the trade to its old channel, though they have of late partially fallen. The conveyance upon the new canal is practicable at times when the other is too shallow for any barge to travel; and as the canal is but a mile and three quarters, whilst by the river the line is five and a half, the difference is considerably in favour of the first: the trade, however, usually passes along the old navigation, except in summer, or in floods. From the end of the canal to Palingham quay, three miles, the river is navigable; but from thence to Newbridge another cut has been made by the same company, at the expense of 15,000*l*.

Timber, plank, and all sorts of convertible underwood, are sent from the Weald, and the barges return with chalk, coal, or lime, at a less expense than what the same articles were purchased for before the navigation was effected; and the roads about Newbridge, and

in

in the line of the canal are much improved, and less
expensive to mend. The barges are three sorts: the
largest carry 30 ton; the second size 25; the smallest
15 ton: the second are the best. The passage from
Little Hampton to Newbridge is two days and a half,
using a horse: the tide flows 17 miles of the way, and
by going through Hardham tunnel, the barges save six
hours of time.

In order to extend the benefit of water-carriage to
other parts of Sussex, the Earl of Egremont lately procured an Act of Parliament, enabling his Lordship, at
his own sole expense, to make the Rother navigable
from its junction with the Arun, as far as Midhurst;
and by a collateral branch to Haslingbourne, within
half a mile of Petworth. The Rother joins the Arun
at Stopham, and is now navigable to the sea, as the
subjoined plan will more fully explain. His Lordship
means to extend it to Hamper's-common, close to the
town: it has eight locks in the line from Midhurst to
its fall into 'the Arun, and five from Haslingbourne,
making 35 feet of lockage from Stopham-bridge to Haslingbourne, which when it comes to be continued to
the common, will add 51 more, altogether 86 feet fall,
and 52 from Midhurst to the Arun.

By this most useful and public spirited undertaking, many thousand acres of land are necessarily rendered more valuable to the proprietors. Timber is
now sent by water. Large falls have been exported
which would scarcely have been felled; and the Government Agents and Contractors have made large pur-
chases, in consequence of a more easy communication to the sea. An additional tract of country is also
supplied with lime, from the Houghton and Bury pits;
and when the collateral branch joins Hamper's-com-
                                                      mon,

mon, the whole country, which is at present supplied
with chalk from Duncton-hill, will take it from Petworth at a cheaper rate. At least 40,000 ton is annually
sent from the Houghton pits, in consequence of Lord
Egremont's improving the navigation of this part of the
county.

At present the farmers in the Weald take their chalk
from Duncton; but it has happened that they have
been thrown out of this manure by the state of their
roads. Having no other, the greater demand is for
this. It is to be observed, that the farmers generally
manure their land for the crop of wheat almost as often
as they sow that grain, and lay upon an acre at the rate
of from 80 to 120 bushels. The vast benefit therefore
of facilitating the transport of so necessary a commodity, is too obvious to dwell upon; and when the waggons can go to Hamper's-common for chalk, each team
will return with three or four ladings, according to the
distance which now takes only one in a day.

Another considerable benefit to the country, arises
from the facility with which coal is freighted and carried through the heart of the Weald, which has been
the means of extending the consumption of this article
in lime-burning, and proportionably lessened the demand for furze; for the generality of the farmers set
apart a few acres for the growth and cultivation of this
plant to feed their kilns, which are giving way to coal-
kilns, as a cheaper and more expeditious method of
burning; and that land which is at present used in cultivating furze, can in future be sown with grain, according to the distinction which nature has drawn, that the
bowels of the earth should warm us, and the surface
feed us.

Let us for a moment reflect upon the advantages
which

which result from the employment of between one and two hundred workmen, all natives of Sussex. In the usual method of cutting canals, these men are a constant nuisance to the neighbourhood, and the terror of all other descriptions of people. But in Lord Egremont's canal, the men are all drawn from amongst his own workmen, and have none of that turbulence and riot with which foreign workmen are inspired; and as these labourers use implements equal to the best navigation diggers, the employment of domestic workmen is an evident advantage: and still farther, the expenses of the job are much less to the employer, whilst the weekly wages of the men in this business, instead of 8s. or 9s. rise up to 14s. or 15s. They are now a set of men who have been so long accustomed to the employment, as to be ready to engage themselves in any work of the sort.

In the navigation of the Rother, the course of the river was adopted in preference to a canal. In this instance, it affords a safer and easier passage to the sea than if a canal had been made, since the fall of water is gradual, and the current gentle at all times of the year. In numberless instances, and especially in the one before us, it would have been a great loss of labour and expense to have cut a canal along the side of the Rother at an immense expense of digging, banking, bridges, sluices, tunnels, &c. when at a much less expense, and to better effect, the river has been made navigable, and without any apprehension of overflowing its banks, which has been effectually provided against.

The great superiority of a canal over a river, for navigation, consists in its not being so subject to the violence of inundations and torrents; but this enemy has in the Rother been converted into a friend, for the

drains

drains can be opened at will, and all the adjoining meadow land irrigated to the great benefit of the proprietors, which is done by stopping the drains. In canal navigations, a disadvantage arises from their cutting asunder, as they always do, one part of an estate or a farm from the other, to the great injury of the owner.

The limits of parishes are not seldom bounded by rivers, and in this respect the advantage in favour of a river is obvious; and by banks and dams that are well constructed, and drains to let off the superabundant waters, it is easier and more expeditious to render a river navigable where the fall of water is gentle, and no particular cataracts or steep descents are in the way of the undertaking.

One great advantage that the county derives from the navigation of the Rother, is, that it is vested in the hands of a single proprietor living upon the spot, and who having a large property in the county, is the more interested in the prosperity of the undertaking, and feels a greater spur in the success of it, than any company of merchants who live at a distance and subscribe their money.

By vesting the undertaking in the hands of an individual, no opposition is likely to be met with; nor is the business liable to be thwarted or counteracted.

A considerable part of the original plan still remains to be carried into execution: it is, to connect London with Sussex, and to lay open that market to the produce of this county, and receiving its goods and merchandize in return. By a direct communication from Petworth to Guildford, by a collateral branch to Horsham, a very considerable proportion of the county would be benefited: the ground has been surveyed, and
the

the levels taken ; and if ever it should be effected, the value of estates would in many places be more than doubled.

From Hamper s-common to Stonebridge-wharf is 23 miles, making 133 feet of lockage : the collateral branch to the town of Horsham, 12 miles, is upon a level. In the intervening country between Duncton-hill and Godalming, which is 20 miles, there is no chalk but what is brought from either of these places, so that the water communication would supply Surrey as well as Sussex, as all the timber, and all the productions of the soil, as well from that part of Surrey as from Sussex, would then go to the London market by a very short inland navigation, which is now sent by a circuitous passage along the coast; and what is of still greater consequence, this cut would take off corn which now goes to Guildford by land at a great expense.

It is impossible not to feel great respect, in contemplating the energy of an individual of the highest rank and fortune, animated with such ideas, and expending his income in so meritorious a manner, forming navigations, rewarding industry in the lower classes, improving the breeds of live-stock by bounties, encouraging all useful and mechanical artisans; setting on foot multiplied experiments to ascertain the comparative merit of different agricultural implements; introducing improvements, by extending the knowledge of new plants, animals, or implements, all of them in so many and various shapes contributing their assistance to national prosperity. The thought of one man having been instrumental in the improvement of his country, and still exerting himself in the same career, must be a constant fund of gratification to every benevolent mind;

and

and that long may he live to enjoy the fruits of his labour in the service of his country, is the wish of every man in the county.

SECT. III.—FAIRS.

MARCH
12. S. Bourne.
14. Seaford.

APRIL
5. Midhurst.
Ditchling.
Lambrust.
Hailsham.
W. Tarring.
14. Cat-street.
18. Gardner-street.
22. Rushlake.
Tolesfield.
25. Crowborough.
Loxwood.
29. Wadhurst.
30. Newick.

MAY
1. Egdean.
Hoo.
Hursterpoint.
3. Cowden.
4. Chichester St. Geo.
Henfield.
Nutley.
Ticehurst.

6. Lewes. Cliff.
8. Crawley.
9. Hoathley.
Horsbridge
Old Tye.
12. Alfriston.
Burwash.
Lindfield.
14. Storrington,
Arundel.
Uckfield.
Winchelsea.
Worley-common,
17. Bolney.
Groombridge.
18. Westfield.
19. Southwick.
20. Rackham.
21. Lambrust.
Whitesmith.
23. Guestling.
24. Woodscor.
25. Hayward's-heath.
27. Horsted Keynes.
29. Cuckfield.
30. Ardingly.
30. May-

30. Mayfield.
   JUNE
3. Hailsham.
   Hurst Gr.
4. Balcomb.
   S. Harting.
6. Bodiham.
9. Steyning.
12. Biues Green.
18. Rotherfield.
22. Broadwater.
   Cross in Hand.
24. Green.
   W. Preston.
   Frantfield.
   Chelwood.
25. Forrest-row.
27. Cat-street.
29. Ashington.
   Wivelsfield.
   JULY
3. Whitesmill.
5. Bognor.
   Clayton.
   Pevensey.
   Sompting.
8. Southwater.
13. E. Grinstead.
14. Hollington.
18. Horsham.
   Pett.
20. Longbridge.
22. Beeding.

25. Seaford.
   Shoreham.
   Danehill.
   Blackboys.
26. Hastings.
29. Chailey.
31. Angmering.
   Buxted.
   Henfield.
   AUGUST
2. Ripe.
5. Chichester(StJohns).
   Ewhurst.
   Lindfield.
10. Hawkhurst.
   Rye
12. Green.
21. Arundel.
29. Uckfield.
   SEPTEMBER
1. Nothiam.
4. Brighton.
   Egdean.
   Maresfield.
   Playden.
9. Crawley.
12. Adversean.
   Horsted Keynes.
14. Finden.
15. Westham.
16. Cuckfield.
17. Wilmington.
19. Selmiston.
   19. Steyning.

19. Steyning.
21. Boreham-street.
25. Groombridge.
    Arundel.
    Robertsbridge.
26. Clayton.
27. Rogate.
29. Horsebridge.

OCTOBER
1. Hastings (Bl. Rock.)
2. Lewes. Cliff.
   Warborne.
   West Tarring.
6. Blackboys.
8. Alfriston.
10. Chichester.
    E. Bourne.
    Withyham.
    Newhaven.
    Rushlake.
    Steyning.
12. Ditchling.
13. Rackham.
15. Ashurst.
16. Turner's-hill.
20. Rotherfield.

20. Chichester (Stow).
21. Shipley.
28. E. Dean.
    South Harting.
29. Scaynes.
    Broadwater.
    Midhurst.

NOVEMBER
1. Wadhurst.
2. Bletchingly.
8. Forrest-row.
   Billinghurst.
11. Storrington.
13. Mayfield.
17. St. Leonard.
18. Hayward's-heath.
    Cuckfield.
19. Cross in Hand.
20. Petworth.
22. Battel.
23. Hastings.
27. Horsham.

DECEMBER
11. Bolney.
    E. Grinstead.
17. Arundel.

Moveable Fairs.

Easter Tuesday.

Brede.
Pulborough.
Slinfold.
Slaugham.

Turner's-hill.

Easter Thursday.

Beckley.
Brighton.
Dicker.

Finden.

Finden.
Petworth.

Monday before Whit-Sunday.

Fletching.
Horsham.

Whit-Monday.

Battel.
Billinghurst.
Chichester.
Rye.
West Hoathley.
Willingdon.

Whit-Tuesday.

Hastings.
Lewes.
Midhurst.
Thakeham.
Warnham.

Whit-Thursday.

Cuckfield.
Jevington.

Trinity Monday.

Rudgwick.

Thursday after Trinity.

Hartfield.
Peasmarsh.

Lewes Wool Fair, July 26.

This fair was first established in 1786, and the county is indebted to the happy thought which suggested to Lord Sheffield the establishment of such an excellent plan. Before this era, the mode of buying and selling wool was entirely left to chance and uncertainty; and by nobody knowing the fair price, every one sold for what he could get, which necessarily left the seller at the mercy of the stapler; but his Lordship, by instituting this fair, collected the flock-masters together, and a proper price has ever since been obtained.

Other counties soon imitated the example: Norfolk and Suffolk, Essex, &c.

Lewes Sheep Fair.

This fair is annually held upon the second day of October; and it is from hence that the South Down flocks are dispersed over various quarters of England, as the buyers come from a great distance to attend Lewes upon this day, where large droves are bought up by commission. From 20' to 30,000 sheep are generally collected upon this occasion.

Previous to this fair, there is one at Selmiston (September 19) upon a much smaller scale. But the principal flocks are drafted and sold previous to either of these fairs, so that a buyer who comes from another county, and examines the sheep upon the day of the fair, is deprived of seeing the finest part of this celebrated stock.

SECT. IV.—MANUFACTURES.

These are, iron, charcoal, gunpowder, paper, &c. &c.

1. *Iron.*

Sussex, in the common acceptation of the term, is not a manufacturing district. Formerly there were very extensive iron-works which flourished in the Weald, but only the remnant of them are at present in existence. The vast woods supplied an inexhaustible fund of fuel in the working of the material the iron-stone pervades the greatest part of the county; but the Scotch, by some late discoveries, work the manufacture so much cheaper than can be afforded in Sussex,

Sussex, that the furnaces have nearly vanished before
the cheapness of the Northern establishments. The
Earl of Ashburnham's extensive forests, before the application to the making lime, were used for the production of iron. At present, to make 13 ton of pig-
iron, takes 50 load of charcoal (two cord of wood makes
one load of coal, and two of these a weighing load),
and 50 load of ironstone (12 bushels in a load).

2. *Charcoal.*

The manufacture of charcoal is an object of some
consequence in such a county as Sussex.   Large quantities are annually sent to London by land-carriage.
The old process in burning has been lately laid aside,
and a new method substituted; as, after various experiments, the powder made upon this new principle,
has, upon proof of its strength, been found much superior to that which was made in the old way.   And
accordingly this ingenious mode has been suggested to
Government, by the Bishop of Llandaff, of making the
charcoal in iron cylinders, of such a construction, as
effectually to exclude the air, and to preserve all the
tar acid which is extracted from the wood in the course
of burning.

Adjoining the turnpike at North Chappel, and
within five miles of Petworth, Government has lately
purchased a small piece of land of Lord Egremont,
and upon it have erected this charcoal manufactory.
The cylinder room is 60 feet in length, and proportionably high and wide: three sets of iron cylinders are
placed in a very thick wall, or bed of brick-work,
built nearly along the centre of the house; each of them
contains three cylinders, each being six feet long and two
feet diameter. To prevent every possibility of air being
admitted,

admitted, iron stops are contrived, 18 inches in length, and the size of the inner circumference of the cylinder, which are placed in the mouth, and are filled and rammed down with sand; besides which, sand-doors (as they call them) are made to project obliquely over the front or opening of the cylinder, and are entirely filled with sand, and the stops covered with it. At the back part of the building are copper-pipes projecting seven feet in length, communicating at one extremity with the far end of the cylinder, and at the other extremity immersed in half-hogshead barrels. These pipes serve to draw off the steam or liquid, which flows in large quantities into the tar barrels during the process of charring. Sea-coal fires are made under them, one to each set; and in order to convey the heat as equally as possible to all parts of the cylinder alike, four flues or cavities equidistant from each other in the brick-work, spirally encircle the cylinders, and conduct the heat over every part. The position of the grate was at first under the centre cylinder. Various alterations have since been made, as it was found that this method did not answer so as to heat all the cylinders equally. The grate is now placed under the outside cylinder in each of the sets; and by the flues being so conveyed, it follows, that the further cylinder is first heated, and that which is nearest the fire, last. Each set holds 5 cwt. of wood; so that when all three are in full work, the daily consumption is 15 cwt. of wood, which makes 4 cwt. of coal; it loses nearly three parts out of four in charring;—and if all the three sets were at constant work, the annual consumption would be nearly 550,000 cwt.—27,500 ton.

The process of this novel and valuable operation may be thus explained: very early in the morning the first thing done by the workmen is to take the doors down, by a pulley suspended at the ceiling, remove the sand, and also to take the sand out of the stops, previous to being drawn out and suspended; large tin coolers are then brought up to the mouth of the cylinder, and the charcoal of the preceding day is then drawn with a rake into the cooler, and shut up close. As soon as the cylinders are emptied, the workmen are employed in recharging them. For this purpose the sorts of wood are various, but withy and elder are the best: the cordwood is about 18 inches long, but before it is placed in the furnace, they cut it into five lengths, and all the black knots are cut away. In the act of filling, the largest pieces are placed in the centre, and the smaller adjoining the rim: when it is charged, the iron stop is let down by the pulley, put into its place, and the sand rammed into the front; the doors are then hung over the mouth, and filled up with sand, and the fire is kindled and fed till the wood is completely charred, which is known by the tar ceasing to flow through the copper-pipes. If the fire is lighted about half after six o'clock in the morning, it will take from two to two hours and a half before the wood is at all heated, and the liquid begins to flow. At this time the fume becomes extremely offensive, and soon after almost intolerable to any but the workmen. The time required is eight hours, but this depends upon the size of the wood. During the operation, attention is paid to the pipes, which are inspected, lest any air might be accidentally admitted, which would infallibly stop the pipes from working. The fires are kept up as strong and as bright as it is possible; though the waste of sea-coal is not considerable.; about eight bushels to each set daily. When the wood ceases to work, and the tar to flow,

the

the fire is gradually extinguished, which concludes the day's work, the furnaces remaining in the same state till the next morning, in order to give them time to cool; and when drawn, they are replenished in the manner before-mentioned, but are always cleaned each day, and the pipes once a month.

The wood for this manufactory comes out of the neighbourhood, and is bought in at 24s. per stack (fell, flaw, and stack), besides the carriage. Large quantities of wood are kept in the yard, and stand about a year before using: the stack is here twelve feet long, three feet ten inches high, and three feet six inches over, and from each is extracted about 55 gallons of tar-liquor. This tar acid they daily draw from the barrels, put into a large tub, and preserve it in hogsheads; but at present it cannot be used, because a patent is out for the monopoly of the sale. It is worth 6d. per gallon. The charcoal goes to Waltham and Faversham.

3. *Gunpowder.*

There is an extensive private manufactory of gun-powder at Battel, and for some time it is reputed to have been a famous place for the excellency of the powder manufactured there. The chief proprietors are Sir Godfrey Webster and Mr. Harvey. Every sportsman knows it; but the Dartford is stronger, and the quality superior.

At Chichester is a small woollen fabric; sacks, blankets, and some other articles, are made in many of the workhouses, and assortments of linen and worsted yarn, cotton and stuff goods; though it deserves in-quiry, whether to promote manufactures in workhouses, is founded in justice to the poor. Husbandry is a more
rational

rational employment, and more congenial to the temper of the people in this county.

Paper is manufactured at Iping and other places. Lord Egremont has established a manufactory of it at Duncton, and a fulling-mill at the same place, besides a mill for grinding oatmeal, which supplies all the neighbourhood with a very useful article, which used to be had from a distance at a greater expense.

Brick-kilns are established in various places. At Little Hampton, white bricks are made. Near Petworth, a kiln has been lately constructed for supplying the West Indies; an open kiln, and a dome-kiln, each holding 28,000 : they take 30 hours burning with 2500 bavins, at 9s. per 100 : three men fill in three days, and draw in three more. If the demand was brisk, the kiln would burn all the year. In 1796, only 300,000 and 100,000 tiles were made; sold at 29s. per thousand on the spot. At Arundel, 34s. To burn 400,000 requires nine men; wages 4s. 6d. per thousand; size, 9 inches, 4, 2¼.

Potash is made at Bricksill-hill, adjoining Petworth, for the soap-masters of the town.

SECT. V.—POOR.

By this term is understood, in a general sense, the labouring poor, and those who at any time of the year seek assistance from the parish.

I shall set out with observing, that the present state of this class of people is in many parts of England inferior to what every humane person would wish, and much

much below that condition which they may reasonably expect in so wealthy a community. Too many of their houses are the residence of filth and vermin; their dress insufficient; their minds uneducated, uninstructed; and their children, from insufficiency of earnings, trained to vice; their daughters to follow the easy road to prostitution, and too many of them at last to become injurious, instead of a blessing. Give each man an interest, a stake in the welfare of his country, and we should no longer hear of so many crimes. The possession of property is so deeply interwoven with earthly considerations, that every country labourer who has strength to labour, ought in a well ordered society to enjoy as much land as he can cultivate. It is this great principle which forms the cement of society, and would establish the perfect security of the country. For example, if each labourer rented as much grazing land as enabled him to support a cow in winter and summer, with a few pigs and poultry, and a garden extensive enough to supply his own family, it would completely do away the frivolous and unfounded complaints of the ignorant, that the price of provision is owing to the size of farms. Each labourer would then be as fully interested in support of the Constitution as the most potent peer of the realm. By such an arrangement, sedition, which once menaced the country, would vanish, as the great mass of the community would naturally feel an affection for their country exactly in the ratio of their own domestic felicity. Such an order of things would signally promote the comfort of life, would improve the understanding of the 'poor, and give them ideas of moral obligation, the rights of society, and the duties of christianity. It will be worth

inquir-

inquiring into the state and price of the following articles :

1. Expenses,
2. Earnings,
3. Cottage,
4. Food,
5. Dress,
6. Friendly Societies,
7. Charity,
8. Houses of Industry.

These circumstances involve their maintenance and support: let us then compare the income with the expenses of industrious labouring families in Sussex, and examine whether the wages are not absolutely inadequate to support them honestly in their calling. For this purpose, the annual expenses and earnings of several labouring families must be stated, and a medium year is the fairest for calculating the account, because during the last three years, the price of provisions has fluctuated too much to strike an even balance in any one of these years ; for if the account had been averaged according to the valuation of the necessaries of life during this period, such a table would have exhibited too severe a picture of distress. The following account was made with great accuracy and correctness.

Expenses

Expenses and Earnings of Six Families of Labourers, by the Week and by the Year, in the Parish of Glynde, Sussex, 1793.

Necessaries.	No. 1.* Eight Persons.			No. 2.* Three Persons.			No. 3.* Six Persons.			No. 4.* Six Persons.			No. 5.* Seven Persons.			No. 6.* Three Persons.		
Expenses per Week:	£.	s.	d.	£.	s.	d.	£.	s.	d.	£.	s.	d.	£.	s.	d.	£.	s.	d.
Bread or flour, ...	0	6	8	0	1	11	0	5	9	0	5	9	0	6	2½	0	2	10½
Yeast and salt, ...	0	0	6	0	0	2	0	0	4	0	0	6	0	0	6	0	0	3
Pork, or other meat, ...	0	2	0	0	0	5	0	1	8	0	1	8	0	1	10	0	0	10
Tea, sugar, butter, ...	0	1	7½	0	0	6	0	1	3	0	1	3	0	1	6	0	0	7
Cheese, ...	0	0	10	0	0	1¼	0	0	6	0	0	6	0	0	9	0	0	3
Soap, starch, blue, ...	0	0	6	0	0	2	0	0	5	0	0	6	0	0	6	0	0	4
Candle, ...	0	0	4½	0	0	2¼	0	0	4½	0	0	4½	0	0	4½	0	0	4½
Thread, worsted, ...	0	0	7	0	0	2	0	0	6	0	0	6	0	0	7	0	0	2
Total, ...	0	13	1	0	3	7½	0	10	9½	0	11	0½	0	12	2¾	0	5	8
Per annum, ...	34	0	4	9	7	6	28	0	2	28	14	2	31	15	11	14	14	8

* No. 1. A man, his wife, and six children; the eldest twelve years old, the youngest two years old.
No. 2. A woman, whose husband has run away, and two small children.
No. 3. A man, his wife, and four children; the eldest fifteen, and the youngest three years old.
No. 4. A man, his wife, and four small children; the eldest not quite five years of age, the youngest an infant.
No. 5. A man, his wife, and five children; the eldest ten years old, and the youngest an infant.
No. 6. A man, his wife, and one child; the man but one leg, his wife lame, but industrious, the child six years of age.

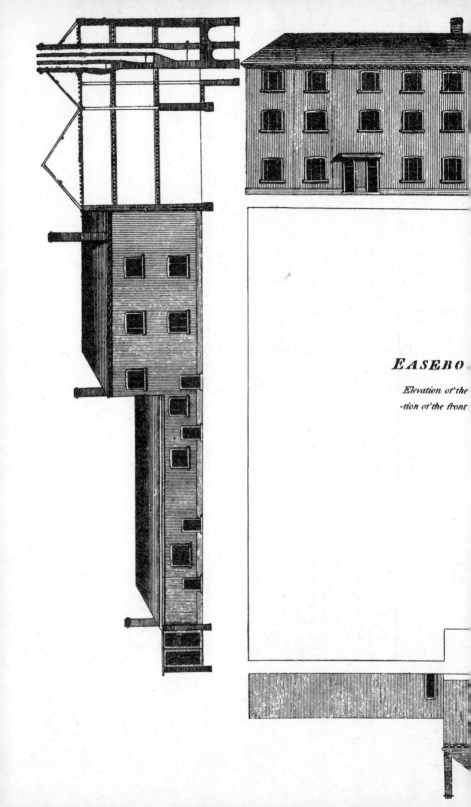

EASEBO

Elevation of the

-tion of the front

Hock sculp.ᵗ 352 Strand

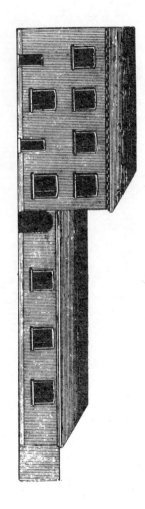

POOR HOUSE

nd the Yard, with Sec-

Committee Room

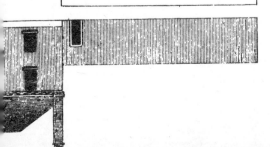

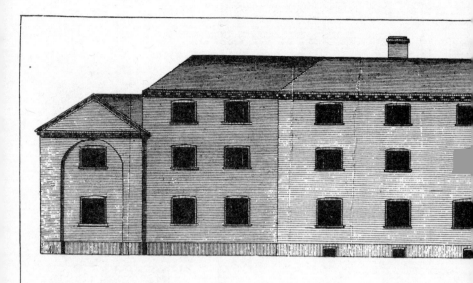

ELEVATION

ONT

	No. 1. Eight Persons.			No. 2. Three Persons.			No. 3. Six Persons.			No. 4. Six Persons.			No. 5. Seven Persons.			No. 6. Three Persons.		
	£.	s.	d.	£.	s.	d.	£.	s.	d.	£.	s.	d.	£.	s.	d.	£.	s.	d.
Earnings per Week:																		
The man earns, at a medium,	0	9	0	0	4*	0	0	9	0	0	9	0	0	10	0	0	4	0
The woman,	0	0	0	0	1	0	0	2	0	0	3	0	0	3	0	0	3	6
Children,	0	2	0	0	0	0	0	3	0	0	0	0	0	1	0	0	1	0
Total,	0	11	0	0	5	0	0	14	0	0	12	0	0	14	0	0	8	6
Per annum,	28	12	0	13	0	0	36	8	0	31	4	0	36	8	0	22	2	0
To the above amount of expenses per ann.	34	0	4	9	7	6	28	0	2	28	14	10	31	15	11	14	14	8
Add rent, fuel, clothes, lying-in,	8	14	0	8	14	0	8	14	0	8	14	0	8	14	0	8	14	0
Total expense per ann.	42	14	4†	18	1	6	36	14	2	37	8	0	40	9	11	23	8	8
Total earnings per ann	28	12	0	13	0	0	36	8	0	31	4	0	36	8	0	22	2	0
Deficiency of earnings,	14	2	2	5	1	6	0	6	2	6	4	10	4	1	11	1	6	8

* Parish pay.

† According to the expenses of the boasted houses of industry in Norfolk and Suffolk, this, instead of 42l. 14s. 4d. would have been 65l. 4s.

Annual Expenses.

Rent of a cottage and garden, ...	£.2	10	0
Fuel, if bought, costs 1*l*. 1*s*. to 1*l*. 4*s*. The labourers are allowed the old wood; their wives pick up sticks, ...	1	1	0
Clothing: the man wears a frock, per annum	0	5	0
Wear of a working waistcoat and breeches,	0	6	0
Two shirts, ...	0	10	0
One pair of stout shoes, nailed, ...	0	9	0
A pair of stockings, ...	0	4	0
Hat, handkerchief, &c.	0	6	0
	5	11	0
The woman wears a gown and petticoat, ...	0	9	0
Two shifts, ...	0	7	0
A pair of strong shoes, ...	0	5	0
Two pair of stockings, ...	0	3	0
Two aprons, ...	0	3	0
Handkerchiefs, caps, &c.	0	4	0
	1	11	0
Lying-in, sickness, and loss of time, ...	1	12	0
Total, ...	£.8	14	0
Price of the half peck loaf of wheaten bread,	0	1	0
Gallon of flour, ...	0	0	11½
A week's labour in winter, ...	0	9	0
Throughout the year, ...	0	10	0
In harvest, ...	0	15	0

The tea used in a family, is from two to four ounces per week, at 3*d*. per ounce.

Moist sugar, half a pound, at 9*d*. or 10*d*. per lb.

Salt butter, quarter of a lb. at 8½*d*. to 9*d*. per lb.

Cheese,

Cheese, from 5*d*. to 6*d*. per lb.

Beer, none.

Soap, 4*d*. per lb.

Many of the women wash for the unmarried labourers.

Those labourers who can rent a cottage and garden, can generally keep poultry, and fatten a hog; and all have frequent and great help from the charitable and considerable farmers, such as milk, broth and inferior meat, which must make the deficiencies of earnings*.

Indeed

---

* This table is truly excellent indeed: and its utility would have been increased by a similar table 40 or 50 years ago, deduced from extracts from farmers' books.—*Rev. John Howlett, Dunmow, Essex.*

These particulars tend to confirm me in my opinion (which I have stated in other Reports), of the necessity of increasing the rate of wages generally, to labourers in husbandry. Their situation ought to be made more comfortable; otherwise we must not be surprised at their being dissatisfied, and wishing for a change.—*Mr. William Dann, Gillingham, Kent.*

The following statement will shew the earnings of the labourers that I employ on my farm, and the family to consist of eight persons:

	£		
The man 26 weeks, at 11*s*. per week, - - - -	£.14	6	0
Ditto 26 ditto, by the great, at 15*s*. - - - -	19	10	0
Wife three weeks in spring, planting, hoeing, &c. at 5*s*. a week,	0	15	0
Ditto hay-harvest three weeks, at 5*s*. per week, - -	0	15	0
Corn harvest, - - - - - - - -	1	16	0
Getting potatoes, three weeks, at 5*s*. - - - -	0	15	0
The oldest child half price of the mother, - - -	0	7	6
Hay-harvest, - - - - - - - -	0	7	6
Corn ditto, - - - - - - - -	0	18	0
Getting potatoes, - - - - - - -	0	7	6
Some little manure they make and gather themselves, and one cart-load that I give them, and give them land to set their potatoes, and work it for them with the plough, and cart their dung for them free of expense, by which they will average 60 bushels of potatoes, at 1*s*. 4*d*. per bush.	4	0	0

The labourer and family's earnings for one year, - - £.43 17 6

Now,

Indeed there was no necessity to demonstrate, that the wages of labour are inadequate to the price of provisions: it is too striking to be controverted by any, but those who think it beneath their dignity to examine the residence of the cottager, especially exemplified in the instance of numerous families: their bed and bedding, blankets, sheets, &c. &c. their little houshold furniture, clothes, fuel, food, garden; the clothing of their children, the rags and nastiness in which too many of them are wrapt up; and last of all, the state of their cottages in general. The architecture for a palace is well enough understood: for a cottage it is unknown: the comment is sufficiently clear. One grand requisite in the amelioration of the poor, is the construction of cottages upon a new principle. In a

---

Now, where the wife is industrious, which is mostly the case with labourers, after working for me 13 weeks, she spins and knits for the family, and often makes a piece of coarse cloth for the family use, and jobs about in the neighbourhood; by which means she will earn a shilling per week, which will be 39 shillings; but that may be allowed for loss of time, &c. and after the oldest child has worked with the mother 18 weeks, it is sent to the school, and all the rest that are qualified to go, except one that the mother may sometimes keep at home to nurse the infant child, as it is mostly the case there is one. And by this means they pay their way well; and both they and their children appear clean and decent, and have change of clothes to go to their devotion, &c.

Labourers in husbandry and their wives are mostly industrious; for both them and their wives have generally been servants to farmers, and have been brought up to industry and economy.

Where there is one labourer in husbandry that becomes needful to his parish, there are ten to one that have been brought up to mechanical trades; for they mostly ramble about, and marry to servants in gentlemen's families, and, always living in plenty and luxuriousness, when either old age or any thing happens to the husband, they do not know how to turn their hands to any kind of industry, but immediately put themselves upon the parish.—*Mr. Harper, Bank-hall, Liverpool.*

northern

northern climate like our own, fuel enters very deeply
into the absolute necessaries of life.

To sustain health and strength, and give the human
frame the means and power of undergoing the burthen
of unremitted labour, hearty and strengthening diet is
absolutely necessary to preserve life to a vigorous old
age. Animal food once in each day, should form a
part of the labourer's diet; not indeed entirely salted
provisions, but fresh meat with a mixture of vegetables;
potatoes, carrots, and parnips, are very nourishing,
particularly the latter. A garden might raise a sufficiency for his own family; an orchard might perhaps
be added for cyder; and the whole lot, including the
field for a cow and hogs, from one to three acres, according to circumstances: these, with a neat brick or
stone built cottage, would supersede every necessity of
resorting to the parish for relief, and would render the
labouring poor happy and flourishing, population multiplied, consumption increased, gross produce augmented, provisions cheap, rates abolished, industry
incessant, morals, education, &c. improved. To the
possession of property would be attached a native dignity of mind, which would excite in them a new train
of ideas, and a more extensive reach of understanding.

In the house of industry at East bourne, the experiment of potatoe puddings was frequently tried: it did
not answer; no saving in the flour: 70 lb. of flour
made into puddings, has dined the whole family; the
same quantity mixed with half a bushel of potatoes,
served a less quantity of paupers than the pudding of
flour only, as some part of the latter generally re-
mained after the dinner was ended, but of the other
there was seldom a sufficiency. Potatoes are a great
digester,

digester, and left the paupers much more hungry for
the next meal than when wheaten flour was used by
itself.

Mr. Williams, of Woolbeding, and Mr. Islip, of
Stedham, have tried potatoe bread, and the effect has
been much the same as the foregoing. This is, in all
probability, to be attributed to the bread being used
soon after having been made; but if these gentlemen
had kept it for a few days before trial of its merit, I
have no doubt but the result would have been the same
with that which was tried at Petworth.

But the right application of potatoes for human food,
is, beyond all question of doubt, *roasted* or *baked;*
boiled, and eaten with salt, is good, but not to be compared with the other. In the former method they are
superior to bread; but in either (simply boiled or
roasted) preferable to any preparation with other com-
posts.

Friendly Societies ought to be encouraged by all
possible means: the utility of them is so obvious, that
it is much to be lamented they are not more general.
If gentlemen and farmers were to encourage them with
their protection and assistance, it would tend to the
support of old age, sickness and infirmity.

But of all the duties which we owe to society, charity is the first. There are various ways in which the
exercise of this virtue may contribute its assistance to
the poor; but certainly the best method is when it operates to the promotion of industry. In this manner
the relief of the poor in Sussex has been taken up by
Lord Egremont. The great utility which results to
the community at large, from holding out rewards to
the poor and industrious among the lower ranks, and
from discouraging, as much as it is possible, every
propen-

propensity to idleness, long since induced his Lordship to distribute to the sober and industrious, bounties in clothes, which are intended to serve for the en-
couragment of active industry, at the same time that
it might operate as a check and discouragement of the
idle, by cutting off all hopes of such being recommended as objects of charity. For such a plan to have
its fullest effect, by rendering it known to the great
mass of the people, it was necessary to circulate the
following certificate, describing the age, sex, &c. of
the person who may be entitled to the bounty, upon
application to two respectable persons (not parish officers), for if it rested with these, it might tend to defeat
the very end for which the bounty was intended, as
the officers might recommend any persons as objects of
the charity, in order that they might not any longer
apply to the parish for relief. Here follows the form
of the certificate.

We the undersigned inhabitants of the parish of Petworth, do recommend *Mary, the wife of William
Ayling*, a poor person resident within the said parish, being of the age of 47 years, and having six
children living at home, under the age of twelve
years, as sober and industrious, and a proper object
to receive the Earl of Egremont's bounty of clothes.
Dated this 25th day of February, 1799.

(Signed) { WILLIAM WOUT,
ISAAC IRELAND.

The above recommendation is to be signed by two
respectable housekeepers living in the same parish with
the party recommended. And to prevent any abuse of
the charity, it is earnestly requested that no person will
sign

sign this paper, unless he perfectly knows such party, and is fully convinced that in all respects he or she is a proper object of the bounty.

The number of families that partook of this bounty for the last year, was as under :

Parishes.	Numbers.	Children under Twelve.
Stopham,	5	25
Fittleworth,	18	68
Petworth,	74	190
Tillington,	28	82
North Chappel,	14	36
Green,	6	22
Kindford,	19	57
Lurgershall,	8	18
Ambersham,	3	11
Sutton,	1	3
Lodsworth,	2	7
Wollavington,	1	5
Bury,	2	8
Egdean,	3	8
Wisborough-green,	1	6
Cold Waltham,	1	0

186

Besides this bounty (which is very much increasing), of the strongest Yorkshire cloth, Lord Egremont constantly distributes three and four times a week, good soup to the poor of the neighbourhood, made of barrelled beef, Scotch barley, and potatoes, besides regaling between three and four hundred families at Christmas with beef and pork pies.

A sur-

A surgeon-apothecary (Mr. Andrew) lives in Petworth-house, for the express purpose of attending upon the poor of that and the neighbouring parishes gratis. All who come under his care, are treated in the best possible manner: if a limb is fractured, it is set; if physic is wanted, it is administered, as there is a complete apothecary's shop and surgical apparatus at hand.

In 1795 his Lordship sent an expert woman to the British Lying-in Hospital, to pass through the qualifications requisite in the business of midwifery, preparatory to her settlement at Petworth.

Another woman has been settled in the neighbourhood, for the sole purpose of inoculating the children of the poor.

Houses of industry, well attended to, are a fair mode of relieving the poor in bodies.

The 22d Geo. III. cap. 83, on this subject, is certainly an excellent Act, as well for the benefit of the poor as for the good of the public. Objections indeed have been raised. Whoever examines the interior economy of any one of them, and then turns into a poor man's cottage in the neighbourhood, cannot fail of being struck with the difference. Examine the regularity, the cleanliness, the industry, wholesome diet, and clothing of the paupers, and then compare them with similar circumstances in a cottage. Which is the better? There is great attention paid by professional men, to sickness in these houses; there is little in a cottage: but the mental improvement, the instruction of the young in the principles of religion, and in the practice of morality, is, I hope, one of the primary concerns in houses of industry. We all know how

they

they have tended to diminish the poor-rates, which may partly be attributed to economy; and they are fed upon a more nutritive diet. It is really astonishing, how very cheap great numbers of people may be fed upon a very hearty food when they mess together, without one morsel of animal food. Count Rumford has explained this fact in the clearest and most decisive manner, by shewing that in the house of industry at Munich, the poor (1200) are fed with a soup of pearl barley, pease, potatoes, wheaten bread, salt and vinegar, prepared in the following proportions.

Expense of this food in the scarcity of 1795 (November).

70 lb.	9 oz.	pearl-barley, at 2d.	£.0	11	9
65 —	10 —	pease, at 1½d.	0	8	2
230 —	4 —	potatoes, at 0½d.	0	13	9
69 —	10 —	bread, at 11$\frac{5}{8}$$d$.	0	16	6
19 —	13 —	salt, at 1½d.	0	2	5¼
Vinegar, 1 gallon, ...			0	1	8
Expenses for fuel, servants, &c.			0	10	4¼

Total, ... £.3 4 7¼

3l. 4s. 7¼d. divided by 1200, gives for each 2¼d. very nearly; so that notwithstanding the uncommonly high prices of provisions at that time, the cheapness of subsistence is most extraordinary. Not, however, that any thing like this management has been exerted in England; but the capability of doing it remains the same.

The house of industry at East-bourne was formed July 5, 1792; house, offices, and conveniences, have been erected under the powers of the 22d Geo. III.

c. 83, and opened for the reception of paupers, October 10, 1794. The parishes united are, Bepton, Cocking, Chithurst, East-bourne, Farnhurst, Iping, Linchmere, Lodworth, Lurgeshall, Selham, Stedham, Tillington, Trayford, Trotton, Woolbeding, Woollavington.

1. *Expenses.*—For the first year: the buildings (computed to contain 180 persons), according to the first proposed plan, were contracted for at 2800*l.* and the expenses of the purchase of the ground, and of those buildings (inclusive of some alterations in the original plans), and of fitting up the premises to receive the furniture, amount to .... £.4005 8 5¼

Manufacture amounts to 948 17 9

Raw materials of manufacture (which will be returned to the undertaking on sale of the produce), amounts to } 75 8 1

Miscellaneous, comprising all other articles under the 15th Schedule (exclusive of the pay of officers, or assistant, not yet due), amounts to } 73 14 9

£.5103 9 0¼

Provisions, and other articles, including the stock of provisions in hand, amount to } 1058 8 9¼

Total of both, ... £.6161 17 10½

2. *Employment.*—So many of the paupers within the house are engaged in the domestic employment of the institution, as are capable and necessary for that service: the males in gardening and out-door work, and the females in the houshold work; and the males are

are let out to labour to the neighbouring farmers who choose to hire them. There are likewise established manufactures of linen and woollen for clothes ; but as no persons are admitted into the house between the ages of 14 and 60 years, except those wanted for the management of domestic affairs, women with infant children, and persons who, from infirmity of mind or body, are incapable of procuring a livelihood ; and as the principal objects of the institution are to obtain a comfortable asylum to persons of a very *advanced* age, and to bring up the children at an early age to habits of industry, and to make them handy and useful servants, it cannot be expected that much profit should arise from the manufactures, especially in their present infant state ; and in fact, that profit has not for this first year been sufficient to reimburse to the house the expenses of the Board, and wages of the persons employed to instruct the children ; the charge of wear of implements ; the interest of the cost of them, and of the raw materials of the fabric.

3. *Earnings.*—The earnings of the paupers from October 10, 1794, to September 26, 1795, are as follows :

By spinning linen, woollen, mops, ...	£.14	4	7¾
Knitting, ...	3	1	0½
Needle-work, ...	3	17	4¼
Husbandry, deducting 2d. in 1s.	12	17	2
	£.34	0	2¼

The value of the labour of the poor within the house or garden, not estimated.

4. *Num-*

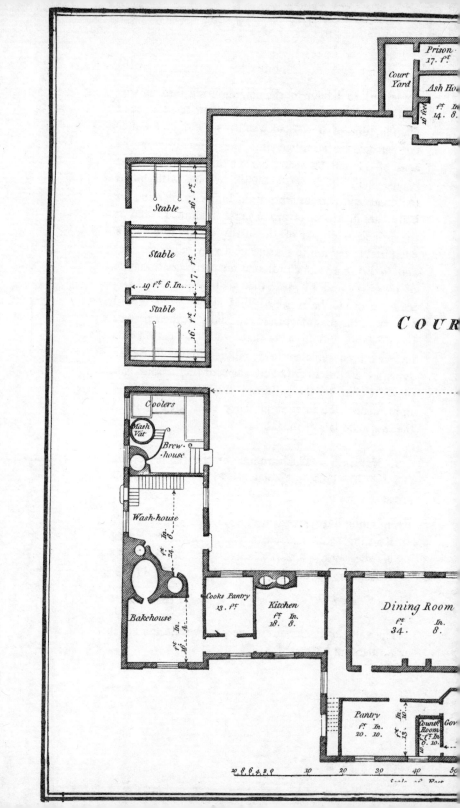

Prison
17. f.t

Court
Yard

Ash Ho...

f.t In.
14. 8.

10 feet

Stable

16. f.t

Stable

17. f.t

...19 f.t 6. In...

Stable

16. f.t

C O U R

Coolers

Mash
Vat

Brew-
house

Wash-house

f.t In.
24. 8.

Bakehouse

f.t In.
40. 4.

Cooks Pantry

13. f.t

Kitchen

f.t In.
18. 8.

Dining Room

f.t   In.
34.   8.

Pantry

f.t In.
20. 10.

f.t In.
13. 10.

oun...
Room
f.t In.
6. 10.

Gov

10. 8. 6. 4. 2. 0.   10   20   30   40   50

Scale of Feet

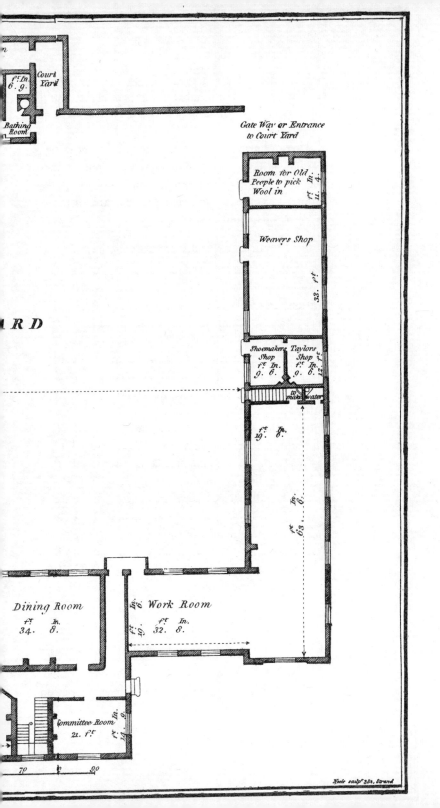

Court
Yard

f.t In.
6. 9.

Bathing
Room

Gate Way or Entrance
to Court Yard

Room for Old
People to pick
Wool in

f.t In.
u. 4.

Weavers Shop

33. f.t

Shoemakers Taylors
Shop Shop

f.t In. f.t In.
9. 6. 9. 6.

to make water

f.t In.
19. 0.

f.t In.
63. 0.

R D

Dining Room

f.t In.
34. 8.

Work Room

f.t In.
32. 8.

Committee Room

21. f.t

f.t In.
14. 9.

70   90

Neele sculp.t 352, Strand

4. *Numbers.*—The number of paupers in the house, October 1795, was 159.

5. *Diet.*—For breakfast, bread and cheese, or water-gruel and milk-pottage, sometimes broth and onion-pottage;—for dinner, pudding, mutton and pork, trimmings, ox-heads and bacon, and coarse beef;—for supper, bread and cheese.

Provisions expended for one month:

	£	s	d
Beef, 72 lb.	1	11	6
Three ox-heads,	0	10	4
Mutton, 22½ lb.	0	10	5
Pork, 453 lb.	15	2	0
Lard, 61 lb.	1	15	7
Flour, 211½ gallons,	16	14	$10\frac{1}{2}$
Bread, 2972 lb.	34	1	1
Cheese, 491 lb.	9	4	$1\frac{1}{2}$
Oatmeal, 3 quarts,	0	1	6
Butter, 19 lb.	0	14	6
Rice and Scotch barley, 3¼ lb.	0	9	11
Sugar, 12 lb.	0	9	0
Milk, 42 gallons,	0	11	4
Beer, 802 gallons,	8	7	1
Soap, 40 lb.	1	1	8
Candles, 6 lb.	0	4	5
Wood and faggot, 500,	5	0	0
Sundries,	0	11	$9\frac{1}{4}$
Small incidental expenses,	3	0	1
£.	100	1	$2\frac{1}{2}$

$157\frac{7}{17}$ paupers, for 31 days, cost 100*l.* 1*s.* 2¼*d.* = 2*s.* 11¼*d.* $\frac{1011}{4681}$ each.

Sutton

Sutton house of industry was effected in 1791, and a manufactory established, for spinning, and making worsted. The original uniting, comprehended eight parishes; three more have been since added, and the undertaking now includes the parishes of Bersted, Bignor, Burton, Bury, Clapham, Coates, Duncton, Egdean, Patching, Slindon, and Sutton. Mr. Samuel Bryan, the governor, contracts to pay the parishes for the labour of the poor within the house, at the rate of 1s. 3d. per head per week, for each pauper of the age of six years and upwards. Previous to October 10, 1794, the food of the poor had been contracted for, and the weekly average expense of each pauper's maintenance during the above year, was 2s. 10¼d.

These poor-houses are too recently established, to draw much instruction from the experiments in this county. It carries in its favour every prospect of lowering the rates, of mending the morals, of instructing the children, and educating them in habits of sobriety and industry: their health is not by any means impaired by the nature of their employment; their diet is wholesome; and the only solid objection is farming them. This is perhaps the consequence of the manufacture. In East-bourne it is unknown; though some employment is absolutely necessary, and husbandry-work for the very old and very young, is utterly impracticable; and after all, a manufacture of some sort is perhaps the best possible way of promoting such industry.

On the subject of the poor laws and management of the poor, Lord Sheffield, who has had upwards of thirty years' experience of their effect, is fully aware of all the difficulties which have arisen from the abuse of them, and the misconception of their great object; and

and he considers all the deviations from the principles
of the law of Elizabeth, as promoting the mischiefs
which now embarrass us. The original law of Elizabeth was excellent in principle, but a false interpreta-
tion, and bad execution of it, and above all, the non-
sense of sentimental *economists*, who never comprehended its spirit, have rendered it a great nuisance,
highly oppressive of the landed interest, and crippling
of the resources of the country. He observes, that a
kind of system has been established, of relieving the
poor, by no means supported by law. The statute of
Elizabeth was well imagined, and answered all the
purposes intended; it merely gave a power to the
parish officers to provide for the lame, blind, and im-
potent, and to set the idle to work; which was peculiarly necessary at that time, as there was a number of
idle, needy, and disorderly people, who used to receive
alms from the monasteries, previously to their dissolu-
tion, and of soldiers and mariners, who were turned
loose after the defeat of the Spanish Armada. But it
has been so much misconstrued and abused, that it has,
in a great degree, destroyed a provident spirit on the
part of the lower ranks, and promoted the neglect of
their families and children, by suggesting notions that
the parish is obliged to maintain, not only their children, but themselves also; thereby leading them to
look to other means of subsistence than their own industry, than which a greater mischief cannot be imagined. Unfortunately an ill-judged conduct on the part
of those who were not aware of the views of this excellent law, and who never had a practical knowledge of
the country, has encouraged these evil consequences,
which are becoming highly calamitous.

Lord Sheffield conceives that all the changes in this
law

law have been for the worse; and among other instances, he mentions a late alteration, which allows the
poor to wander out of the reach and observance of
those who are obliged to maintain them: that those
who promoted this measure, did not perceive that the
principal object in not permitting the poor to ramble,
without a certificate, where they pleased, was with a
moral view; it was in part intended to prevent their
intrusion into places where they were unknown, and
might introduce much mischief. This license greatly
lessens the necessity of supporting a good character;
and it became less necessary to recommend themselves
by their good conduct, to the principal persons of the
parish; it told them they had nothing to do but to run
away to another parish, when they had transgressed;
and they became a severe scourge to those among
whom they intruded, from whence they cannot now be
removed. It gives an opportunity to smugglers and to
the greatest villains, to assemble where they please,
from all parts, to the ruin of the morals of a parish,
before orderly and well regulated. In answer to the
common-place observation, that it is very hard, not to
suffer the poor to seek work wherever they suppose they
can best find it, it should be observed, that the law
provided a power to give certificates to those who did
not readily meet with employment in their own parish,
which parish officers will of course grant, rather than
relieve them; and it is but reasonable that there should
be something reciprocal in the connexion, and that
those who are liable to maintain them, should have the
advantage of their work, if they can find employment
for them.

The rental of the Weald of Sussex is much affected
by the extravagance of the poor-rates; and, comparatively

tively with the intrinsic value of the land, there is no part of the island where it is lett at so low a price : in common years the rate through a considerable district, is at ten shillings in the pound, rack-rents ; and during late years of scarcity, they a,.*ounted to 25s., and even in some parishes to 35s. in the pound, at rack-rents.

The miserable parish workhouses seem principally intended *in terrorem*, and without them, the parishes would be overwhelmed by the demands of the paupers ; they are, in general, the vilest establishments (if they are worthy of such a description), devoid of any thing like tolerable superintendance ; some of them, however, make feeble attempts to employ the poor that are lodged in them, but in great part, there is no attempt at any work. In a few instances, the county affords examples of a certain number of parishes having united to form houses of industry, in imitation of those established in several parts of England ; but there is little prospect of its ever becoming, by any means, general, notwithstanding the evident good effects of the system.

Lord Sheffield has, at different times, represented the great benefits arising, not only in respect to the management of the poor, but also in the reduction of an enormous expense, from such institutions: the most obstinate prejudice, and want of intelligence, however, renders every attempt to persuade any number of parishes to agree to a measure of the kind, so troublesome and disagreeable, that nothing but an obligatory law is likely to bring about so desirable a purpose. The sum paid for rent of houses for the poor, would more than pay the interest of the money that would be necessary for building a large house of industry, and other habitations, for the poor.

In

In all management, this district seems to be behind other parts of England : attempts have been made to encourage the poor to keep a cow, but without success; and it is found most expedient to lett small grass farms for the purpose of a dairy only, stipulating that the milk shall be sold by retail : the necessity of such stipulation arises from the circumstance, that it is not worth the while of a considerable farmer to sell milk in small quantities to the poor, who too often evade payment where they possibly can.

On the whole, Lord Sheffield is of opinion, that the evils in respect to the poor are so deeply rooted, and the abuses so inveterate, that the most intelligent man will find it very difficult to please or satisfy himself, in regard to the correction of those abuses ; and he ascribes part of the difficulty to the incompetency of the mass of those who are necessarily to be consulted in vestry or otherwise : yet he says, if some great measure is not soon adopted, the extent of the mischief must become ruinous in various shapes.

SECT. VI.—POPULATION.

By reference to the parish registers, and abstracting the state of births and burials at different periods in different parts of the county, and comparing the former with the present state, a tolerable judgment may be drawn of the state of population in Sussex. A great and almost uniform increase of the births over the burials, is apparent in every register that has been examined. The improvement of the Weald, by growing more corn and cattle, has in some degree contributed

to

to the health of the inhabitants; and the drainage of the marshes has had its effect, as the stagnated waters have been let off, and these fenny ditches made more salutary.

HORSHAM.

	B.	C.		B.	C.
1579	32	65	1592	58	78
1580	47	60	1593	36	47
1581	34	74	1594	32	54
1582	35	51	1595	58	60
1583	35	63	1596	64	40
1584	37	61	1597	66	40
1585	37	65	1598	61	44
1586	37	55	1599	45	42
1587	41	68	1600	42	53
1588	44	59			
1589	34	67		957	1300
1591	38	68			

Excess of christenings 343 in the 22 last years of the 16th century.

Decennially in a century and a half.

	B.	C.		B.	C.
1610	81	62	1700	54	63
1620	59	69	1710	44	35
1630	73	63	1720	63	48
1640	80	53	1730	63	53
1650	47	60	1740	56	53
1660	51	68	1750	59	49
1670	91	62	1760	33	51
1680	56	57			
1690	65	67		975	915

Excess of burials 60.

1783

	B.		C.			B.		C.
1783	....	44	....	82	1789	....	58	....	86
1784	....	82	....	72	1790	....	45	....	75
1785	....	41	....	90	1791	....	42	....	90
1786	....	50	....	92	1792	....	57	....	83
1787	....	70	....	83			530		841
1788	....	41	....	88					

Excess of christenings 311 in ten last years.

EAST-BOURNE.

	B.	C.
24 years, 1648 to 1671	808	....	676
Ditto 1769 to 1792	588	....	1062
Excess of burials in the first period		132
Christenings in the latter		474

WESTHAM.

Decennially 1570 to 1670, and 1690 to 1780 :
    1571 to 1671, excess of burials   70
    1690 to 1780, ditto   7
In the last ten years, excess of christenings 120

When the present vicar came to the parish of Pevensey, he was at first troubled with the ague, but since 1783 or 1784 this complaint has disappeared, which is attributed to the better drainage of the levels : the dykes are now opened, and the waters no longer remain in a state of stagnation, which was found to be so extremely unfavourable to the healthiness of the place.

HAILSHAM.

HAILSHAM.

	Burials.	Christenings.	Excess of Burials.
1599 to 1608	254	202	52
1611 1620	423	262	161
1630 1640	381	272	109

In every register that I examined, the winter months were invariably found to be by far the most unfavourable to human life.

	Burials.	Christenings.	Excess of Christenings.
1752 to 1762	98	163	65
1762 1771	130	155	25
1772 1781	142	239	97
1783 1792	140	280	just double.

RYE.

	Burials.	Christenings.	Excess of Burials.	Excess of Christenings.
1610 to 1620	783	834	—	49
1630 1640	824	666	158	—
1650 1660	618	484	134	—
1670 1680	402	496	—	92
1690 1700	419	342	77	—
1713 1722	478	336	142	—
1730 1740	497	493	4	—
1750 1760	380	498	—	118
1770 1780	602	646	—	44
1781 1792	562	742	—	180

BEDINGHAM.

In this parish are 25 families. Mr. Carr's farm maintains 25 labourers from the parish, all but three married,

married, having 41 children. Mr. Martin's farm maintains two labourers, both married, eight children. Mr. Taylor and Mr. Davis upon both their farms maintain 16 labourers; 15 of them married, 38 children. The poor-house maintains three men, four women, and 18 children. Mr. Davis has in his family 11 souls; Mr. Taylor five; Mr. Carr 24; Mr. Martin 20 : total in the parish 252.

FIRLE.

Males 223; females 198; total 421.—Families 77; houses 66; farmers 9; poor 100.

RINGMER.

Males 467; females 436; total 903.—Families 170; houses 105; farmers 20.

TARRING NERILL.

Males 43; females 37; total 80.—Families 14; houses 7; farmers 3; poor 9.

GLYNDE.

Males 96; females 126; total 222.—Families 35; houses 25; farmers 8; labourers 21; poor 19.

			Burials.	Christenings.
1729 to	1738	....	27	 33
1739	1748	....	34	 35
1749	1758	....	40	 40
1759	1768	....	36	 53
1769	1778	... ...	37	 73
1779	1788	....	47	.... 65*

ARUNDEL.

---

* The number of houses in the parish are 24, but as six of them are double, we reckon 30 families in the parish. The present number
of

ARUNDEL.

	Burials.	Christen-ings.	Excess of Burials.	Excess of Christen-ings.
1560 to 1578	433	 349	... 84	.... —
1580 1592	232	 200	 32	.... —
1581 1792	403	 638	.... —	 235

In 1780 the population was 1200: in 1786, it was 1753: in 1792 it was 1926.

PETWORTH.

	Burials.	Christen-ings.	Excess of Christen-ings.
1774 to 1783	417	 616	 199
1783 1794	447	 745	 298

From these few extracts, and many others, it appears beyond all question, that the people in this county are very rapidly multiplying, and increase faster than they are able to feed themselves.

---

of inhabitants, men, women, and children, is 212, which gives seven to a family and two over, one with the other. We have no tradesmen or artificers, but what are connected with and dependent on agriculture, as carpenters, wheelwrights, blacksmiths, bricklayers, &c. Our land-tax is high, very near 4s. in the pound, owing, it is said, to the too warm zeal of some friends to the Revolution, who rated their money with the land, and now the personal estate is gone, the burden of the tax remains on the land. The poor-tax we think rather moderate : it was last year about 2s. in the pound, rack rate. Most parishes about us are much higher.—*John Ellman.*

CHAP.

CHAP. XVI.

OBSTACLES TO IMPROVEMENT:
WOODS, AND COMMON RIGHTS.

THIS obstruction to tillage reigns over two-thirds of this county. The timber, woods, coppices and shaws, are sufficiently mischievous to grazing land: to the growth of corn they are ruinous and destructive to the last degree. The enclosures are so small, and the soil so wet and binding, that to lay such land dry is no easy task, though the action of wind and sun is more necessary here than elsewhere: yet each hedge-row is a nursery for timber, and so enormously thick, as to be perfectly impervious to the rays of the sun, and so tall as to convert the country into the appearance of a forest: the consequence to the corn is indeed deplorable*.

Common

---

* It is a great pity, that the injury done to the corn and grass in most parts of this county, by the growing trees and large belts of underwoods surrounding almost every enclosure in the country, is not remedied. For in the autumn and winter seasons the grounds, by the exclusion of sun and air, are rendered torpid, greatly perishing the seeds and roots of the vegetating corn: whereas in the other seasons, the shade, and even attractive qualities of these woods, prevent any grain or vegetables which happen to spring, from coming to maturity. It will be found in chemical trials, that grain raised in small enclosures does not produce the same virtues as the corn in large or open fields do, where air and sun have free communication. A second evil: these woods are nurseries for

insects

Common Rights.

These are unexceptionably the most perfect nuisance that ever blasted the improvement of a country; and till they are done away, no tolerable husbandry will flourish in those districts where they are in force.

Tithes.

These are certainly in some measure an impediment to improvement, though not to that degree which has

---

insects and vermin, if the expression be allowed. I have seen and felt this matter severely, where fields of turnips and pease have been destroyed by such insects, and particularly by the slug: even the Windsor bean could not, after laying one night in the ground, resist their devouring every part but the husk. But when too late to remedy the evil, I found that saw-dust, and the shavings cut by the hatcher, in preparing the bark for the tanner, were preventatives. Hence I conclude, that a liquid, or a preparation made from bark, or the branches, &c, of trees of a bitter and acid quality, and sprinkled over the ground, might prevent the mischief done by such insects, when saw-dust and bark could not be had. A third evil: the profits made by these underwoods, when maturely considered, are of consequence to the farmer. Suppose that each acre of the growth of ten years were worth 10*l.* 10*s.* and it seldom exceeds that money; when the loss in interest of money is calculated, *half* the sum is thereby lost at least. Now the other half would be gained in two years in the crops of potatoes and turnips that might be raised: for I look upon the roots to be of greater value than the expense laid out in grubbing. Nay, I have found the ashes of roots equal to the expenses of grubbing, were they only applied to dressing meadow lands: and the losses by insects, and the seclusion of sun and air from the enclosures, as already pointed out, far exceeded the expense that would ensue to a farmer, however small his tenure, in the article of other fuel to his family being brought from a distance. Last spring I laid some wood-ashes in a wet part of a meadow, where only benty-grass and bushes grew. It has nearly extirpated them, and brought a fine foliage of natural grasses, so that the horses can now be scarcely kept from forcing their way to that part which formerly they would not visit, owing to the quantity of salts the present grass contains.—*F.*

been

been alleged. If any alteration takes place in the mode
of paying the Clergy, the general opinion seems in favour of a corn-rent; but perhaps land would be better.
Land in mortmain is undoubtedly cultivated to much
disadvantage; but if a fair compensation was allowed
for the improvement which the Clergy made in their
respective parishes, it would obviate the disadvantage
attending this mode of settling the business. Where
the land has been in a state of cultivation for many
years, it would of itself be an improvement to put it
into the hands of able and intelligent men, and without
the least reimbursement, four-fifths would feel their
own interest too much implicated, not to exert themselves to keep it in a state of progressive improvement.
Surely there is much good sense in the idea of each parish having a resident clergyman who is a good farmer,
and much more rational, than to witness that idle life
which too many of them lead.

CHAP. XVII.

MISCELLANEOUS OBSERVATIONS.

SECT. I.—AGRICULTURAL SOCIETIES.

IN 1772 a Society was proposed and established at Lewes, for the Encouragement of Agriculture, Manufactures, and Industry, by John Baker Holroyd, Esq. now Lord Sheffield, and Premiums were offered; but on the breaking out of the war in 1778, it was dropped.

Petworth Fair.

A fair has been yearly held upon the 20th of November at this place, but was not remarkable till the Earl of Egremont, with a view of promoting the improvement of cattle, by animating the neighbouring breeders to exertions before unheard of, excited a rivalship among them by offering premiums.

In 1795, Lord Egremont offered a premium of a silver cup to the finest bull that was shown at the fair.

In consequence of this encouragement nine bulls appeared, and the prize was adjudged to Mr. Thomas Coppard, of Woodmancote.

This first experiment was so very satisfactory to the farmers, that they agreed amongst themselves to show their stock of bulls and heifers at the next Storrington fair (December 5).

The cow stock produced on this day was very good, especially Mr. Coppard's, and Mr. Upfold's.

The

The Sussex breed of cattle that are reared upon the borders of Kent, have been very generally praised, as exceeding in beauty all the rest of the county. But Petworth-fair has unequivocally proved, that the true-bred stock is not confined to any local habitation, but that it pervades the whole county.

The first year of the show of cattle turned out so much to the satisfaction of Lord Egremont, that in the following year his Lordship offered the following premiums.

Petworth Fair, November 20, 1796.

A show of three and four-year old bulls, and a show of three-year old heifers which have had a calf.

A *silver cup* will be given to the proprietor of the best bull; and *ten pounds* to the proprietor of the second best bull.

Fifteen guineas will be given to the proprietor of the best heifer; and *five guineas* to the proprietor of the next best heifer.

For these prizes the bulls of the under-mentioned owners appeared :

John White Parsons's bull, three years old, West Camel, Somerset; breed, Devon.

John Ellman's, ditto ditto, Glynde, Sussex; breed, Sussex.

John Ireland's, ditto ditto, Rudgwick; ditto.

Thomas Coppard, ditto ditto, Henfield; ditto.

Henry Colgate, ditto ditto, Frantfield; ditto.

John Upperton, ditto ditto, Rackham; ditto.

Thomas Holman, ditto ditto, Henfield; ditto.

The silver cup was adjudged to Mr. Colgate, for his Sussex bull; the second prize to Mr. Parsons, for his Devon.

Mr.

Mr. Parsons's heifer, three years old; breed, Devon.
Lord Stawell's   ditto,   ditto.
John Ellman's   ditto, Glynde; breed, Sussex.
Mr. Hainse's   ditto, Kindford, ditto.

The first prize for the best heifer was adjudged to Mr. Ellman; Lord Stawell's Devon gained the second.

After the above decision, was a sweepstakes (fifteen subscribers) for the best two-year old heifer; won by Mr. John Salter, of Fittleworth.

1797.

This year there was a great meeting at Petworth; and in order the more to stimulate the farmers to exertions of such importance to their own welfare, as well as to the public good, and to give a larger range to the sphere of their ideas by that collision of opinion which takes place in large companies composed of men of all ranks, he has on these occasions filled his capacious mansion with the most celebrated breeders, graziers, and farmers, from various parts of the kingdom.

Lord Egremont's silver cup, which was this year of the value of fifty guineas, was adjuged to Mr. Harrington; and the sweepstakes for the best heifer was given to Mr. Marchant.

In the year 1797, the Earl of Egremont set on foot a Society at Lewes, for the improvement of cattle and sheep; rewarding industry among the labouring poor, and distributing prizes to the best ploughmen: and the effect has been such, that although the Society has been established only six years, it has in that short space very materially tended to improve the objects for which it was instituted, and it promises still greater success, by the support it is continually receiving from every quarter of the county.

SECT.

SECT. II.—WEIGHTS AND MEASURES.

The confusion which reigns in the weights and measures of this kingdom, has been more than once proposed to be remedied by substituting an universal standard; but the great difficulty seems to consist in the apparent impossibility of fixing upon any substance in nature subject to no impressions, and liable to no decay from climate, or length of time. It cannot be doubted but that the thing is feasible, and that an equalization might be effected, to be extended to other countries, and all Europe enjoy the benefits of such useful regulations, which tend so strongly to cement a good understanding between different nations, and unite them in friendship.

The weights and measures which are more commonly used in Sussex, are the acre, pound, stone, load, bushel, &c. There are several sorts of acres, a great source of perplexity and confusion—the short acre, the statute acre, the forest acre, and various others: the forest acre is nine score rods; the statute acre eight score; the short acre six score in some places, in others five score.

A stranger, unaware of the variations that prevail, is liable to fall into mistakes in every step he takes. The eight-gallon measure only is used; the load of wheat is 40 bushels; of oats, 80. The stone of meat is eight pounds. The tod of wool, 32 pounds; and both troy and avoirdupoise are in use.

Until a radical reform is brought to bear, the present confusion in buying and selling must prevail, and the honest and unsuspecting will be taken in by the crafty and designing.

CONCLU-

CONCLUSION.

———

MEANS OF IMPROVEMENT,

AND MEASURES CALCULATED FOR THAT PURPOSE.

THE two grand improvements required are, 1*st*, the enclosure of the waste lands, commons, and common rights; and, 2*dly*, a better distribution of arable and woodland in the Weald. Subordinate to these are others: a more extended culture of sainfoin upon the Downs; the annihilation of the husbandry of the old school upon clay land (a fallow and two crops of corn), by substituting tares, rape, rye, cabbages, beans, potatoes, and where the soil is lighter, pease, carrots, turnips. These meliorating crops answer far better the purpose for which fallowing was intended, a dead loss, and no profit. But the fallow crops, either fed upon the ground or soiled in the yards, will contribute their assistance for a crop of wheat something better than the present mode, not only of a whole year's expenses of rent, taxes, and labour, but at least four or five guineas more in lime, and this, moreover, to raise 20 bushels of wheat. How then is it possible, under such a.system, to look for the sunshine which animates the exertions of other districts?

Hollow-draining is far the most capital improvement ever worked upon wet land. But to improve the Weald, corn is not an object. Grass upon wet land, corn upon dry, and both where it is temperate. The

Weald

Weald of Sussex should be a grazing district. Large dairies, with butter, cheese, and hogs ; with beef and mutton for Smithfield.

It is clear, that such an arrangement must be pursued, if the right application of the soil ever comes to be an object of attention.

The quantity of waste land is very great, and affords a most striking proof how much the public encouragement is required, in order to bring those neglected lands under some system of improvement. They are most decidedly capable of being converted to profit. Skill and capital are the main springs for such an undertaking. Judgment to plan, and perseverance in the execution, will eventually triumph over the most untractable desert. It is a cause that is not to be starved.

If this soil was properly treated, something like the following arrangement might be adopted.

If the forest be broken up for the first time, the furze, ling, broom, heath, with all other rubbish covering the surface, should be burnt as it stands, and then pared and burnt, and rye sown the same year; or, if done sufficiently early, a crop of turnips may be first obtained. On this poor sandy soil, care should be taken that the turnips be sown in good time, or they will not arrive to any size : if, therefore, the turnips be not in the ground before or by Midsummer, rye should then take place, to be spring fed with sheep, and succeeded by turnips, and then with oats, laid with artificial grasses, to remain so long as the layer continues good, and the longer that is, the better will the land be for it, as such a soil is more profitable under pasture than it ever can be in a state of tillage. A method somewhat similar to the above ought to be adopted--a hint is sufficient

APPEN-

APPENDIX.

WEALD OF SUSSEX.

NOTHING can be more various than the soil of the Weald. In the range of black mountainous land which stretches from the neighbourhood of Tunbridge Wells, under the names of Waterdown, Ashdown, Tilgate and St. Leonard's forests, the soil is generally bad ; a considerable part incorrigible at any expense that will repay the cultivator, and would be most profitable for the growth of birch. But the country between that range and the South Downs, contains much good land, rich sandy loam, and fertile clay, generally mixed with some sand : capable of producing every kind of crop.

Lord Sheffield's estate, which is the largest, is nearly in the centre of the country just described, and answers to the above description. As he is the largest farmer, the particulars of the management of the land he has had above 30 years in hand (about 1400 acres, which I have often visited), will best describe what may be done in that soil. The arable bears no proportion to the pasture, meadow, and woodland ; the park being between five and six hundred acres, and the woodland between four and five hundred acres. He has tried every mode and every instrument of agriculture ; but observes, that as a gentleman cannot attend markets and fairs, and as he buys dearer and sells cheaper than a mere farmer, it is

not

not wise to undertake the most complicated, although
the most profitable, therefore he pursues the most
simple, plan. The breeding of sheep and cattle is the
chief object of his farming, the redundancy of which
is disposed of at two or three principal fairs. His
course of crops is, turnips, oats, seeds, and wheat, on
one ploughing. He has often sown rape when the land
was not in time, or in a fit state for turnips, or when
the latter have failed. He uses it always as food for
cattle; sometimes fed it, towards Michaelmas, with
the old ewes, and fatting sheep; then shut it up for
the ewes and lambs in the spring: sometimes it is
mown, and given to the cattle, because he observed the
frost was apt to kill the rape when much jagged by the
sheep-bite; and he has also mown it as food for cows
in winter, and found it answer well. Heavy snow is
the greatest enemy to this practice, when the crop is
luxuriant, which it often is; and it cannot be mown
in frost without endangering the spring sprout, the
most valuable for ewes and lambs. He plants a few
acres with cabbages, for cows and other cattle; and
also two or three acres with potatoes, for hogs and
poultry. He has sown what is called in Sussex, *winter
barley*, and also *rye*, on a stubble, as soon as the crop
has been carried, as early food for sheep in the spring,
when the turnips are gone. At first, he was disposed
to consider the winter barley as the best for the purpose, because it matted well on the ground; but the
rye is not only earlier, but bears repeated feeding better
than the barley, therefore, is most useful and productive.

The best white wheat succeeds well on his land, and
produces from three to five quarters per acre; but in
the neighbourhood, red wheat is generally most hardy
and

and productive; he has frequently drilled the wheat, but never dibbled any, as there are no persons in the county experienced in that practice, and he found the business could not be dispatched; he did not find any great advantage from drilling, except as to the saving of seed, which he rather imputes to imperfect hoeing. His land would bear good barley, but he sows little, as oats are most wanted; he has grown good beans, but as it is not customary to give them to horses in that neighbourhood, there is very little demand. He urges all gentlemen living in the country not to yield to the common notion, that farming will not answer to them ; he strongly recommends farming, so far as may be necessary to supply the family, observing, that the team is the heaviest part of the expense, and as no gentleman can reside with convenience in the country without a team, he of course incurs the principal expense of farming; and he argues, that by a proper management of the team, it may easily be made to maintain itself, at the same time that it will do all the business necessary to the convenience and other supply of the family.

Cattle.

He has tried every breed of cattle: he thinks he never had a finer bull than one of the long-horned kind from Craven; he has also had a bull from Mr. Bakewell, which he conceives to have been originally of the same kind. He has had a very good Cleveland and a very good Herefordshire bull; but he prefers the Devonshire breed, which is found very kindly, to any other ; yet, after such experience, considering to how much greater advantage the breed of the country, when it happens to be good, goes to fair or market, he now endeavours

to

to raise cattle of the best Sussex breed, of which he
conceives there are two kinds: the coarser resembles
the Herefordshire, except there being no mixture of
white; the lighter breed resembles very much the
North Devon, and seem to have been originally the
same breed, and not to be inferior to the cattle of any
country. It has been suggested, that some of the
lighter breed may have been produced by a cross some
time past with the best Norman or Alderney breed.

In respect to colour, the deep chesnut red seems to
be preferred by many; the yellow red is very kindly,
but least hardy, and most apt to scour. But Lord
Sheffield prefers the blood bay, such as he has seen of
the Devonshire breed.

A principal reason for preferring the cattle which he
now rears is, that they make the best working oxen.
The coarsest kind of Sussex grow too heavy for work
soon after six years old, and are very slow; but the
kind which Lord Sheffield raises, step out better and
faster than horses, and do the same work, for he never employs more than four for any purpose. He has
only two cart-horses and eighteen working oxen, who
are harnessed like horses, have bridles, and are accustomed to be led; they have never any food but grass or
straw, until they begin to work hard in the spring, when
they have hay cut with their straw: he often works them,
when they are hardy and do the business well, till they
are upwards of twelve years old. He has proved the
fallacy of the notion, that if worked hard to that age,
they will not fat well; he used two of his largest oxen
beyond that age, without ever sparing them, and
within one year they were fatted with oil-cake to the
great weight of nearly 210 stone each. Such of his
oxen or steers as prove clumsy or short-legged, he sells

                                                        or

or fats; but he seldom fats any stock except with grass, and for his own use, consequently he has no dealings with salesmen.

Lord Sheffield supposes that he keeps 18 oxen as cheap as eight horses are usually kept; the oxen eat no oats, and comparatively very little hay, and there is no blacksmith's bill, on their account; it not being usual to shoe them in the Weald. He admits that this breed is not the best for milking; but as the business of the dairy is ill understood in that part of Sussex, he does not impute the deficiency entirely to the breed. He has seen very fine black cattle, bulls, cows, and oxen, the breed having been long in Sussex, which he thinks may have been a cross some time past with the Welsh; but they have the shape and size of the best Sussex. If there is any white, or other colour than red, it is not allowed to be the true Sussex: the least appearance of white is considered as a stain in the breed.

Above 30 years ago Lord Sheffield gave 50 guineas, a large price at that time, to Mr. Bakewell, for the use of one of his rams; the ewes were of the South Down breed, and the cross appeared at first to have answered well; but he soon found that he had sheep of no character; the lambs appeared larger, but weighed little more than South Down, and the wool was very indifferent, being neither long nor short. He has had the Hereford breed of sheep, the Ryeland, and Urchingfield: he liked them very much in respect to shape and wool; but they did not drive to fold so well as South Down; and because they had not grey faces instead of white when they went to fair, nobody would look at them. The same principle, therefore, which induced him to adhere to the Sussex breed of oxen, induced him to return to the South Down, especially as he found that

that they weighed as well on the same food as the others. At the same time it should be observed, that about that period the South Down breed of sheep began to improve very much. The prices which are given, and the great spirit of improvement which has arisen from the exertions and attention of gentlemen, who have spared no expense in obtaining the best breed of cattle, have and will prove highly advantageous to the country. Every farmer begins to feel the advantage of raising good stock; and instead of receiving 6s. or 7s. for head lambs, which was the case when Lord Sheffield first became a farmer, he now receives from 20s. to 30s. The large prices which are given for some rams and ewes, have little to do with the price of mutton; the price of the ram is soon recovered by obtaining a breed that fattens more speedily, or attains more weight on the same quantity of food.

Lord Sheffield's flock consists of about 1000 sheep, and during the greater part of the year he works two folds; but he is clearly of opinion, where a sufficiency of manure can be obtained, that sheep which are not folded, thrive much better, and especially when they are kept in small parcels in different enclosures. He found that those ewes which were not folded, but kept in small parcels, brought twins, and the lambs, though twins, were much larger and better than those of the ewes which were folded and kept in the flock.

Lord Sheffield has improved the breed of hogs in his neighbourhood, by introducing the most approved sorts, and he is now endeavouring to ascertain which is the best, Mr. Western's Essex, or Mr. Ashley's Leicestershire. The former seems to be much more prolific. He has crossed with the best China, and also with the wild kind.

An

An engraved plan of Lord Sheffield's farm-yard is annexed: it is very commodious, and contains every thing necessary for a considerable farm: although some attention has been paid to symmetry and appearance, he in general rejects every improvement that cannot come within the reach of the common farmer; and his endeavour is to simplify every construction and implement, knowing how much greater the expense, and how much more liable to be out of repair, if complicated.

Perhaps there is no object in the Weald of Sussex so worthy of observation as the growth of timber; there is no region of the earth where trees of all kinds thrive better, particularly oak and ash. The district called the Weald has formerly been covered with trees, and was called the forest of Anderida; and even now, if a field is neglected, it will become a wood, principally of oak and birch, intermixed with hazel, some kinds of willow, and dogwood. Lord Sheffield has paid particular attention to this subject, and there is no estate in the county on which there is so great a stock of fine young oak. The increased value will promote general attention, and more than ordinary care for the preservation of that most useful and necessary article. Within little more than 20 years the value is more than doubled. The Navy Board has relinquished the bad policy of endeavouring to avail itself of a kind of monopoly of large timber: the price was kept down so much, that it became a maxim, on account of the debased price, that it answered better, for the sake of the quickness of the return, to cut down a tree before it reached the value of 40s. than to suffer it to remain till it acquired a large size.

The highest price for the largest timber 30 years

ago,

Elevation of Lord Sheffields Farm Yard. Sheffield Park.

SUSSEX.

Made at Prescot.

Scale of Feet.

Lord Sheffield's FARMYARD Sheffield Park, Sussex.

STACK YARD

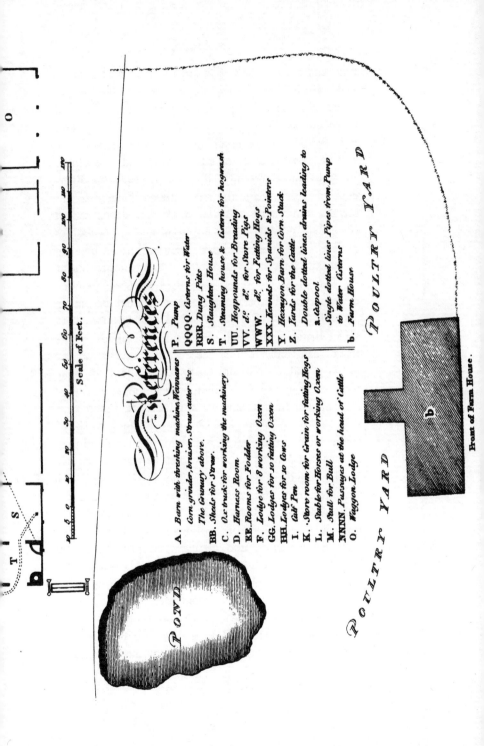

Scale of Feet.

Références

A. Barn with threshing machine, Winnowers
Corn grinder, bruiser, Straw cutter &c
The Granary above.
BB. Sheds for Straw.
C. Ox truck for working the machinery
D. Harness Room
EE. Rooms for Fodder
F. Lodge for 8 working Oxen
GG. Lodges for 10 fatting Oxen
HH. Lodges for 10 Cows
I. Calf Pen
K. Store room for Grain for fatting Hogs
L. Stable for Horses or working Oxen
M. Stall for Bull
NNNN. Passages at the head of Cattle
O. Waggon Lodge

P. Pump
QQQQ. Cisterns for Water
RRR. Dung Pits
S. Slaughter House
T. Steaming house & Cistern for hogwash
UU. Hogpounds for Breading
VV. dº dº for Store Pigs
WWW. dº for Fatting Hogs
XXX. Kennels for Spaniels & Pointers
Y. Hexagon Barn for Corn Stack
Z. Yards for the Cattle
Double dotted lines. drains leading to
a cesspool.
Single dotted lines Pipes from Pump
to Water Cisterns
b. Farm House.

POULTRY YARD

POULTRY YARD

POULTRY YARD

b

Front of Farm House.

PLAN

POND

ago, did not exceed five guineas per load; the same, and even inferior timber, would now sell for 15*l.* per load. The increased price of oak-bark may also tend to encourage the growth. A good price being now obtained, there is a much better prospect that timber will be allowed to reach a large size; and it would be promoted more certainly, if a more considerable difference of price were allowed between timber of the largest dimensions and the next size, which is not the case at present. The improvidence and necessities of families of landed property, and of those who have only a life-interest in an estate without impeachment of waste, must however always prevent the practice of preserving timber to a good size, from becoming general: otherwise, Lord Sheffield is of opinion, that Great Britain could with ease furnish the utmost quantity of oak-timber which can ever be required for her military and commercial shipping, without any material interference with other branches of agriculture. At the same time it must be admitted, that more care is taken to preserve and encourage the growth of timber than formerly; it is much to be lamented that the management of woods has not been more scientifically attended to; that little has been written, and that little instruction can be derived from books.

Lord Sheffield observes, that a good system of setting out the fellows or saplings, and of preserving them when young, and during their growth, would have rendered the growing timber in this island infinitely more valuable to the public, and consequently to the individual; and he conceives that the timber on his estate would be worth many thousands more, if he had earlier attended to the pruning and management of the woods. It is not sufficient merely to leave

a great

a great number of young trees, they require regular care and training: if they are left too thin on the first setting out, they will not thrive, nor become clean lengthy plank timber; but it is absolutely necessary, as they grow up to thin them properly, leaving at last after the rate of from 40 to 50 trees on an acre.

Ash timber is become highly valuable, the best growing in Sussex and Kent, for the use of coach-makers; it now sells for upwards of 3s. per foot; and it should be observed, that it attains that value comparatively in a short time; and as it is an article principally used by the makers of all carriages and husbandry implements, there must always be a great demand.

The underwoods in this part of Sussex, are converted into hop-poles, hoops and cordwood; the principal part of the latter goes to London in the shape of charcoal; the spray or small branches are made into faggots for houshold use, and burning lime and bricks. If the woodland be good, it will produce from 12l. to 20l. per acre, at 10 to 14 years growth.

THE END.

Printed by B. M'Millan,
Bow Street, Covent Garden.

SUSSEX.]

DIRECTIONS TO THE BINDER,

FOR PLACING THE PLATES.

HARROWS.—See page 57.

A, the first harrow.

B, C, D, E, the bulls, or ledges of the harrow.

F, F, F, the iron for strengthening and connecting the bulls.

G, cross bar moveable on two pivots, for connecting the harrow.

H, the chain for ditto.

I, chain which connects the whipple tree with the corner of the harrow.

K, the eye through which the hook acts.

N, a harrow tooth.

P, ropes.

Q, the collar.

R, ropes connecting the horses.

S, collar band.

T, tooth collar.

A NIDEOT,

AN ENGINE TO CLEAR LAND OF WEEDS.

A, A, the frame.

B, B, the two outside shares.

C, C, C, C, C, C, the standers fixing the shares

D, D, the wheel.

F, F, fixture of the wheel.

G, G, axis point of wheel.

H, H, front fixtures of shares.

I, shar. fixture in frame.

K, K, ties to frame.